AF720119

Fourier Transform: Algorithms and Applied Principles

Fourier Transform: Algorithms and Applied Principles

Edited by
May Hooper

MURPHY & MOORE
www.murphy-moorepublishing.com

Published by Murphy & Moore Publishing,
1 Rockefeller Plaza,
New York City, NY 10020, USA

Copyright © 2022 Murphy & Moore Publishing

This book contains information obtained from authentic and highly regarded sources. Copyright for all individual chapters remain with the respective authors as indicated. All chapters are published with permission under the Creative Commons Attribution License or equivalent. A wide variety of references are listed. Permission and sources are indicated; for detailed attributions, please refer to the permissions page and list of contributors. Reasonable efforts have been made to publish reliable data and information, but the authors, editors and publisher cannot assume any responsibility for the validity of all materials or the consequences of their use.

Trademark Notice: Registered trademark of products or corporate names are used only for explanation and identification without intent to infringe.

ISBN: 978-1-63987-239-8

Cataloging-in-Publication Data

Fourier transform : algorithms and applied principles / edited by May Hooper.
p. cm.
Includes bibliographical references and index.
ISBN 978-1-63987-239-8
1. Fourier transformations. 2. Transformations (Mathematics).
3. Fourier analysis. I. Hooper, May.

QC20.7.F67 F68 2022
515.723--dc23

For information on all Murphy & Moore Publications
visit our website at www.murphy-moorepublishing.com

MURPHY & MOORE

Contents

Preface

The world is advancing at a fast pace like never before. Therefore, the need is to keep up with the latest developments. This book was an idea that came to fruition when the specialists in the area realized the need to coordinate together and document essential themes in the subject. That's when I was requested to be the editor. Editing this book has been an honour as it brings together diverse authors researching on different streams of the field. The book collates essential materials contributed by veterans in the area which can be utilized by students and researchers alike.

Fourier transformations are mathematical transformations which decompose functions dependent on space or time into functions which depend on temporal or spatial frequency. Fourier transform of a time function is a complex-valued function whose magnitude shows the frequency of the original function and argument is the phase offset of that frequency. Inverse Fourier transforms are used to obtain the original mathematical function from its frequency domain representation. They find extensive applications in the fields of signal processing, spectroscopy and quantum mechanics. This book presents researches and studies performed by experts across the globe. The various advancements in Fourier transformations are glanced at and their applications as well as ramifications are looked at in detail. This book is a vital tool for all researching or studying Fourier transforms as it gives incredible insights into emerging trends and concepts of this field.

Each chapter is a sole-standing publication that reflects each author's interpretation. Thus, the book displays a multi-facetted picture of our current understanding of application, resources and aspects of the field. I would like to thank the contributors of this book and my family for their endless support.

Editor

1

Fourier Transformation Method for Computing NMR Integrals over Exponential Type Functions

Hassan Safouhi
Mathematical Section, Campus Saint-Jean,
University of Alberta, Alberta
Canada

Introduction

Experimental methods based on magnetic resonance are among the most used techniques for investigating molecular and electronic structure. Nuclear magnetic resonance (NMR) is mostly applied to closed-shell molecules and can be used for structural research of matter in solid, liquid and gaseous form. The computation of NMR parameters, which are of a great interest not only in chemistry but also in biology and solid-state physics, presents severe analytical and numerical difficulties [Dickson & Ziegler (1996); Ditchfield (1974); Fukui et al. (2004); Helgaker et al. (1999); Ishida (2003); London (1937); Pople et al. (1968); Pyykkö (1988); Schreckenbach & Ziegler (1995); Vaara (2007); Watson et al. (2004)]. The computation of NMR parameters for any of the standard models of quantum chemistry constitute an important challenge [Helgaker et al. (1999)]. Calculations involving a magnetic field should preserve gauge invariance. This is conveniently accomplished by using a gauge including atomic orbitals (GIAO) [London (1937)], which is based on atom-centered basis functions with an explicit field dependence. Magnetic properties are sensitive to the quality of the basis sets due to many contributing physical phenomena arising from both the vicinity of the nucleus and from the valence region. A good atomic orbital basis should decay exponentially for large distances [Agmon (1985)] and should also satisfy Kato's conditions for exact solutions of the appropriate Schrödinger equation [Kato (1957)]. Exponential type functions (ETFs) are better suited than Gaussian functions (GTFs) [Boys (1950a;b)] to represent electron wave functions near the nucleus and at long range. Among the ETFs, Slater type functions (STFs) [Slater (1932)], have a dominating position due to their simple analytic expression, but their multi-center integrals are extremely difficult to evaluate for polyatomic molecules, particularly bi-electronic terms. We note that many researchers hope that the next generation of ab initio programs will be based on the usage of ETFs. Indeed much effort is being made to develop efficient molecular algorithms for integrals over conventional ETFs (STFs or B functions) [Barnett (1990); Fernández et al. (2001); Kutzelnigg (1988); Niehaus et al. (2008); Ozdogan & (Editors); Rico et al. (1998; 1999; 2001); Steinborn et al. (2000); Weatherford & Jones (1982)].

Various studies focussed on the use of B functions. The use of B functions was proposed by Shavitt [Shavitt (1963)], since reduced Bessel functions possess a representation in terms of a remarkably simple Gauss transform. Detailed discussions of the mathematical properties of reduced Bessel functions and of their anisotropic generalizations can be found in [Weniger (1982)]. Furthermore, B functions have much more appealing properties applicable to multi-center integral problems, compared to other exponentially decaying functions [Filter & Steinborn (1978a;b); Steinborn & Filter (1975); Weniger (2005); Weniger & Steinborn (1983a)]. The multi-center molecular integrals over B functions can be computed much more easily than the corresponding integrals of other exponentially decaying functions. This can be explained in terms of the Fourier transform of B functions, which is of exceptional simplicity among exponentially decaying functions [Niukkanen (1984); Weniger (1982); Weniger & Steinborn (1983b)]. Moreover, the Fourier transforms of STFs, of hydrogen eigenfunctions, or of other functions based on the generalized Laguerre polynomials can all be expressed as finite linear combinations of Fourier transforms of B functions [Weniger (1985); Weniger & Steinborn (1983b)]. The basis set of B functions is well adapted to the Fourier transform method [Geller (1962); Grotendorst & Steinborn (1988); Prosser & Blanchard (1962); Trivedi & Steinborn (1983)], which allowed analytic expressions to be developed for molecular multi-center integrals over B functions [Grotendorst & Steinborn (1988); Trivedi & Steinborn (1983)].

Of the NMR parameters, the nuclear shielding tensor is of a great importance. The computation of the shielding tensor presents severe analytical and numerical difficulties especially when using ETFs as a basis set of atomic orbitals. The main difficulty arises from the operators associated with these parameters. An example of such operators is $3\,r_{jN,\beta}\left(\vec{r}_{jN}\cdot\vec{\sigma}(j)\right)/r_{jN}^{5}$, where β represents a cartesian coordinates, $\vec{r}_{jN}$ is the vector separating the jth electron and the Nth nuclei and σ stands for Pauli spin matrix. These operators lead to extremely complicated integrals. Analytic treatment of the NMR parameters over GTFs was a subject of many articles (see for example [Ishida (2003)] and references therein). Although, the interest of using ETFs in the computation of NMR parameters is increasing [see the pioneer work by Dickson & Ziegler (1996); Schreckenbach & Ziegler (1995)] and Watson et al. (2004)], no effort was dedicated to their analytic treatment over ETFs. Straightforward numerical integration was used for the computation of integrals associated with these parameters.

The analytical development of NMR integrals can be obtained using the Fourier transform method combined with B functions as a basis set of atomic orbitals [Berlu & Safouhi (2008); Safouhi (2010b); Slevinsky et al. (2010)]. The obtained analytic expressions turned out to be similar to those obtained for the so-called three-center nuclear attraction integrals (zeroth order integrals). The latter were the subject of significant research [Berlu & Safouhi (2003); Duret & Safouhi (2007); Fernández et al. (2001); Grotendorst & Steinborn (1988); Homeier & Steinborn (1993); Niehaus et al. (2008); Rico et al. (1998; 1999); Safouhi (2001b; 2004); Slevinsky & Safouhi (2009)]. In our research, we used techniques based on extrapolation methods combined with numerical quadratures to compute the analytic expressions of the NMR integrals. Numerical tables are listed and we refer the interested reader to [Safouhi (2010b); Slevinsky et al. (2010)] for an extensive list of numerical tables as well as detailed numerical discussions.

Molecular integrals in the absence of magnetic fields

In the absence of magnetic fields, the molecular electronic Hamiltonian operator $\mathcal{H}_{e}$ corresponding to total energy E_{e} for a molecule of N nuclei and n_e electron is given by:

$$\mathcal{H}_e = -\frac{1}{2}\sum_{j=1}^{n_e} \nabla_j^2 - \sum_{j=1}^{n_e}\sum_{K=1}^{N} \frac{Z_K}{r_{jK}} + \sum_{i=1}^{n_e}\sum_{j>i} \frac{1}{r_{ij}}, \tag{1}$$

where:

- Z_K is the atomic number of the Kth nucleus whose mass is M_K.
- r_{jK} is the distance that separates the Kth nucleus from the jth electron.
- r_{ij} is the distance that separates the ith electron from the jth electron.
- ∇_j^2 is the Laplacian operators for the coordinates of electron j.

The stationary Schrödinger equation that needs to be solved is:

$$\mathcal{H}_e \, \Psi(\boldsymbol{r}, \boldsymbol{R}) = E_e \, \Psi(\boldsymbol{r}, \boldsymbol{R}).$$

The above Schrödinger equation is solved only in the case of hydrogen-like atom. The solutions are one-electron functions and are referred to as hydrogen-like atomic orbitals. These atomic orbitals form a complete and orthonormal basis. The use of hydrogen-like atomic orbitals was prevented due to the fact that their molecular multi-center integrals are extremely difficult to evaluate analytically and numerically. Linear combinations of the hydrogen-like atomic orbitals lead to STFs, which form the most popular basis set of atomic orbitals. Unormalized STFs are given by [Slater (1932)]:

$$\chi_{n,l}^{m}(\zeta, \vec{r}) = r^{n-1} \, \mathrm{e}^{-\zeta r} \, Y_l^m(\theta_{\vec{r}}, \varphi_{\vec{r}}), \tag{2}$$

where $Y_l^m(\theta, \varphi)$ is the surface spherical harmonic [Condon & Shortley (1951)] and where n is the principal quantum number, l is the orbital angular momentum number and m is the magnetic quantum number.

The B functions are given by [Filter & Steinborn (1978a); Steinborn & Filter (1975)]:

$$B_{n,l}^{m}(\zeta, \vec{r}) = \frac{(\zeta r)^l}{2^{n+l}(n+l)!} \, \hat{k}_{n-\frac{1}{2}}(\zeta r) \, Y_l^m(\theta_{\vec{r}}, \varphi_{\vec{r}}), \tag{3}$$

where $\hat{k}_{n-\frac{1}{2}}(\zeta r)$ stands for the reduced spherical Bessel function of the second kind [Shavitt (1963); Steinborn & Filter (1975)]:

$$\hat{k}_{n+\frac{1}{2}}(z) = z^n \, e^{-z} \sum_{j=0}^{n} \frac{(n+j)!}{j!\,(n-j)!} \, \frac{1}{(2\,z)^j}. \tag{4}$$

STFs can be expressed as finite linear combinations of B functions [Filter & Steinborn (1978a)]:

$$\chi_{n,l}^{m}(\zeta, \vec{r}) = \frac{1}{\zeta^{n-1}} \sum_{p=\tilde{p}}^{n-l} \frac{(-1)^{n-l-p} \, 2^{2p+2l-n} \, (l+p)!}{(2p-n+l)! \, (n-l-p)!} \, B_{p,l}^{m}(\zeta, \vec{r}), \tag{5}$$

where $\tilde{p} = \dfrac{n-l}{2}$ if $n-l$ is even or $\tilde{p} = \dfrac{n-l+1}{2}$ if $n-l$ is odd.

Fourier transform in molecular multi-center integrals calculation

The Fourier transform of B functions, which is of exceptional simplicity among exponentially decaying functions, is given by [Niukkanen (1984); Weniger (1982); Weniger & Steinborn (1983b)]:

$$\bar{B}^m_{n,l}(\zeta,\vec{p}) = \sqrt{\frac{2}{\pi}}\,\zeta^{2n+l-1}\,\frac{(-i\,|p|)^l}{(\zeta^2+|p|^2)^{n+l+1}}\,Y^m_l(\theta_{\vec{p}},\varphi_{\vec{p}}). \tag{6}$$

In [Trivedi & Steinborn (1983)], the Fourier transform method is used in combination with equation (6) to derive analytic expressions for the following integrals:

$$\mathcal{T} = \int_{\vec{r}} \left[B^{m_1}_{n_1,l_1}(\zeta_1,\vec{r})\right]^* e^{-i\vec{k}\cdot\vec{r}}\,B^{m_2}_{n_2,l_2}(\zeta_2,\vec{r}-\vec{R}_2)\,\mathrm{d}\vec{r}. \tag{7}$$

The main idea of the Fourier integral transformation is given by:

$$\begin{aligned}
\int [f(\vec{r})]^*\,e^{-i\vec{x}.\vec{r}}\,g(\vec{r}-\vec{R})\,\mathrm{d}\vec{r} &= (2\pi)^{-3/2}\int\left[\int [\bar{f}(\vec{q})]^*\,e^{-i\vec{q}.\vec{r}}\,e^{-i\vec{x}.\vec{r}}\,g(\vec{r}-\vec{R})\,\mathrm{d}\vec{q}\right]\mathrm{d}\vec{r}\\
&= e^{-i\vec{x}.\vec{R}}\int [\bar{f}(\vec{q})]^*\,e^{-i\vec{q}.\vec{R}}\,(2\pi)^{-3/2}\left[\int e^{-i(\vec{q}+\vec{x}).(\vec{r}-\vec{R})}\,g(\vec{r}-\vec{R})\,\mathrm{d}\vec{r}\right]\mathrm{d}\vec{q}\\
&= e^{-i\vec{x}.\vec{R}}\int \bar{f}^*(\vec{q})\,e^{-i\vec{q}.\vec{R}}\,\bar{g}(\vec{q}+\vec{x})\,\mathrm{d}\vec{q},
\end{aligned} \tag{8}$$

where $\bar{f}(\vec{k})$ stands for the Fourier transform of $f(\vec{r})$. The function $f(\vec{r})$ and its Fourier transform $\bar{f}(\vec{k})$ are connected by the symmetric relationships:

$$\bar{f}(\vec{k}) = (2\pi)^{-3/2}\int_{\vec{r}} e^{-i\vec{k}\cdot\vec{r}}\,f(\vec{r})\,\mathrm{d}\vec{r} \qquad \text{and} \qquad f(\vec{r}) = (2\pi)^{-3/2}\int_{\vec{k}} e^{\,i\vec{k}\cdot\vec{r}}\,\bar{f}(\vec{k})\,\mathrm{d}\vec{k}. \tag{9}$$

Replacing f by $B^{m_1}_{n_1,l_1}(\zeta_1,\vec{r})$ and g by $B^{m_2}_{n_2,l_2}(\zeta_2,\vec{r}-\vec{R})$ we obtain:

$$\int \left[B^{m_1}_{n_1,l_1}(\zeta_1,\vec{r})\right]^* e^{-i\vec{x}.\vec{r}}\,B^{m_2}_{n_2,l_2}(\zeta_2,\vec{r}-\vec{R})\,\mathrm{d}\vec{r} = e^{-i\vec{x}.\vec{R}}\int \left[\bar{B}^{m_1}_{n_1,l_1}(\zeta_1,\vec{q})\right]^* e^{-i\vec{q}.\vec{R}}\,\bar{B}^{m_2}_{n_2,l_2}(\zeta_2,\vec{q}+\vec{x})\,\mathrm{d}\vec{q}. \tag{10}$$

Equations (10) and (6) led to an analytic expression for the integral $\mathcal{T}$ in equation (7). This analytic expression is given by [Trivedi & Steinborn (1983)]:

$$\begin{aligned}
\mathcal{T} &= \frac{(4\pi)^3\,(2l_1+1)!!\,(2l_2+1)!!\,(n_1+l_1+n_2+l_2+1)!\,\zeta_1^{2n_1+l_1-1}\,\zeta_2^{2n_2+l_2-1}}{(n_1+l_1)!\,(n_2+l_2)!\,2^{n_1+n_2+l_1+l_2+1}}\\
&\times \sum_{l_1'=0}^{l_1}(-i)^{l_1+l_1'}\sum_{m_1'=-l_1'}^{l_1'}\frac{\langle l_1m_1\,|l_1'm_1'|\,l_1-l_1'm_1-m_1'\rangle}{(2l_1'+1)!![2(l_1-l_1')+1]!!}\\
&\times \sum_{l_2'=0}^{l_2}(-i)^{l_2+l_2'}\sum_{m_2'=-l_2'}^{l_2'}\frac{\langle l_2m_2\,|l_2'm_2'|\,l_2-l_2'm_2-m_2'\rangle}{(2l_2'+1)!![2(l_2-l_2')+1]!!}\\
&\times \sum_{l'=l'_{\min},2}^{l_1'+l_2'}(-1)^{l_1'}\,\langle l_2'm_2'\,|l_1'm_1'|\,l'm_2'-m_1'\rangle\,R_2^{l'}\,Y_{l'}^{m_2'-m_1'}(\theta_{\vec{R}_2},\varphi_{\vec{R}_2})\\
&\times \sum_{l_{12}=l_{12min},2}^{l_1-l_1'+l_2-l_2'}\langle l_2-l_2'm_2-m_2'\,|l_1-l_1'm_1-m_1'|\,l_{12}m_{12}\rangle\,Y_{l_{12}}^{m_{12}}(\theta_{\vec{k}},\varphi_{\vec{k}})
\end{aligned}$$

$$\times \sum_{j=0}^{\Delta l} \frac{(-2)^j \binom{\Delta l}{j}}{(n_1+n_2+l_1+l_2-j+1)!}$$
$$\times \int_{s=0}^{1} s^{n_{22}} (1-s)^{n_{11}} k^{l_1-l_1'+l_2-l_2'} \frac{\hat{k}_\nu [R_2 \gamma(s,k)]}{[\gamma(s,k)]^{n_\gamma}} e^{-i(1-s)\vec{k}\cdot\vec{R}_2} \, ds, \tag{11}$$

where:

$$\begin{aligned} n_\gamma &= 2(n_1+n_2+l_1+l_2)-(l_1'+l_2'+l')+1 \\ \gamma(s,k) &= \sqrt{(1-s)\,\zeta_1^2 + s\,\zeta_2^2 + s\,(1-s)\,k^2} \\ \nu &= n_1+n_2+l_1+l_2-l'-j+\tfrac{1}{2} \\ m_{12} &= (m_2-m_2')-(m_1-m_1') \\ n_{11} &= n_1+l_1+l_2-l_2' \\ n_{22} &= n_2+l_2+l_1-l_1' \\ \Delta l &= \tfrac{l_1'+l_2'-l'}{2}, \end{aligned} \tag{12}$$

Gaunt coefficients $\langle l_1 m_1 | l_2 m_2 | l_3 m_3 \rangle$ are defined by [Gaunt (1929)]:

$$\langle l_1 m_1 | l_2 m_2 | l_3 m_3 \rangle = \int_{\theta=0}^{\pi} \int_{\varphi=0}^{2\pi} [Y_{l_1}^{m_1}(\theta,\varphi)]^* \, Y_{l_2}^{m_2}(\theta,\varphi) \, Y_{l_3}^{m_3}(\theta,\varphi) \, \sin(\theta) \, d\theta \, d\varphi. \tag{13}$$

Equation (11) led to analytical expressions for all molecular multi-center integrals over B functions or STFs [Grotendorst & Steinborn (1988); Safouhi (2001a); Trivedi & Steinborn (1983)].

Relativistic formulation of NMR shielding tensor

In the presence of an external uniform magnetic field $\vec{B}_0$, the electronic non-relativistic Hamiltonian is given by:

$$\mathcal{H} = \sum_{i=1}^{n} \left[\frac{1}{2} \vec{p}_i^{\,2} + V(i) + \sum_{i<j}^{n} \frac{1}{r_{ij}} \right], \tag{14}$$

where the electronic impulsion $\vec{p}_j$ is given by:

$$\vec{p}_i = \left[-i\,\vec{\nabla}_i + e\,\vec{A}_i \right] \qquad \text{where} \qquad \vec{A}_i = \frac{1}{2}\left(\vec{B}_0 \wedge \vec{r}_{i0}\right) + \frac{\mu_0}{4\pi} \sum_N \frac{\vec{\mu}_N \wedge \vec{r}_{iN}}{r_{iN}^3}, \tag{15}$$

where $\vec{A}_i$ stands for the vector potential induced by the nuclear moments $\vec{\mu}_N$ and the external uniform magnetic field $\vec{B}_0$. μ_0 stands for dielectric permittivity. r_{ij} is the modulus of the vector $\vec{r}_{ij}$ separating the electrons i and j. $\vec{r}_{iN}$ is the vector separating the electron i and the nuclei N.

The relativistic effects are important for the fourth and fifth rows in the periodic table and for transitions metals [Pyykkö (1988)]. In terms of perturbations with respect to $\mu_{N,\alpha}$ and $B_{0,\beta}$ where α and β stand for cartesian coordinates $(\alpha, \beta \in (x,y,z))$, the electronic relativistic Hamiltonian is given by:

$$\mathcal{H} = \mathcal{H}^{(0)} + \mathcal{H}^{(r)} + \mu_{N,\alpha} \mathcal{H}_\alpha^{(0,1)} + B_{0,\beta} \mathcal{H}_\beta^{(1,0)} + \mu_{N,\alpha} B_{0,\beta} \mathcal{H}_{\alpha\beta}^{(1,1)} + \cdots, \tag{16}$$

where $\mathcal{H}^{(0)}$ is the zeroth-order hamiltonian (1) and $\mathcal{H}^{(r)}$ is the relativistic perturbation term, which is independent of the magnetic perturbations and is given by [Fukui & Baba (1998)]:

$$\mathcal{H}^{(r)} = \sum_{j<k}^{n} \left[\mathcal{H}_{1,j}^{(r)} + \mathcal{H}_{2,j}^{(r)} + \mathcal{H}_{3,j}^{(r)} + \mathcal{H}_{4,j}^{(r)} + \mathcal{H}_{5,j}^{(r)} \right], \tag{17}$$

which include the contributions; $\mathcal{H}_{1,j}^{(r)}$: two-electron Darwin term, $\mathcal{H}_{2,j}^{(r)}$: two-electron spin-orbit term, $\mathcal{H}_{3,j}^{(r)}$: the retarded orbit-orbit term, $\mathcal{H}_{4,j}^{(r)}$: spin-other-orbit term and $\mathcal{H}_{5,j}^{(r)}$: spin-spin term.

The perturbations $\mathcal{H}_{\alpha}^{(0,1)} = \left(\frac{\partial \mathcal{H}}{\partial \mu_{N,\alpha}}\right)_{\{\vec{\mu}_N=\vec{0},\vec{B}_0=\vec{0}\}}$, $\mathcal{H}_{\beta}^{(1,0)} = \left(\frac{\partial \mathcal{H}}{\partial B_{0,\beta}}\right)_{\{\vec{\mu}_N=\vec{0},\vec{B}_0=\vec{0}\}}$ and $\mathcal{H}_{\alpha\beta}^{(1,1)} = \left(\frac{\partial^2 \mathcal{H}}{\partial \mu_{N,\alpha} \partial B_{0,\beta}}\right)_{\{\vec{\mu}_N=\vec{0},\vec{B}_0=\vec{0}\}}$ are given by [Fukui & Baba (1998)]:

$$\mathcal{H}_{\alpha}^{(0,1)} = \frac{\mu_0}{2\pi} \sum_{j=1}^{n} \frac{\vec{l}_{jN,\beta}}{r_{jN}^3} + \frac{\mu_0}{4\pi} \sum_{j=1}^{n} \left[\frac{8\pi}{3} \delta(\vec{r}_{jN}) \sigma_\beta(j) - \frac{\sigma_\beta(j)}{r_{jN}^3} + 3\, r_{jN,\beta} \frac{\vec{r}_{jN} \cdot \vec{\sigma}(j)}{r_{jN}^5} \right] \tag{18}$$

$$\mathcal{H}_{\beta}^{(1,0)} = \frac{1}{2} \sum_{j=1}^{n} \vec{l}_{j,\beta} + \frac{1}{2} \sum_{j=1}^{n} \sigma_\beta(j) \tag{19}$$

$$\mathcal{H}_{\alpha\beta}^{(1,1)} = \frac{\mu_0}{8\pi} \sum_{j=1}^{n} \frac{\vec{r}_j \cdot \vec{r}_{jN} \delta_{\alpha\beta} - r_{jN,\alpha}\, r_{j,\beta}}{r_{jN}^3}, \tag{20}$$

where $\vec{l}_{jX} = -i \left(\vec{r}_{jX} \wedge \vec{\nabla}_j\right)$, $\vec{l}_j = -i \left(\vec{r}_j \wedge \vec{\nabla}_j\right)$ and $\vec{\sigma}_j$ stands for the Pauli spin matrix of the electron j and its cartesian coordinates are given by:

$$\sigma_{j,x} = \begin{pmatrix} 0 & 1 \\ 1 & 0 \end{pmatrix}, \quad \sigma_{j,y} = \begin{pmatrix} 0 & i \\ i & 0 \end{pmatrix} \quad \text{and} \quad \sigma_{j,z} = \begin{pmatrix} 1 & 0 \\ 0 & -1 \end{pmatrix}.$$

Fourier transformation for the analytic development of NMR integrals

The operators involved in equations (18), (19) and (20) lead to very complicated integrals. The analytic development of the these NMR integrals is difficult due to the presence of the operator involving $1/r^n$, which is not the case of the usual three-center molecular integrals (zeroth order molecular integrals) where the Coulomb operator $1/r$ is involved.

In this review, we present the method based on Fourier transform that led to analytic expressions for first and second order integrals of the shielding tensor. Among the operators involved in integrals of the shielding tensors are $\frac{\vec{r}_j \cdot \vec{r}_{jN} \delta_{\alpha\beta} - r_{jN,\alpha}\, r_{j,\beta}}{r_{jN}^3}$ in equation (20) for the second order terms and $3 r_{jN,\beta} \left[\frac{\vec{r}_{jN} \cdot \vec{\sigma}_j}{r_{jN}^5}\right]$ in equation (18) in the case of first order relativistic terms.

The integrals induced by the above operators are given by:

$$\mathcal{I}_{13} = \int_{\vec{r}_j} \left[\chi_{n_1,l_1}^{m_1}(\zeta_1, \vec{r}_{jA}) \right]^* 3\, r_{jN,\beta} \frac{\vec{r}_{jN} \cdot \vec{\sigma}_j}{r_{jN}^5} \chi_{n_2,l_2}^{m_2}(\zeta_2, \vec{r}_{jB})\, d\vec{r}_j, \tag{21}$$

$$\mathcal{I}_{23} = \int_{\vec{r}_j} \left[\chi^{m_1}_{n_1,l_1}(\zeta_1,\vec{r}_{jA})\right]^* \frac{\vec{r}_j\cdot\vec{r}_{jN}\delta_{\alpha\beta} - r_{jN,\alpha}\, r_{j,\beta}}{r^3_{jN}}\, \chi^{m_2}_{n_2,l_2}(\zeta_2,\vec{r}_{jB})\,\mathrm{d}\vec{r}_j, \tag{22}$$

where $\vec{r}_{jA} = \vec{r}_j - \overrightarrow{OA}$, $\vec{r}_{jB} = \vec{r}_j - \overrightarrow{OB}$ and $\vec{r}_{jN} = \vec{r}_j - \overrightarrow{ON}$. A, B and N are three arbitrary points of the Euclidean space and 0 is the origin of the fixed coordinate system.

First order integrals

After expanding the operator in the integrals (21), we can write $\mathcal{I}_{13}$ as follows:

$$\mathcal{I}_{13} = -\sum_{\alpha}\sigma_{j,\alpha}\int_{\vec{r}_j}\left[\chi^{m_1}_{n_1,l_1}(\zeta_1,\vec{r}_{jA})\right]^* r_{jN,\beta}\,\frac{\partial}{\partial r_{jN,\alpha}}\left(\frac{1}{r^3_{jN}}\right)\chi^{m_2}_{n_2,l_2}(\zeta_2,\vec{r}_{jB})\mathrm{d}\vec{r}_j. \tag{23}$$

Let $\mathcal{I}^{(\alpha,\beta)}_{13}$ represent the integrals in summation of the RHS of the above equation. Using equation (5), the integrals $\mathcal{I}^{(\alpha,\beta)}_{13}$ are expressed as linear combinations of integrals ${}_B\mathcal{I}^{(\alpha,\beta)}_{13}$ over B functions which are given by:

$${}_B\mathcal{I}^{(\alpha,\beta)}_{13} = \int_{\vec{r}_j}\left[B^{m_1}_{n_1,l_1}(\zeta_1,\vec{r}_{jA})\right]^* r_{jN,\beta}\,\frac{\partial}{\partial r_{jN,\alpha}}\left(\frac{1}{r^3_{jN}}\right) B^{m_2}_{n_2,l_2}(\zeta_2,\vec{r}_{jB})\mathrm{d}\vec{r}_j. \tag{24}$$

Using the Fourier transform method, we obtain:

$${}_B\mathcal{I}^{(\alpha,\beta)}_{13} = (2\,\pi)^{-3/2}\int_{\vec{k}}\overline{r_{jN,\beta}\,\frac{\partial}{\partial r_{jN,\alpha}}\left(\frac{1}{r^3_{jN}}\right)}\left[\int_{\vec{r}_j}\left[B^{m_1}_{n_1,l_1}(\zeta_1,\vec{r}_{jA})\right]^* e^{-i\vec{k}\cdot\vec{r}_{jN}}\,B^{m_2}_{n_2,l_2}(\zeta_2,\vec{r}_{jB})\,\mathrm{d}\vec{r}_j\right]\mathrm{d}\vec{k}. \tag{25}$$

In the case where α and β represent two different cartesian coordinates, the Fourier transform of the operator is given by [Safouhi (2010b)]:

$$\overline{r_\beta\,\frac{\partial}{\partial r_\alpha}\left(\frac{1}{r^3}\right)} = \sqrt{\frac{2}{\pi}}\,\frac{k_\alpha\,k_\beta}{k^2}. \tag{26}$$

In the case where α and β represent the same cartesian coordinate, the calculations leads to the potential $1/r^3$, which poses serious difficulties because of the singularity and its Fourier transform does not exist in a sense of classical analysis. This case is a part of ongoing research where the theory of generalized functions will be used in order to derive the Fourier transform of the operator in the generalized function sense.

Using the analytic expression (11) obtained by Trivedi and Steinborn [Trivedi & Steinborn (1983)] for the integrals over $\vec{r}_j$ involved in equation (25) and with the help of equation (26), one can derive the following analytic expression for the integrals ${}_B\mathcal{I}^{(\alpha,\beta)}_{13}$ [Safouhi (2010b)]:

$$\begin{aligned}{}_B\mathcal{I}^{(\alpha,\beta)}_{13} &= \frac{4^3\,\pi\,(2l_1+1)!!\,(2l_2+1)!!\,(n_1+l_1+n_2+l_2+1)!\,\zeta_1^{2n_1+l_1-1}\,\zeta_2^{2n_2+l_2-1}}{(n_1+l_1)!\,(n_2+l_2)!\,2^{n_1+n_2+l_1+l_2+2}}\\ &\times\sum_{l_1'=0}^{l_1}(-i)^{l_1+l_1'}\sum_{m_1'=-l_1'}^{l_1'}\frac{\langle l_1m_1|l_1'm_1'|l_1-l_1'm_1-m_1'\rangle}{(2l_1'+1)!![2(l_1-l_1')+1]!!}\end{aligned}$$

$$
\begin{aligned}
&\times \sum_{l_2'=0}^{l_2} (-i)^{l_2+l_2'} \sum_{m_2'=-l_2'}^{l_2'} \frac{\langle l_2 m_2 \,|l_2' m_2'|\, l_2 - l_2' m_2 - m_2' \rangle}{(2l_2'+1)!![2(l_2-l_2')+1]!!} \\
&\times \sum_{l'=l'_{\min},2}^{l_1'+l_2'} (-1)^{l_1'} \langle l_2' m_2' \,|l_1' m_1'|\, l' m_2' - m_1' \rangle \; R_2^{l'} \, Y_{l'}^{m_2'-m_1'}(\theta_{\vec{R}_2}, \varphi_{\vec{R}_2}) \\
&\times \sum_{l_{12}=l_{12min},2}^{l_1-l_1'+l_2-l_2'} \langle l_2 - l_2' m_2 - m_2' \,|l_1 - l_1' m_1 - m_1'|\, l_{12} m_{12} \rangle \\
&\times \sum_{m_3=-1}^{1} \sum_{m_4=-1}^{1} (-1)^{m_4} \, c_{\alpha,m_3} \, c_{\beta,m_4} \sum_{l''=l''_{\min,2}}^{2} \langle 1\, m_4 | 1\, m_3 | l''\, m_4 - m_3 \rangle \\
&\times \sum_{\lambda=\lambda_{\min,2}}^{l''+l_{12}} (-i)^{\lambda} < l_{12} m_{12} | l'' m_3 - m_4 | \lambda\, \mu > \\
&\times \sum_{j=0}^{\Delta l} \frac{(-2)^j \binom{\Delta l}{j}}{(n_1+n_2+l_1+l_2-j+1)!} \\
&\times \int_{s=0}^{1} s^{n_{22}} (1-s)^{n_{11}} \, Y_\lambda^{\mu}(\theta_{\vec{v}}, \varphi_{\vec{v}}) \\
&\times \left[\int_{x=0}^{+\infty} x^{n_x} \, \frac{\hat{k}_\nu\, [R_2\, \gamma(s,x)]}{[\gamma(s,x)]^{n_\gamma}} \, j_\lambda(v\, x) \, \mathrm{d}x \right] \mathrm{d}s, \qquad (27)
\end{aligned}
$$

where:

$$
\begin{aligned}
\vec{R}_1 &= \overrightarrow{AN}, \vec{R}_2 = \overrightarrow{AB} \quad \text{and} \quad \vec{v} = (1-s)\, \vec{R}_2 - \vec{R}_1 \\
\gamma(s,x) &= \sqrt{(1-s)\, \zeta_1^2 + s\, \zeta_2^2 + s\,(1-s)\, x^2} \\
n_\gamma &= 2(n_1+n_2+l_1+l_2) - (l_1'+l_2'+l') + 1 \\
n_{11} &= n_1 + l_1 + l_2 - l_2', \; n_{22} = n_2 + l_2 + l_1 - l_1' \\
\nu &= n_1 + n_2 + l_1 + l_2 - l' - j + \tfrac{1}{2} \qquad (28) \\
m_{12} &= (m_2 - m_2') - (m_1 - m_1') \\
n_k &= l_1 - l_1' + l_2 - l_2' + 2 \\
\mu &= m_{12} - m_3 + m_4 \\
\Delta l &= \tfrac{l_1'+l_2'-l'}{2}.
\end{aligned}
$$

In the case of one- and two-center integrals ${}_B\mathcal{I}_{11}^{(\alpha,\beta)}$ and ${}_B\mathcal{I}_{12}^{(\alpha,\beta)}$, corresponding to $A = B = N$ and $A = B \neq N$ respectively, we derived analytical expressions in [Slevinsky et al. (2010)].

For the one-center integrals ($A = B = N$):

$$
\begin{aligned}
{}_B\mathcal{I}_{11}^{(\alpha,\beta)} &= -\frac{\zeta_1^{l_1}\, \zeta_2^{l_2}\, \zeta_s^{2-l_1-l_2}}{\sqrt{\pi}\, 2^{2\,n_1+l_1+2\,n_2+l_2+3}\, (n_1+l_1)!\, (n_2+l_2)!} \\
&\times \sum_{m_3=-1}^{1} \sum_{m_4=-1}^{1} (-1)^{m_3} \, c_{\alpha,m_3} \, c_{\beta,m_4}
\end{aligned}
$$

$$\times \sum_{l=l_{\min},2}^{2} 2^{-l} \langle 1\, m_4 | 1\, m_3 | l\, m_4 - m_3 \rangle \langle l_1 m_1 | l_2 m_2 | l m_1 - m_2 \rangle \delta_{m_1-m_2, m_4-m_3}$$
$$\times \sum_{\tau=2}^{n_1+n_2} \sum_{\varsigma=\tau_1}^{\tau_2} \frac{2^{\tau} \zeta_1^{\varsigma-1} \zeta_2^{\tau-\varsigma-1}}{\zeta_s^{\tau}} \frac{(2n_1-\varsigma-1)!\,(2n_2-\tau+\varsigma-1)!\,(\tau+l_1+l_2)_{l+1}}{(\varsigma-1)!\,(n_1-\varsigma)!\,(\tau-\varsigma-1)!\,(n_2-\tau+\varsigma)!\,\Gamma(l+\frac{3}{2})}$$
$$\times \sum_{r=0}^{\eta'} \frac{(\frac{\eta}{2})_r (\frac{\eta+1}{2})_r (-r-\frac{l+1}{2})_{l_1+l_2+\tau-1}}{(l+\frac{3}{2})_r\, r!}, \tag{29}$$

where:

$$\begin{aligned} \tau_1 &= \max(1, \tau - n_2) \\ \tau_2 &= \min(n_1, \tau - 1) \\ \zeta_s &= \zeta_1 + \zeta_2 \\ \eta &= l - \tau - l_1 - l_2 + 1 \\ \eta' &= -\tfrac{\eta}{2} \quad \text{if} \quad \eta \quad \text{is even otherwise} \quad \eta' = -\tfrac{\eta+1}{2}. \end{aligned} \tag{30}$$

For the two-center integrals ($A = B \neq N$):

$${}_B\mathcal{I}_{12}^{(\alpha,\beta)} = \frac{\sqrt{\pi}\, \zeta_1^{l_1} \zeta_2^{l_2} \zeta_s^{l_1+l_2-1} R^{2l_1+2l_2-3}}{4^{n_1+l_1+n_2+l_2} (n_1+l_1)!\,(n_2+l_2)!} \sum_{l=l_{min},2}^{l_1+l_2} 2^{-l} \langle l_1 m_1 | l_2 m_2 | l m_1 - m_2 \rangle$$
$$\times \sum_{m_3=-1}^{1} \sum_{m_4=-1}^{1} (-1)^{m_3} c_{\alpha,m_3} c_{\beta,m_4} \sum_{l'=l'_{\min},2}^{2} \langle 1\, m_4 | 1\, m_3 | l'\, m_4 - m_3 \rangle$$
$$\times \sum_{\lambda=\lambda_{\min},2}^{l+l'} (R\,\zeta_s)^{-\lambda} \langle l' m_4 - m_3 | l m_1 - m_2 | \lambda m_4 - m_3 - m_1 + m_2 \rangle Y_{\lambda}^{m_4-m_3-m_1+m_2}(\theta_{\vec{R}}, \varphi_{\vec{R}})$$
$$\times \sum_{\tau=2}^{n_1+n_2} \sum_{\varsigma=\tau_1}^{\tau_2} \frac{R^{2\tau} (2n_1-\varsigma-1)!\,(2n_2-\tau+\varsigma-1)!\,\zeta_1^{\varsigma-1} \zeta_2^{\tau-\varsigma-1} \zeta_s^{\tau} (\tau+l_1+l_2)_{l+1}}{(\varsigma-1)!\,(n_1-\varsigma)!\,(\tau-\varsigma-1)!\,(n_2-\tau+\varsigma)!\,\Gamma(l+\frac{3}{2})}$$
$$\times \sum_{r=0}^{\eta'} \frac{(\frac{\eta}{2})_r (\frac{\eta+1}{2})_r}{(l+\frac{3}{2})_r\, r!} \sum_{s=0}^{r+\frac{l-\lambda}{2}} \binom{r+\frac{l-\lambda}{2}}{s} \frac{2^s (1-l_1-l_2-\tau)_s}{(R\,\zeta_s)^{2s}} \hat{k}_{\lambda-l_1-l_2-\tau+s+\frac{3}{2}}(R\,\zeta_s). \tag{31}$$

Second order integrals

The second order integrals $\mathcal{I}_{23}$ are given by (22). For simplicity and without loss of generality, we assume $A = O$ in equation (25).

The operator involved in (22) is given by:

$$\frac{\vec{r}_j \cdot \vec{r}_{jN} \delta_{\alpha\beta} - r_{jN,\alpha}\, r_{j,\beta}}{r_{jN}^3} = \begin{cases} \dfrac{-r_{jN,\alpha}\, r_{j,\beta}}{r_{jN}^3} & \text{when } \alpha \neq \beta \\[2ex] \dfrac{r_{jN,u}\, r_{j,u}}{r_{jN}^3} + \dfrac{r_{jN,v}\, r_{j,v}}{r_{jN}^3} & \text{when } \alpha = \beta \quad \text{and} \quad u, v \neq \alpha \in \{x, y, z\}. \end{cases} \tag{32}$$

From the above equation, it is obvious that the integrals $\mathcal{I}_{23}$ can be expressed in terms of integrals of the form:

$$\mathcal{I}_{23}^{\alpha\beta} = \int_{\vec{r}_j} \left[\chi_{n_1,l_1}^{m_1}(\zeta_1, \vec{r}_j) \right]^* \frac{r_{jN,\alpha}\, r_{j,\beta}}{r_{jN}^3} \chi_{n_2,l_2}^{m_2}(\zeta_2, \vec{r}_{jB})\, \mathrm{d}\vec{r}_j. \tag{33}$$

The cartesian coordinate $r_{jN,\alpha}$ and $r_{j,\beta}$ can be expressed in terms of spherical harmonics and r_{jN} and r_j respectively as follows:

$$r_{jN,\alpha} = r_{jN} \sum_{\mu_1=-1}^{1} c_{\alpha,\mu_1} Y_1^{\mu_1}(\theta_{\vec{r}_{jN}}, \phi_{\vec{r}_{jN}}) \quad \text{and} \quad r_{j,\beta} = r_j \sum_{\mu_2=-1}^{1} c_{\beta,\mu_2} Y_1^{\mu_2}(\theta_{\vec{r}_j}, \phi_{\vec{r}_j}), \tag{34}$$

where the coefficients $c_{\alpha,\mu}$ are given as follows:

$$\begin{cases} c_{x,-1} = \sqrt{\frac{2\pi}{3}}, & c_{y,-1} = i\sqrt{\frac{2\pi}{3}} & \text{and } c_{z,-1} = 0 \\ c_{x,0} \;\;= 0, & c_{y,0} \;\;= 0 & \text{and } c_{z,0} \;\;= \sqrt{\frac{4\pi}{3}} \\ c_{x,1} \;\;= -\sqrt{\frac{2\pi}{3}}, & c_{y,1} \;\;= i\sqrt{\frac{2\pi}{3}} & \text{and } c_{z,1} \;\;= 0. \end{cases} \tag{35}$$

Using the analytic expression of the Unormalized STFs (2), one can obtain:

$$\begin{aligned} r_{j,\beta} \left[\chi_{n_1,l_1}^{m_1}(\zeta_1, \vec{r}_j)\right]^* &= \left[r_j \sum_{\mu_2=-1}^{1} c_{\beta,\mu_2} Y_1^{\mu_2}(\theta_{\vec{r}_j}, \phi_{\vec{r}_j})\right] r_j^{n_1-1} \mathrm{e}^{-\zeta_1 r_j} \left[Y_{l_1}^{m_1}(\theta_{\vec{r}_j}, \varphi_{\vec{r}_j})\right]^* \\ &= \sum_{\mu_2=-1}^{1} c_{\beta,\mu_2} r_j^{n_1} \mathrm{e}^{-\zeta_1 r_j} \left[Y_{l_1}^{m_1}(\theta_{\vec{r}_j}, \varphi_{\vec{r}_j})\right]^* Y_1^{\mu_2}(\theta_{\vec{r}_j}, \phi_{\vec{r}_j}). \end{aligned} \tag{36}$$

The product of two spherical harmonics can be linearized by Gaunt coefficients as follows:

$$\left[Y_{l_1}^{m_1}(\theta, \varphi)\right]^* Y_{l_2}^{m_2}(\theta, \varphi) = \sum_{l=l_{\min},2}^{l_1+l_2} \langle l_2\, m_2 | l_1\, m_1 | l\, m_2 - m_1 \rangle\, Y_l^{m_2-m_1}(\theta, \varphi), \tag{37}$$

where the summation index l runs in steps of 2 from $l_{\min}$ to $l_1 + l_2$. The constant $l_{\min}$ is given by [Weniger & Steinborn (1982)]:

$$l_{\min} = \begin{cases} \max(|l_1 - l_2|, |m_2 - m_1|) & \text{if } l_1 + l_2 + \max(|l_1 - l_2|, |m_2 - m_1|) \text{ is even} \\ \max(|l_1 - l_2|, |m_2 - m_1|) + 1 & \text{if } l_1 + l_2 + \max(|l_1 - l_2|, |m_2 - m_1|) \text{ is odd.} \end{cases} \tag{38}$$

From equation (37), it follows that:

$$\begin{aligned} r_{j,\beta} \left[\chi_{n_1,l_1}^{m_1}(\zeta_1, \vec{r}_j)\right]^* &= \sum_{\mu_2=-1}^{1} c_{\beta,\mu_2} r_j^{n} \mathrm{e}^{-\zeta r_j} \sum_{l=l_{\min},2}^{l_1+1} \langle 1\, \mu_2 | l_1\, m_1 | l\, \mu_2 - m_1 \rangle\, Y_l^{\mu_2-m_1}(\theta_{\vec{r}_j}, \varphi_{\vec{r}_j}) \\ &= \sum_{\mu_2=-1}^{1} \sum_{l=l_{\min},2}^{l_1+1} c_{\beta,\mu_2} \langle 1\, \mu_2 | l_1\, m_1 | l\, \mu_2 - m_1 \rangle\, (-1)^{\mu_2-m_1} \left[\chi_{n_1+1,l}^{m_1-\mu_2}(\zeta_1, \vec{r}_j)\right]^*. \end{aligned} \tag{39}$$

Using equations (34) and (39), we obtain the following relation:

$$\begin{aligned} \mathcal{I}_{23}^{\alpha\beta} = &\sum_{\mu_2=-1}^{1} \sum_{l=l_{\min},2}^{l_1+1} \sum_{\mu_1=-1}^{1} (-1)^{\mu_2-m_1} c_{\beta,\mu_2} c_{\alpha,\mu_1} \langle 1\, \mu_2 | l_1\, m_1 | l\, \mu_2 - m_1 \rangle \\ &\times \int_{\vec{r}_j} \left[\chi_{n_1+1,l}^{m_1-\mu_2}(\zeta_1, \vec{r}_j)\right]^* \frac{Y_1^{\mu_1}(\theta_{\vec{r}_{jN}}, \phi_{\vec{r}_{jN}})}{r_{jN}^2} \chi_{n_2,l_2}^{m_2}(\zeta_2, \vec{r}_{jB})\, \mathrm{d}\vec{r}_j. \end{aligned} \tag{40}$$

Let $\tilde{\mathcal{I}}_{23}^{\alpha\beta}$ represent the integrals involved in the above equation. Using equation (5), these integrals can be expressed as linear combinations of integrals ${}_B\hat{\mathcal{I}}_{23}^{\alpha\beta}$ over B functions of the form:

$${}_B\hat{\mathcal{I}}_{23}^{\alpha\beta} = \int_{\vec{r}_j} \left[B_{\tilde{n}_1,l}^{\tilde{m}_1}(\zeta_1, \vec{r}_j)\right]^* \frac{Y_1^{\mu_1}(\theta_{\vec{r}_{jN}}, \phi_{\vec{r}_{jN}})}{r_{jN}^2} B_{n_2,l_2}^{m_2}(\zeta_2, \vec{r}_{jB})\, d\vec{r}_j, \tag{41}$$

where $\tilde{m}_1 = m_1 - \mu_2$ and $\tilde{n}_1 = n_1 + 1$.

Using the Fourier transform method, we obtain:

$${}_B\hat{\mathcal{I}}_{23}^{\alpha\beta} = (2\,\pi)^{-3/2} \int_{\vec{k}} \overline{\frac{Y_1^{\mu_1}(\theta_{\vec{r}_{jN}}, \phi_{\vec{r}_{jN}})}{r_{jN}^2}} \left[\int_{\vec{r}_j} \left[B_{\tilde{n}_1,l}^{\tilde{m}_1}(\zeta_1, \vec{r}_j)\right]^* e^{-i\,\vec{k}\cdot\vec{r}_{jN}} B_{n_2,l_2}^{m_2}(\zeta_2, \vec{r}_{jB})\, d\vec{r}_j\right] d\vec{k}. \tag{42}$$

The Fourier transform of the operator involved in the above integrals is given by [Berlu & Safouhi (2008)]:

$$\overline{\left(\frac{Y_1^{\mu}(\theta_{\vec{r}_{jN}}, \phi_{\vec{r}_{jN}})}{r_{jN}^2}\right)}(\vec{k}) = -i\sqrt{\frac{2}{\pi}}\, \frac{Y_1^{\mu}(\theta_{\vec{k}}, \phi_{\vec{k}})}{k}. \tag{43}$$

Using the analytic expression (11) obtained by Trivedi and Steinborn [Trivedi & Steinborn (1983)] for the integrals over $\vec{r}_j$ involved in equation (42) and equation (43), one can derive the following analytic expression for the integrals ${}_B\hat{\mathcal{I}}_{23}^{\alpha\beta}$ [Berlu & Safouhi (2008)]:

$$\begin{aligned}
{}_B\hat{\mathcal{I}}_{23}^{\alpha\beta} &= 8\,(4\pi)^2\,(2l_1+1)!!\,(2l_2+1)!!\,\frac{(\tilde{n}_1+n_2+l_1+l_2+1)!}{(\tilde{n}_1+l_1)!\,(n_2+l_2)!}\,\frac{\zeta_1^{2\tilde{n}_1+l_1-1}\,\zeta_2^{2n_2+l_2-1}}{2^{\tilde{n}_1+n_2+l_1+l_2+1}} \\
&\times \sum_{l_1'=0}^{l_1}\sum_{m_1'=-l_1'}^{l_1'} (-i)^{l_1+l_1'}\frac{\langle l_1, \tilde{m}_1 | l_1', m_1' | l_1-l_1', \tilde{m}_1-m_1'\rangle}{(2l_1'+1)!![2(l_1-l_1')+1]!!} \\
&\times \sum_{l_2'=0}^{l_2}\sum_{m_2'=-l_2'}^{l_2'} (-i)^{l_2+l_2'}\frac{\langle l_2, m_2 | l_2', m_2' | l_2-l_2', m_2-m_2'\rangle}{(2l_2'+1)!![2(l_2-l_2')+1]!!} \\
&\times \sum_{l'=l'_{\min},2}^{l_1'+l_2'} (-1)^{l_1'}\,\langle l_2', m_2' | l_1', m_1' | l' m_2', -m_1'\rangle\, R_2^{l'}\, Y_{l'}^{m_2'-m_1'}(\theta_{\overrightarrow{OB}}, \varphi_{\overrightarrow{OB}}) \\
&\times \sum_{l_{12}=l_{12\min},2}^{l_1-l_1'+l_2-l_2'} \langle l_2-l_2', m_2-m_2' | l_1-l_1', \tilde{m}_1-m_1' | l_{12}, m_{12}\rangle \\
&\times \sum_{\lambda=\lambda_{\min},2}^{1+l_{12}} (-i)^{\lambda+1}\,\langle \lambda, \mu | l_{12}, m_{12} | 1, \mu_1\rangle \\
&\times \sum_{j=0}^{\Delta l}\binom{\Delta l}{j}\frac{(-2)^j}{(\tilde{n}_1+n_2+l_1+l_2-j+1)!} \\
&\times \int_{s=0}^{1} s^{n_2+l_2+l_1-l_1'}(1-s)^{\tilde{n}_1+l_1+l_2-l_2'}\,Y_\lambda^{\mu}(\theta_{\vec{v}}, \varphi_{\vec{v}}) \\
&\times \left[\int_{x=0}^{+\infty} x^{n_x}\,\frac{\hat{k}_\nu[\gamma(s,x)\,R_2]}{\gamma(s,x)^{n_\gamma}}\,j_\lambda(vx)\,dx\right] ds,
\end{aligned} \tag{44}$$

where:

$$
\begin{aligned}
&m_{12} = (m_2 - m_2') - (\tilde{m}_1 - m_1') \text{ and } \mu = \mu_1 + m_{12} \\
&n_\gamma = 2(\tilde{n}_1 + n_2 + l_1 + l_2) - (l_1' + l_2' + l') + 1 \\
&\gamma(s,x) = \sqrt{s\zeta_2^2 + (1-s)\zeta_1^2 + s(1-s)x^2} \\
&\vec{v} = (1-s)\,\overrightarrow{OB} - \overrightarrow{ON} \quad \text{and} \quad \text{and } v = |\vec{v}| \\
&\nu = \tilde{n}_1 + n_2 + l_1 + l_2 - l' - j + \frac{1}{2} \\
&n_x = l_1 - l_1' + l_2 - l_2' + 1 \\
&\Delta l = (l_1' + l_2' - l')/2.
\end{aligned}
\tag{45}
$$

n_1	l_1	m_1	ζ_1	n_2	l_2	m_2	ζ_2	${}_B\mathcal{I}_2^{(x,y)}$
2	1	-1	2.0	2	1	-1	1.5	.844938515752529(-3)
2	1	0	2.0	2	1	0	1.5	.197895122059999(-3)
3	2	-1	2.0	2	1	-1	1.5	-.183961338177373(-4)
3	2	1	2.0	3	2	1	1.5	.314124180426832(-4)
3	2	2	2.0	3	2	2	1.5	.132485365811557(-3)
4	2	1	2.0	2	1	1	1.5	-.361569492726092(-4)
4	2	1	2.0	3	2	1	1.5	.764186724377223(-4)
4	2	2	2.0	3	2	2	1.5	.331525812542680(-3)
4	2	0	2.0	4	2	0	1.5	-.297315128733290(-4)
4	2	1	2.0	4	2	1	1.5	.215564810358993(-3)
4	2	2	2.0	4	2	2	1.5	.951453780055960(-3)

Table 1. Evaluation of ${}_B\mathcal{I}_1^{(x,y)}$ (29).

n_1	l_1	m_1	ζ_1	n_2	l_2	m_2	ζ_2	${}_B\mathcal{I}_2^{(y,z)}$
2	1	1	1.0	2	1	-1	1.5	-.980852389239356(-3)
3	2	1	1.0	2	1	-1	1.5	-.319537014904337(-1)
3	2	1	1.0	2	1	1	1.5	.858668805557491(-2)
3	2	1	1.0	3	2	-1	1.5	-.219369423462985(-1)
3	2	2	1.0	3	2	-2	1.5	.391684472826858(-4)
3	2	2	1.0	3	2	-1	1.5	-.204743121872522(-1)
3	2	2	1.0	3	2	1	1.5	.163009587528461(2)
4	2	1	1.0	2	1	-1	1.5	-.414819029773496(-1)
4	2	2	1.0	3	2	-2	1.5	.441086792217064(-4)
4	2	2	1.0	3	2	-1	1.5	-.298416140021138(-1)
4	2	2	1.0	3	2	1	1.5	.291083584089616(2)

Table 2. Evaluation of ${}_B\mathcal{I}_2^{(y,z)}$ (31). $\vec{R} = (1.5, 75^o, 0^o)$ in spherical coordinates.

n_1	l_1	m_1	ζ_1	n_2	l_2	m_2	ζ_2	${}_B\mathcal{I}_3^{(x,z)}$
2	1	-1	2.0	2	1	1	1.0	.769304745002408(-4)
2	1	1	2.0	2	1	1	1.0	-.620498371396534(-4)
3	2	1	2.0	2	1	0	1.0	-.146444907098019(-5)
3	2	1	2.0	2	1	1	1.0	-.576608712417475(-5)
3	2	1	2.0	3	2	1	1.0	.137450124735083(-6)
3	2	1	2.0	3	2	-1	1.0	-.117106174581176(-6)
3	2	2	2.0	3	2	-2	1.0	.287818321676003(-6)
3	2	2	2.0	3	2	1	1.0	.143900856736229(-6)
3	2	2	2.0	3	2	2	1.0	-.121032718682770(-5)
4	2	-1	2.0	3	2	1	1.0	-.359169492307861(-6)
4	2	1	2.0	3	2	1	1.0	.343245699707943(-6)

Table 3. Evaluation of $\mathcal{I}_3^{(x,z)}$ (27). $\vec{R}_1 = (12.0, 90^o, 0^o)$ and $\vec{R}_2 = (2.0, 90^o, 0^o)$ in spherical coordinates.

Conclusion

In the present review, we showed how the Fourier transformation method allowed the derivation of compact formulae for one of the most challenging integrals, namely molecular multi-center integrals and NMR multi-center integrals.

Analytic expressions are obtained for integrals of the paramagnetic contribution in the relativistic calculation of the shielding tensor as well as integrals of second order in the non-relativistic calculation of the shielding tensor. The basis set of ETFs is used and it is well known that these functions are better suited than GTFs.

The obtained analytic expressions for the one- and two-center integrals can be computed easily and no quadrature rule is required. In the case of the three-center integrals, we need to compute semi-infinite integrals involving oscillatory functions. These oscillatory integrals can be computed to a high pre-determined accuracy using existing methods and algorithms based on extrapolation methods and numerical quadrature [Berlu & Safouhi (2003); Duret & Safouhi (2007); Safouhi (2001b; 2004; 2010a); Slevinsky & Safouhi (2009)].

Numerical tables for the NMR integrals of interest can be found in [Safouhi (2010b); Slevinsky et al. (2010)].

Acknowledgment

The author acknowledges the financial support for this research by the Natural Sciences and Engineering Research Council of Canada (NSERC).

References

Agmon, S. (1985). *Bounds on exponential decay of eigenfunctions of Schrödinger operators, in S. Graffi (editor), Schrödinger operators,* Springer-Verlag, Berlin.

Barnett, M. (1990). Molecular integrals over Slater orbitals, *Chem. Phys. Lett.* 166: 65–70.

Berlu, L. & Safouhi, H. (2003). An extremely efficient and rapid algorithm for a numerical evaluation of three-center nuclear attraction integrals over Slater type functions, *J. Phys. A: Math. Gen.* 36: 11791–11805.

Berlu, L. & Safouhi, H. (2008). Analytical treatment of nuclear magnetic shielding tensor integrals over exponential type functions, *J. Theor. Comp. Chem.* 7: 1215–1225.

Boys, S. (1950a). Electronic Wave Functions. I. A General Method of Calculation for the Stationary States of Any Molecular System , *Proc. R. Soc. Lond. Series A, Math. & Phys. Sciences.* 200: 542–554.

Boys, S. (1950b). Electronic Wave Functions. II. A Calculation for the Ground State of the Beryllium Atom, *Proc. R. Soc. Lond. Series A, Math. & Phys. Sciences.* 201: 125–137.

Condon, E. & Shortley, G. (1951). *The theory of atomic spectra*, Cambridge University Press, Cambridge, England.

Dickson, R. & Ziegler, T. (1996). NMR spin-spin coupling constants from density functional theory with Slater-type basis functions, *J. Phys. Chem.* 100: 5286–5290.

Ditchfield, R. (1974). Self-consistent perturbation-theory of diamagnetism. I. Gauge-invariant LCAO method for NMR chemical-shifts, *Molecular Physics* 27: 789–807.

Duret, S. & Safouhi, H. (2007). The W algorithm and the $\bar{D}$ transformation for the numerical evaluation of three-center nuclear attraction integrals, *Int. J. Quantum Chem.* 107: 1060–1066.

Fernández, J., López, R., Aguado, A., Ema, I. & Ramírez, G. (2001). SMILES Slater Molecular Integrals for Large Electronic Systems: New Program for Molecular Calculations with Slater Type Orbitals, *Int. J. Quant. Chem.* 81: 148–153.

Filter, E. & Steinborn, E. (1978a). Extremely compact formulas for molecular one-electron integrals and Coulomb integrals over Slater-type orbitals, *Phys. Rev. A.* 18: 1–11.

Filter, E. & Steinborn, E. (1978b). The three-dimensional convolution of reduced Bessel functions of physical interest, *J. Math. Phys.* 19: 79–84.

Fukui, H. & Baba, T. (1998). Calculation of nuclear magnetic shieldings XII. Relativistic no-pair equation, *J. Chem. Phys.* 108: 3854–3862.

Fukui, H., Baba, T., Shiraishi, Y., Imanishi, S., Kudo, K., Mori, K. & Shimoji, M. (2004). Calculation of nuclear magnetic shieldings: infinite-order Foldy-Wouthuysen transformation, *Mol. Phys.* 102: 641–648.

Gaunt, J. (1929). The triplets of helium, *Phil. Trans. Roy. Soc.* A. 228: 151–196.

Geller, M. (1962). Two-center, nonintegral, Slater-orbital calculations: Integral formulation and application to the Hydrogen molecule-ion, *J. Chem. Phys.* 36: 2424–2428.

Grotendorst, J. & Steinborn, E. (1988). Numerical evaluation of molecular one- and two-electron multicenter integrals with exponential-type orbitals via the Fourier-transform method, *Phys. Rev. A.* 38: 3857–3876.

Helgaker, T., Jaszuński, M. & Ruud, K. (1999). Ab initio methods for the calculation of NMR shielding and indirect spin-spin coupling constants, *Chem. Rev.* 99: 293–352.

Homeier, H. & Steinborn, E. (1993). Programs for the evaluation of nuclear attraction integrals with B functions, *Comput. Phys. Commun.* 77: 135–151.

Ishida, K. (2003). Molecular integrals over the gauge-including atomic orbitals. II. The Breit-Pauli interaction, *J. Comput. Chem.* 24: 1874–1890.

Kato, T. (1957). On the eigenfunctions of many-particle systems in quantum mechanics, *Commun. Pure Appl. Math.* 10: 151–177.

Kutzelnigg, W. (1988). Present and futur trends in quantum chemical calculations, *J. Mol. Struc (THEOCHEM)* 50: 33–54.

London, F. (1937). The quantic theory of inter-atomic currents in aromatic combinations, *Journal de Physique et le Radium* 8: 397–409.

Niehaus, T., López, R. & Rico, J. (2008). Efficient evaluation of the Fourier transform over products of Slater-type orbitals on different centers, *J. Phys. A: Math. Theor.* 41: 485205–485219.

Niukkanen, A. (1984). Fourier transforms of atomic orbitals. I. Reduction to fourdimensional harmonics and quadratic transformations, *Int. J. Quantum Chem.* 25: 941–955.

Ozdogan, T. & (Editors), M. R. (2008). *Recent Advances in Computational Chemistry. Molecular Integrals over Slater Orbitals*, Published by Transworld Research Network., Kerala, India.

Pople, J., McIver Jr, J. & Ostlund, N. (1968). Self-consistent perturbation theory II. Nuclear spin coupling constants, *J. Chem. Phys.* 49: 2960–2965.

Prosser, F. & Blanchard, C. (1962). On the evaluation of two-center integrals, *J. Chem. Phys.* 36: 1112–1112.

Pyykkö, P. (1988). Relativistic effects in structural chemistry, *Chem. Rev.* 88: 563–594.

Rico, F., López, R., Aguado, A., Ema, I. & Ramírez, G. (1998). Reference program for molecular calculations with Slater-type orbitals, *Int. J. Quantum Chem.* 19: 1284–1293.

Rico, J., Fernández, J., Ema, I., López, R. & Ramírez, G. (1999). Master formulas for two-amd three-center one electron integrals involving cartesian GTO, STO, and BTO, *Int. J. Quantum Chem.* 78: 83–93.

Rico, J., Fernández, J., Ema, I., López, R. & Ramírez, G. (2001). Four-center integrals for gaussian and exponential functions, *Int. J. Quant. Chem.* 81: 16–18.

Safouhi, H. (2001a). An extremely efficient approach for accurate and rapid evaluation of three- center two-electron Coulomb and hybrid integral over *B* functions, *J. Phys. A: Math. Gen.* 34: 881–902.

Safouhi, H. (2001b). The properties of sine, spherical Bessel and reduced Bessel functions for improving convergence of semi-infinite very oscillatory integrals: The evaluation of three-center nuclear attraction integrals over B functions, *J. Phys. A: Math. Gen.* 34: 2801–2818.

Safouhi, H. (2004). Highly accurate numerical results for three-center nuclear attraction and two-electron Coulomb and exchange integrals over Slater type functions, *Int. J. Quantum Chem.* 100: 172–183.

Safouhi, H. (2010a). Bessel, sine and cosine functions and extrapolation methods for computing molecular multi-center integrals, *Numer. Algor.* 54: 141–167.

Safouhi, H. (2010b). Integrals of the paramagnetic contribution in the relativistic calculation of the shielding tensor, *J. Math. Chem.* 48: 601–616.

Schreckenbach, G. & Ziegler, T. (1995). Calculation of NMR shielding tensors using gauge-including atomic orbitals and modern density functional theory, *J. Phys. Chem.* 99: 606–611.

Shavitt, I. (1963). *The Gaussian function in calculation of statistical mechanics and quantum mechanics, Methods in Computational Physics.2. Quantum.Mechanics*, edited by B. Alder, S. Fernbach, M. Rotenberg, Academic Press, New York.

Slater, J. (1932). Analytic atomic wave functions, *Phys. Rev.* 42: 33–43.

Slevinsky, R. & Safouhi, H. (2009). The *S* and *G* transformations for computing three-center nuclear attraction integrals, *Int. J. Quantum Chem.* 109: 1741–1747.

Slevinsky, R., Temga, T., Mouattamid, M. & Safouhi, H. (2010). One- and two-Center ETF-integrals of first order in relativistic calculation of NMR parameters, *J. Phys. A: Math. Theor.* 43: 225202 (14pp).

Steinborn, E. & Filter, E. (1975). Translations of fields represented by spherical-harmonics expansions for molecular calculations. III. Translations of reduced Bessel functions, Slater-type s-orbitals, and other functions, *Theor. Chim. Acta.* 38: 273–281.

Steinborn, E., Homeier, H., Ema, I., López, R. & Ramírez, G. (2000). Molecular calculations with *B* functions, *Int. J. Quantum Chem.* 76: 244–251.

Trivedi, H. & Steinborn, E. (1983). Fourier transform of a two-center product of exponential-type orbitals. Application to one- and two-electron multicenter integrals, *Phys. Rev. A.* 27: 670–679.

Vaara, J. (2007). Theory and computation of nuclear magnetic resonance parameters, *Phys. Chem. Chem. Phys.* 9: 5399–5418.

Watson, M., Handy, N., Cohen, A. & Helgaker, T. (2004). Density-functional generalized-gradient and hybrid calculations of electromagnetic properties using Slater basis sets, *J. Chem. Phys.* 120: 7252–7261.

Weatherford, C. & Jones, H. (1982). *ETO Multicenter Molecular Integrals*, Reidel, Dordrecht.

Weniger, E. (1982). *Reduzierte Bessel-Funktionen als LCAO-Basissatz: Analytische und numerische Untersuchungen, publisher = Ph.D.thesis, Universität Regensburg.*

Weniger, E. (1985). Weakly convergent expansions of a plane wave and their use in Fourier integrals, *J. Math. Phys.* 26: 276–291.

Weniger, E. (2005). The spherical tensor gradient operator, *Collect. Czech. Chem. Commun.* 70: 1125–1271.

Weniger, E. & Steinborn, E. (1982). Programs for the coupling of spherical harmonics, *Comput. Phys.Commun.* 25: 149–157.

Weniger, E. & Steinborn, E. (1983a). Numerical properties of the convolution theorems of B functions, *Phys. Rev. A.* 28: 2026–2041.

Weniger, E. & Steinborn, E. (1983b). The Fourier transforms of some exponential-type functions and their relevance to multicenter problems, *J. Chem. Phys.* 78: 6121–6132.

2

Application of Hankel Transform for Solving a Fracture Problem of a Cracked Piezoelectric Strip Under Thermal Loading

Sei Ueda
Osaka Institute of Technology
Japan

Introduction

In this chapter, an example of the application of Hankel transform for solving a fracture problem will be explained. In discussing axisymmetric problems, it is advantageous to use polar coordinates, and the Hankel transform method is powerful to solve the general equations in polar coordinates. A brief account of the Hankel transform will be given. Here f is a function of r, its transform is indicated by a capital F, J_ν is the ν th order Bessel function of the first kind, and the nature of the transformation either by a suffix or by a characteristic new variable s. It will be assumed without comment that the integrals in question exist, and that, if necessary, the functions and their derivatives tend to zero as the variable tends to infinity. The Hankel transform of order $\nu > -1/2$, $H_\nu[f(r)]$ or $F_\nu(s)$, of a function $f(r)$ is defined as

$$H_\nu[f(r)] \equiv F_\nu(s) = \int_0^\infty r J_\nu(sr) f(r) dr$$

and its inversion formula is

$$f(r) = \int_0^\infty s J_\nu(sr) F_\nu(s) ds$$

Also, integrating by parts twice gives

$$H_\nu\left[\frac{d^2 f}{dr^2} + \frac{1}{r}\frac{df}{dr} - \frac{\nu^2}{r^2} f\right] = -s^2 F_\nu(s)$$

provided that $rf(r)$ and $rdf(r)/dr$ tend to zero as $r \to 0$ and as $r \to \infty$.

The piezoelectric materials have attracted considerable attention recently. Owing to the coupling effect between the thermo-elastic and electric fields in piezoelectric materials, thermo-mechanical disturbances can be determined form measurement of the induced electric potential, and the ensuing response can be controlled through application of an appropriate electric field (Rao & Sunar, 1994). For successful and efficient utilization of piezoelectric as sensors and actuators in intelligent systems, several researches on piezo-thermo-elastic behavior have been reported (Tauchert, 1992).

Moreover a better understanding of the mechanics of fracture in piezoelectric materials under thermal load conditions is needed for the requirements of reliability and lifetime of these systems. Using the Fourier transform, the present author studied the thermally induced fracture of a piezoelectric strip with a two-dimensional crack (Ueda, 2006a, 2006b).

Here the mixed-mode thermo-electro-mechanical fracture problem for a piezoelectric material strip with a penny-shaped crack is considered. It is assumed that the strip is under the thermal loading. The crack faces are supposed to be insulated thermally and electrically. By using the Hankel transform (Sneddon & Lowengrub, 1969), the thermal and electro-mechanical problems are reduced to a singular integral equation and a system of singular integral equations (Erdogan & Wu, 1996), respectively, which are solved numerically (Sih, 1972). Numerical calculations are carried out, and detailed results are presented to illustrate the influence of the crack size and the crack location on the stress and electric displacement intensity factors. The temperature, stress and electric displacement distributions are also presented.

Formulation of the problem

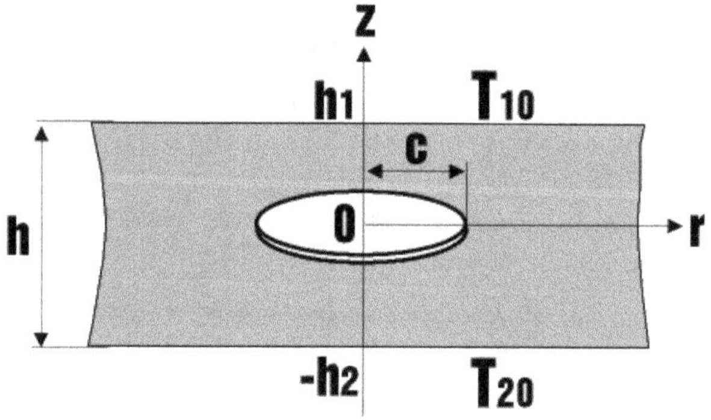

Fig. 1. Penny-shaped crack in a piezoelectric strip

A penny-shaped crack of radius c is embedded in an infinite long piezoelectric strip of thickness $h = h_1 + h_2$ as shown in Figure 1. The crack is located parallel to the boundaries and at an arbitrary position in the strip, and the crack faces are supposed to be insulated thermally and electrically. The cylindrical coordinate system is denoted by (r,θ,z) with its origin at the center of the crack face and the plane $r-\theta$ along the crack plane, where z is the poling axis. It is assumed that uniform temperatures T_{10} and T_{20} are maintained over the stress-free boundaries. In the following, the subscripts r,θ,z will be used to refer to the direction of coordinates. The material properties, such as the elastic stiffness constants, the piezoelectric constants, the dielectric constants, the stress-temperature coefficients, the coefficients of heat conduction, and the pyroelectric constant, are denoted by c_{kl}, e_{kl}, ε_{kk}, λ_{kk} $(k,l=1,2,3)$, κ_r, κ_z, and p_z, respectively.

The constitutive equations for the elastic field are

$$\left.\begin{aligned}\sigma_{rri} &= c_{11}\frac{\partial u_{ri}}{\partial r} + c_{12}\frac{u_{ri}}{r} + c_{13}\frac{\partial u_{zi}}{\partial z} + e_{31}\frac{\partial \phi_i}{\partial z} - \lambda_{11}T_i,\\ \sigma_{\theta\theta i} &= c_{12}\frac{\partial u_{ri}}{\partial r} + c_{11}\frac{u_{ri}}{r} + c_{13}\frac{\partial u_{zi}}{\partial z} + e_{31}\frac{\partial \phi_i}{\partial z} - \lambda_{11}T_i,\\ \sigma_{zzi} &= c_{13}\frac{\partial u_{ri}}{\partial r} + c_{13}\frac{u_{ri}}{r} + c_{33}\frac{\partial u_{zi}}{\partial z} + e_{33}\frac{\partial \phi_i}{\partial z} - \lambda_{33}T_i,\\ \sigma_{zri} &= c_{44}\left(\frac{\partial u_{zi}}{\partial r} + \frac{\partial u_{ri}}{\partial z}\right) + e_{15}\frac{\partial \phi_i}{\partial r}\end{aligned}\right\} \quad (i=1,2) \tag{1}$$

where $T_i(r,z)$ is the temperature, $\phi_i(r,z)$ is the electric potential, $u_{ri}(r,z)$, $u_{zi}(r,z)$ are the displacement components, $\sigma_{rri}(r,z)$, $\sigma_{\theta\theta i}(r,z)$, $\sigma_{zzi}(r,z)$, $\sigma_{zri}(r,z)\,(i=1,2)$ are the stress components. The subscript $i=1,2$ denotes the thermo-electro-elastic fields in $0 \le z \le h_1$ and $-h_2 \le z \le 0$, respectively. For the electric field, the constitutive relations are

$$\left.\begin{aligned}D_{ri} &= e_{15}\left(\frac{\partial u_{zi}}{\partial r} + \frac{\partial u_{ri}}{\partial z}\right) - \varepsilon_{11}\frac{\partial \phi_i}{\partial r},\\ D_{zi} &= e_{31}\frac{\partial u_{ri}}{\partial r} + e_{31}\frac{u_{ri}}{r} + e_{33}\frac{\partial u_{zi}}{\partial z} - \varepsilon_{33}\frac{\partial \phi_i}{\partial z} + p_z T_i\end{aligned}\right\} \quad (i=1,2) \tag{2}$$

where $D_{ri}(r,z)$, $D_{zi}(r,z)\,(i=1,2)$ are the electric displacement components.

The governing equations for the thermo-electro-elastic fields of the medium may be expressed as follows:

$$\kappa^2\left(\frac{\partial^2 T_i}{\partial r^2} + \frac{1}{r}\frac{\partial T_i}{\partial r}\right) + \frac{\partial^2 T_i}{\partial z^2} = 0 \quad (i=1,2) \tag{3}$$

$$\left.\begin{aligned}&c_{11}\left(\frac{\partial^2 u_{ri}}{\partial r^2} + \frac{1}{r}\frac{\partial u_{ri}}{\partial r} - \frac{\partial u_{ri}}{r^2}\right) + c_{44}\frac{\partial^2 u_{ri}}{\partial z^2} + (c_{13}+c_{44})\frac{\partial^2 u_{zi}}{\partial r \partial z} + (e_{31}+e_{15})\frac{\partial^2 \phi_i}{\partial r \partial z} = \lambda_{11}\frac{\partial T_i}{\partial r},\\ &c_{44}\left(\frac{\partial^2 u_{zi}}{\partial r^2} + \frac{1}{r}\frac{\partial u_{zi}}{\partial r}\right) + c_{33}\frac{\partial^2 u_{zi}}{\partial z^2} + (c_{13}+c_{44})\left(\frac{\partial^2 u_{ri}}{\partial r \partial z} + \frac{1}{r}\frac{\partial u_{ri}}{\partial z}\right) + e_{15}\left(\frac{\partial^2 \phi_i}{\partial r^2} + \frac{1}{r}\frac{\partial \phi_i}{\partial r}\right) + e_{33}\frac{\partial^2 \phi_i}{\partial z^2} = \lambda_{33}\frac{\partial T_i}{\partial z},\\ &e_{15}\left(\frac{\partial^2 u_{zi}}{\partial r^2} + \frac{1}{r}\frac{\partial u_{zi}}{\partial r}\right) + e_{33}\frac{\partial^2 u_{zi}}{\partial z^2} + (e_{15}+e_{31})\left(\frac{\partial^2 u_{ri}}{\partial r \partial z} + \frac{1}{r}\frac{\partial u_{ri}}{\partial z}\right) - \varepsilon_{11}\left(\frac{\partial^2 \phi_i}{\partial r^2} + \frac{1}{r}\frac{\partial \phi_i}{\partial r}\right) - \varepsilon_{33}\frac{\partial^2 \phi_i}{\partial z^2} = -p_z\frac{\partial T_i}{\partial z}\end{aligned}\right\} (i=1,2) \tag{4}$$

where $\kappa^2 = \kappa_r / \kappa_z$.

The boundary conditions can be written as

$$\left.\begin{aligned}&\frac{\partial}{\partial z}T_1(r,0) = 0 \quad (0 \le r < c)\\ &T_1(r,0) = T_2(r,0) \quad (c \le r < \infty)\end{aligned}\right\} \tag{5}$$

$$\left.\begin{aligned} T_1(r,h_1) &= T_{10}, \\ \frac{\partial}{\partial z}T_1(r,0) &= \frac{\partial}{\partial z}T_2(r,0), \\ T_2(r,-h_2) &= T_{20} \end{aligned}\right\} \quad (0 \le r < \infty) \tag{6}$$

for thermal loading conditions and

$$\left.\begin{aligned} \sigma_{zz1}(r,0) &= 0 & (0 \le r < c) \\ u_{z1}(r,0) &= u_{z2}(r,0) & (c \le r < \infty) \end{aligned}\right\} \tag{7}$$

$$\left.\begin{aligned} \sigma_{zr1}(r,0) &= 0 & (0 \le r < c) \\ u_{r1}(r,0) &= u_{r2}(r,0) & (c \le r < \infty) \end{aligned}\right\} \tag{8}$$

$$\left.\begin{aligned} D_{z1}(r,0) &= 0 & (0 \le r < c) \\ \phi_1(r,0) &= \phi_2(r,0) & (c \le r < \infty) \end{aligned}\right\} \tag{9}$$

$$\left.\begin{aligned} &\sigma_{zz1}(r,0) = \sigma_{zz2}(r,0), \quad \sigma_{zz1}(r,h_1) = 0, \quad \sigma_{zz2}(r,-h_2) = 0, \\ &\sigma_{zr1}(r,0) = \sigma_{zr2}(r,0), \quad \sigma_{zr1}(r,h_1) = 0, \quad \sigma_{zr2}(r,-h_2) = 0, \\ &D_{z1}(r,0) = D_{z2}(r,0), \quad D_{z1}(r,h_1) = 0, \quad D_{z2}(r,-h_2) = 0 \end{aligned}\right\} \quad (0 \le r < \infty) \tag{10}$$

for the electromechanical conditions.

Temperature field

For the problem considered here, it is convenient to represent the temperature as the sum of two functions.

$$T_i(r,z) = T^{(1)}(z) + T_i^{(2)}(r,z) \quad (i = 1,2) \tag{11}$$

where $T^{(1)}(z)$ satisfies the following equation and boundary conditions:

$$\frac{d^2T^{(1)}}{dz^2} = 0 \tag{12}$$

$$\left.\begin{aligned} T^{(1)}(h_1) &= T_{10}, \\ T^{(1)}(-h_2) &= T_{20} \end{aligned}\right\} \tag{13}$$

and $T_i^{(2)}(r,z)\,(i = 1,2)$ is subjected to the relations:

$$\kappa^2\left(\frac{\partial^2 T_i^{(2)}}{\partial r^2} + \frac{1}{r}\frac{\partial T_i^{(2)}}{\partial r}\right) + \frac{\partial^2 T_i^{(2)}}{\partial z^2} = 0 \quad (i = 1,2) \tag{14}$$

$$\left.\begin{aligned} \frac{\partial}{\partial z}T_1^{(2)}(r,0) &= -\frac{d}{dz}T^{(1)}(0) & (0 \le r < c) \\ T_1^{(2)}(r,0) &= T_2^{(2)}(r,0) & (c \le r < \infty) \end{aligned}\right\} \tag{15}$$

$$\left.\begin{aligned} T_1^{(2)}(r,h_1) &= 0, \\ \frac{\partial}{\partial z}T_1^{(2)}(r,0) &= \frac{\partial}{\partial z}T_2^{(2)}(r,0), \\ T_2^{(2)}(r,-h_2) &= 0 \end{aligned}\right\} \quad (0 \le r < \infty) \tag{16}$$

It is easy to find from Eqs.(12) and (13) that

$$T^{(1)}(z) = \frac{1}{h_1 + h_2}\{(T_{10} - T_{20})z + T_{10}h_2 + T_{20}h_1\} \tag{17}$$

By applying the Hankel transform to Eq.(14) (Sneddon & Lowengrub, 1969), we have

$$T_i^{(2)}(r,z) = \sum_{j=1}^{2} \int_0^\infty D_{ij}(s) J_0(sr) \exp(s\tau_{ij}z) ds \quad (i = 1,2) \tag{18}$$

where $D_{ij}(s)\ (i,j = 1,2)$ are unknown functions to be solved and $\tau_{ij}\ (i,j = 1,2)$ are given by

$$\left.\begin{aligned} \tau_{11} = \tau_{22} &= -\kappa, \\ \tau_{12} = \tau_{21} &= \kappa \end{aligned}\right\} \tag{19}$$

Taking the second boundary condition (15) into consideration, the problem may be reduced to a singular integral equation by defining the following new unknown function $G_0(r)$ (Erdogan & Wu, 1996):

$$G_0(r) = \left\{\begin{array}{lr} \dfrac{\partial}{\partial r}\{T_1^{(2)}(r,0) - T_2^{(2)}(r,0)\} & (0 \le r < c) \\ 0 & (c \le r < \infty) \end{array}\right\} \tag{20}$$

Making use of the first boundary condition (15) with Eqs.(16), we have the following singular integral equation for the determination of the unknown function $G_0(t)$:

$$\int_0^c t\left\{M_0^{(1)}(t,r) + M_0^{(2)}(t,r)\right\}G_0(t)dt = -\frac{2}{\kappa}\frac{T_{10} - T_{20}}{h_1 + h_2} \qquad (0 \le r < c) \tag{21}$$

In Eq.(21), the kernel functions $M_0^{(1)}(t,r)$ and $M_0^{(2)}(t,r)$ are given by

$$M_0^{(1)}(t,r) = \left\{\begin{array}{lr} \dfrac{2}{\pi}\dfrac{1}{t^2 - r^2}E\left(\dfrac{r}{t}\right) & (r < t), \\ \dfrac{2}{\pi}\left\{\dfrac{r}{t(t^2 - r^2)}E\left(\dfrac{t}{r}\right) + \dfrac{1}{rt}K\left(\dfrac{t}{r}\right)\right\} & (r > t) \end{array}\right\} \tag{22}$$

$$M_0^{(2)}(t,r) = -\int_0^\infty s\left\{\frac{2\rho_1(s)\rho_2(s)}{\kappa\rho_0(s)} + 1\right\}J_0(sr)J_1(st)ds \tag{23}$$

where K and E are complete elliptic integrals of the first and second kind, and $\rho_k(s)\ (k = 0,1,2)$ are given by

$$\left.\begin{aligned} \rho_0(s) &= \rho_2(s)\{1-\exp(-2s\kappa h_1)\} - \rho_1(s)\{1-\exp(-2s\kappa h_2)\}, \\ \rho_i(s) &= \tau_{i1} - \tau_{i2}\exp(-2s\kappa h_i) \quad (i=1,2) \end{aligned}\right\} \tag{24}$$

Once $G_0(t)$ is obtained from Eq.(21), the temperature field can be easily calculated as follows:

$$T_i^{(2)}(r,z) = \sum_{j=1}^{2} T_{ij}^{(2)}(r,z) \quad (i=1,2) \tag{25}$$

where

$$T_{ij}^{(2)}(r,z) = \int_0^\infty R_{ij}(s)R_0(s)J_0(sr)\exp(s\tau_{ij}z)ds \quad (i,j=1,2) \tag{26}$$

with

$$\left.\begin{aligned} &R_0(s) = \int_0^c tG_0(t)J_1(st)dt, \\ &R_{11}(s) = -\frac{\rho_2(s)}{\rho_0(s)}, \qquad R_{12}(s) = \frac{\rho_2(s)}{\rho_0(s)}\exp(-2s\kappa h_1), \\ &R_{21}(s) = -\frac{\rho_1(s)}{\rho_0(s)}, \qquad R_{22}(s) = \frac{\rho_1(s)}{\rho_0(s)}\exp(-2s\kappa h_2) \end{aligned}\right\} \tag{27}$$

On the plane $z=0$, the temperatures $T_i^{(2)}(r,0)\ (i=1,2)$ are reduced to

$$T_i^{(2)}(r,0) = \frac{(-1)^i}{2}\int_r^c G_0(t)dt + \int_0^\infty \left\{\sum_{j=1}^{2} R_{ij}(s) - \frac{(-1)^i}{2}\right\} R_0(s)J_0(sr)ds \quad (i=1,2) \tag{28}$$

Thermally induced elastic and electric fields

The non-disturbed temperature filed $T^{(1)}(z)$ given by Eq.(17) does not induce the stress and electric displacement components, which affect the singular field. Thus, we consider the elastic and electric fields due to the disturbed temperature distribution $T_i^{(2)}(r,z)\ (i=1,2)$ only. It is convenient to represent the solutions $u_{zi}(r,z)$, $u_{ri}(r,z)$ and $\phi_i(r,z)\ (i=1,2)$ as the sum of two functions, respectively.

$$\left.\begin{aligned} u_{zi}(r,z) &= u_{zi}^{(1)}(r,z) + u_{zi}^{(2)}(r,z), \\ u_{ri}(r,z) &= u_{ri}^{(1)}(r,z) + u_{ri}^{(2)}(r,z), \\ \phi_i(r,z) &= \phi_i^{(1)}(r,z) + \phi_i^{(2)}(r,z) \end{aligned}\right\} \quad (i=1,2) \tag{29}$$

where $u_{zi}^{(1)}(r,z)$, $u_{ri}^{(1)}(r,z)$, $\phi_i^{(1)}(r,z)\ (i=1,2)$ are the particular solutions of Eqs.(4) replaced T_i by $T_i^{(2)}$, and $u_{zi}^{(2)}(r,z)$, $u_{ri}^{(2)}(r,z)$, $\phi_i^{(2)}(r,z)\ (i=1,2)$ are the general solutions of homogeneous equations obtained by setting $T_i = 0\ (i = 1,2)$ in Eqs.(4). In the following, the superscripts (1) and (2) indicate the particular and general solutions of Eqs.(4). Substituting Eqs.(29) into Eqs.(1) and (2), one obtains stress $\sigma_{rri}(r,z)$, $\sigma_{\theta\theta i}(r,z)$, $\sigma_{zzi}(r,z)$, $\sigma_{zri}(r,z)$ and electric displacement $D_{ri}(r,z)$, $D_{zi}(r,z)\ (i = 1,2)$ expressions.

Using the displacement potential function method (Ueda, 2006a), the particular solutions can be obtained as follows:

$$\left.\begin{aligned}
\sigma_{zzi}^{(1)}(r,z) &= \sum_{j=1}^{2} \int_0^\infty p_{1ij}^{(1)} R_{ij}(s) R_0(s) J_0(sr) \exp(s\tau_{ij} z) ds, \\
\sigma_{zri}^{(1)}(r,z) &= \sum_{j=1}^{2} \int_0^\infty p_{2ij}^{(1)} R_{ij}(s) R_0(s) J_1(sr) \exp(s\tau_{ij} z) ds, \\
D_{zi}^{(1)}(r,z) &= \sum_{j=1}^{2} \int_0^\infty p_{3ij}^{(1)} R_{ij}(s) R_0(s) J_0(sr) \exp(s\tau_{ij} z) ds, \\
u_{zi}^{(1)}(r,z) &= \sum_{j=1}^{2} \int_0^\infty \frac{1}{s} p_{4ij}^{(1)} R_{ij}(s) R_0(s) J_0(sr) \exp(s\tau_{ij} z) ds, \\
u_{ri}^{(1)}(r,z) &= \sum_{j=1}^{2} \int_0^\infty \frac{1}{s} p_{5ij}^{(1)} R_{ij}(s) R_0(s) J_1(sr) \exp(s\tau_{ij} z) ds, \\
\phi_i^{(1)}(r,z) &= -\sum_{j=1}^{2} \int_0^\infty \frac{1}{s} p_{6ij}^{(1)} R_{ij}(s) R_0(s) J_0(sr) \exp(s\tau_{ij} z) ds
\end{aligned}\right\} (i = 1,2) \tag{30}$$

where the constants $p_{kij}^{(1)}\ (i,j = 1,2, k = 1,2,...,6)$ are given in Appendix A. The general solutions are obtained by using the Hankel transform technique (Sneddon & Lowengrub, 1969):

$$\left.\begin{aligned}
\sigma_{zzi}^{(2)}(r,z) &= \sum_{j=1}^{6} \int_0^\infty s p_{1ij}^{(2)} A_{ij}(s) J_0(sr) \exp(s\gamma_{ij} z) ds, \\
\sigma_{zri}^{(2)}(r,z) &= \sum_{j=1}^{6} \int_0^\infty s p_{2ij}^{(2)} A_{ij}(s) J_1(sr) \exp(s\gamma_{ij} z) ds, \\
D_{zi}^{(2)}(r,z) &= \sum_{j=1}^{6} \int_0^\infty s p_{3ij}^{(2)} A_{ij}(s) J_0(sr) \exp(s\gamma_{ij} z) ds, \\
u_{zi}^{(2)}(r,z) &= \sum_{j=1}^{6} \int_0^\infty p_{4ij}^{(2)} A_{ij}(s) J_0(sr) \exp(s\gamma_{ij} z) ds, \\
u_{ri}^{(2)}(r,z) &= \sum_{j=1}^{6} \int_0^\infty p_{5ij}^{(2)} A_{ij}(s) J_1(sr) \exp(s\gamma_{ij} z) ds, \\
\phi_i^{(2)}(r,z) &= -\sum_{j=1}^{6} \int_0^\infty p_{6ij}^{(2)} A_{ij}(s) J_0(sr) \exp(s\gamma_{ij} z) ds
\end{aligned}\right\} (i = 1,2) \tag{31}$$

where $A_{ij}(s)\ (i = 1,2, j = 1,2,...,6)$ are the unknown functions to be solved, and the constants γ_{ij} and $p_{kij}^{(2)}\ (i = 1,2, j,k = 1,2,...,6)$ are given in Appendix B.

Similar to the temperature analysis, the problem may be reduced to a system of singular integral equations by taking the second boundary conditions (7)-(9) into consideration and by defining the following new unknown functions $G_l(r)\ (l=1,2,3)$:

$$G_1(r)=\begin{Bmatrix} \dfrac{\partial}{\partial r}\left\{u_{z1}^{(2)}(r,0)-u_{z2}^{(2)}(r,0)\right\} & (0\le r<c) \\ 0 & (c\le r<\infty) \end{Bmatrix} \tag{32}$$

$$G_2(r)=\begin{Bmatrix} r\dfrac{\partial}{\partial r}\left[\dfrac{1}{r}\left\{u_{r1}^{(2)}(r,0)-u_{r2}^{(2)}(r,0)\right\}\right] & (0\le r<c) \\ 0 & (c\le r<\infty) \end{Bmatrix} \tag{33}$$

$$G_3(r)=\begin{Bmatrix} -\dfrac{\partial}{\partial r}\left\{\phi_1^{(2)}(r,0)-\phi_2^{(2)}(r,0)\right\} & (0\le r<c) \\ 0 & (c\le r<\infty) \end{Bmatrix} \tag{34}$$

Making use of the first boundary conditions (7)-(9) with Eqs.(10), we have the following system of singular integral equations for the determination of the unknown functions $G_l(t)\ (l=1,2,3)$:

$$\begin{aligned} &\int_0^c t[\left\{Z_{11}^{\infty}M_0^{(1)}(t,r)+M_{11}(t,r)\right\}G_1(t)+M_{12}(t,r)G_2(t)+ \\ &+\left\{Z_{13}^{\infty}M_0^{(1)}(t,r)+M_{13}(t,r)\right\}G_3(t)]dt=\sigma_{zz0}(r)\quad (0\le r<c) \end{aligned} \tag{35}$$

$$\begin{aligned} &\int_0^c t[M_{21}(t,r)G_1(t)+\left\{Z_{22}^{\infty}M_1^{(1)}(t,r)+M_{22}(t,r)\right\}G_2(t)+ \\ &+M_{23}(t,r)G_3(t)]dt=\sigma_{zr0}(r)\quad (0\le r<c) \end{aligned} \tag{36}$$

$$\begin{aligned} &\int_0^c t[\left\{Z_{31}^{\infty}M_0^{(1)}(t,r)+M_{31}(t,r)\right\}G_1(t)+M_{32}(t,r)G_2(t)+ \\ &+\left\{Z_{33}^{\infty}M_0^{(1)}(t,r)+M_{33}(t,r)\right\}G_3(t)]dt=D_{z0}(r)\quad (0\le r<c) \end{aligned} \tag{37}$$

where the kernel functions $M_1^{(1)}(t,r)$, $M_{kl}(t,r)$ and the constants $Z_{kl}^{\infty}\ (k,l=1,2,3)$ are given by

$$M_1^{(1)}(t,r)=\begin{Bmatrix} \dfrac{4}{\pi rt}\left\{K\left(\dfrac{r}{t}\right)-E\left(\dfrac{r}{t}\right)\right\}+\dfrac{2}{\pi rt}\left\{\dfrac{t^2}{t^2-r^2}E\left(\dfrac{r}{t}\right)-K\left(\dfrac{r}{t}\right)\right\} & (r<t), \\ \dfrac{4}{\pi t^2}\left\{K\left(\dfrac{t}{r}\right)-E\left(\dfrac{t}{r}\right)\right\}+\dfrac{2}{\pi}\dfrac{1}{t^2-r^2}E\left(\dfrac{t}{r}\right) & (r>t) \end{Bmatrix} \tag{38}$$

$$M_{kl}(t,r)=\begin{Bmatrix} \int_0^\infty s\left\{Z_{kl}(s)-Z_{kl}^{\infty}\right\}J_0(sr)J_1(st)ds & (k=1,3,l=1,3), \\ \int_0^\infty s\left\{Z_{kl}(s)-Z_{kl}^{\infty}\right\}J_0(sr)J_2(st)ds & (k=1,3,l=2), \\ \int_0^\infty s\left\{Z_{kl}(s)-Z_{kl}^{\infty}\right\}J_1(sr)J_1(st)ds & (k=2,l=1,3), \\ \int_0^\infty s\left\{Z_{kl}(s)-Z_{kl}^{\infty}\right\}J_1(sr)J_2(st)ds & (k=2,l=2) \end{Bmatrix} \tag{39}$$

$$Z_{kl}(s)=\sum_{j=1}^{6}p_{k1j}^{(2)}d_{1jl}(s),\quad Z_{kl}^{\infty}=\lim_{s\to\infty}Z_{kl}(s)\quad (k,l=1,2,3) \tag{40}$$

In Eq.(40), the functions $d_{1jl}(s)\ (j=1,2,...,6, l=1,2,3)$ are given in Appendix C.

The functions $\sigma_{zz0}(r)$, $\sigma_{zr0}(r)$ and $D_{z0}(r)$, which correspond to the stress and electric displacement components induced by the disturbed temperature field $T_i^{(2)}(r,z)\ (i=1,2)$ on the r-axis in the plate without crack, are obtained as follows:

$$\left.\begin{aligned}
\sigma_{zz0}(r)&=\int_0^{\infty}R_0(s)\left\{\sum_{j=1}^{6}p_{11j}^{(2)}d_{1j}^{T}(s)+\sum_{j=1}^{2}p_{11j}^{(1)}R_{1j}(s)\right\}J_0(sr)ds,\\
\sigma_{zr0}(r)&=\int_0^{\infty}R_0(s)\left\{\sum_{j=1}^{6}p_{21j}^{(2)}d_{1j}^{T}(s)+\sum_{j=1}^{2}p_{21j}^{(1)}R_{1j}(s)\right\}J_1(sr)ds,\\
D_{z0}(r)&=\int_0^{\infty}R_0(s)\left\{\sum_{j=1}^{6}p_{31j}^{(2)}d_{1j}^{T}(s)-\sum_{j=1}^{2}p_{31j}^{(1)}R_{1j}(s)\right\}J_0(sr)ds
\end{aligned}\right\} \tag{41}$$

where the functions $d_{1j}^{T}(s)\quad (j=1,2,...,6)$ are also given in Appendix C. These components are superficial quantities and have no physical meaning in this analysis. However, they are equivalent to the crack face tractions in solving the crack problem by a proper superposition.

To solve the singular integral equations (21) and (35)- (37) by using the Gauss-Jacobi integration formula (Sih, 1972), we introduce the following functions $\Phi_l(t)\ (l=0,1,2,3)$:

$$G_l(t)=\left(\frac{c+t}{c-t}\right)^{1/2}\Phi_l(t)\quad (l=0,1,2,3) \tag{42}$$

Then the stress intensity factors K_{I}, K_{II} and the electric displacement intensity factor K_{D} may be defined and evaluated as:

$$\left.\begin{aligned}
K_{\mathrm{I}}&=\lim_{r\to c^+}\{2\pi(c-r)\}^{1/2}\sigma_{zz1}(r,0)=(\pi c)^{1/2}\left\{Z_{11}^{\infty}\Phi_1(c)+Z_{13}^{\infty}\Phi_3(c)\right\},\\
K_{\mathrm{II}}&=\lim_{r\to c^+}\{2\pi(c-r)\}^{1/2}\sigma_{zr1}(r,0)=(\pi c)^{1/2}Z_{22}^{\infty}\Phi_2(c),\\
K_{\mathrm{D}}&=\lim_{r\to c^+}\{2\pi(c-r)\}^{1/2}D_{z1}(r,0)=(\pi c)^{1/2}\left\{Z_{31}^{\infty}\Phi_1(c)+Z_{33}^{\infty}\Phi_3(c)\right\}
\end{aligned}\right\} \tag{43}$$

Numerical results and discussion

For the numerical calculations, the thermo-electro-elastic properties of the plate are assumed to be ones of cadmium selenide with the following properties (Ashida & Tauchert, 1998).

The values of the coefficients of heat conduction for cadmium selenide could not be found in the literature. Since the values of them for orthotropic Alumina (Al_2O_3) are $\kappa_r=21.25[\mathrm{W/mK}]$ and $\kappa_z=29.82[\mathrm{W/mK}]$ (Dag, 2006), the value $\kappa^2=\kappa_r/\kappa_z=1/1.5$ is assumed. To examine the effects of the normalized crack size c/h and the normalized crack location h_1/h on the stress and electric displacement intensity factors, the solutions of the system of the singular integral equations have been computed numerically.

$$\left.\begin{array}{ll} c_{11}=74.1\times10^9[\mathrm{N/m^2}], & c_{12}=45.2\times10^9[\mathrm{N/m^2}], \\ c_{13}=39.3\times10^9[\mathrm{N/m^2}], & c_{33}=83.6\times10^9[\mathrm{N/m^2}], \\ c_{44}=13.2\times10^9[\mathrm{N/m^2}], & \\ e_{31}=-0.16[\mathrm{C/m^2}], & e_{33}=0.347[\mathrm{C/m^2}], \\ e_{15}=-0.138[\mathrm{C/m^2}], & \\ \varepsilon_{11}=82.6\times10^{-12}[\mathrm{C/Vm}], & \varepsilon_{33}=90.3\times10^{-12}[\mathrm{C/Vm}], \\ \lambda_{11}=0.621\times10^6[\mathrm{N/Km^2}], & \lambda_{33}=0.551\times10^6[\mathrm{N/Km^2}], \\ p_z=-2.94\times10^{-6}[\mathrm{CK^{-1}m^{-2}}]. & \end{array}\right\} \tag{44}$$

In the first set of calculations, we consider the temperature field and the electro-elastic fields without crack. Figure 2 shows the normalized temperature $(T_i(x)-T_{20})/T_0\ (i=1,2)$ on the crack faces $(0\le r<c, z\to 0^{\pm})$ and the crack extended line $(c\le r\le 2c, z=0)$ for $h_1/h=0.25$ and $c/h=0.5$, where $T_0=T_{10}-T_{20}$. The maximum local temperature difference across the crack occurs at the center of the crack.

Figure 3 exhibits the normalized stress components $(\sigma_{zz0}(r),\sigma_{zr0}(r))/\lambda_{33}T_0$ and the electric displacement component $D_{z0}(r)/p_zT_0$ on the r-axis in the strip without crack due to the temperature shown in Figure 2. The maximum absolute values of $\sigma_{zz0}(r)$ and $D_{z0}(r)$ occur at the center of the crack $(r/c=0.0)$, whereas the maximum value of $\sigma_{zr0}(r)$ occurs at the crack tip $(r/c=1.0)$.

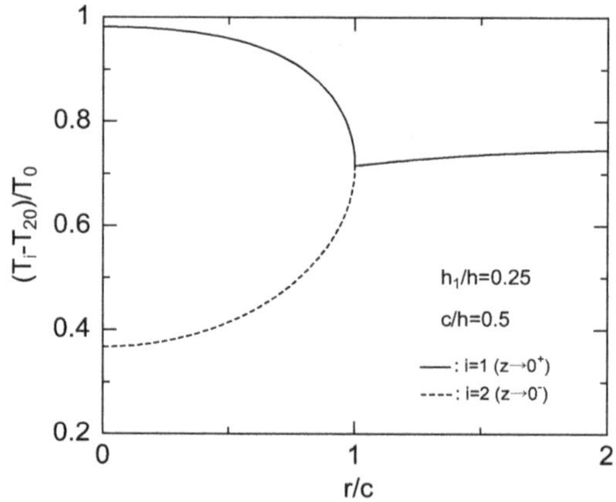

Fig. 2. The temperature on the crack faces and the crack extended line for $c/h=0.5$ and $h_1/h=0.25$

In the second set of calculations, we study the influence of the crack size on the stress and electric displacement intensity factors. Figures 4(a)-(c) show the plots of the normalized stress and electric displacement intensity factors $(K_{\mathrm{I}},K_{\mathrm{II}})/\lambda_{33}T_0(\pi c)^{1/2}$, $K_{\mathrm{D}}/p_zT_0(\pi c)^{1/2}$ versus c/h for $h_1/h=0.25$, 0.5 and 0.75. Because of symmetry, the values of K_{I} and K_{D}

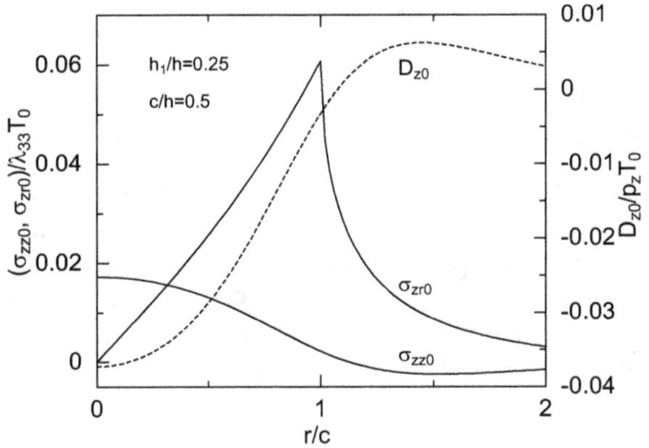

Fig. 3. The stress components σ_{zz0}, σ_{zr0} and the electric displacement component D_{z0} on the r-axis without crack due to the temperature shown in Fig. 2

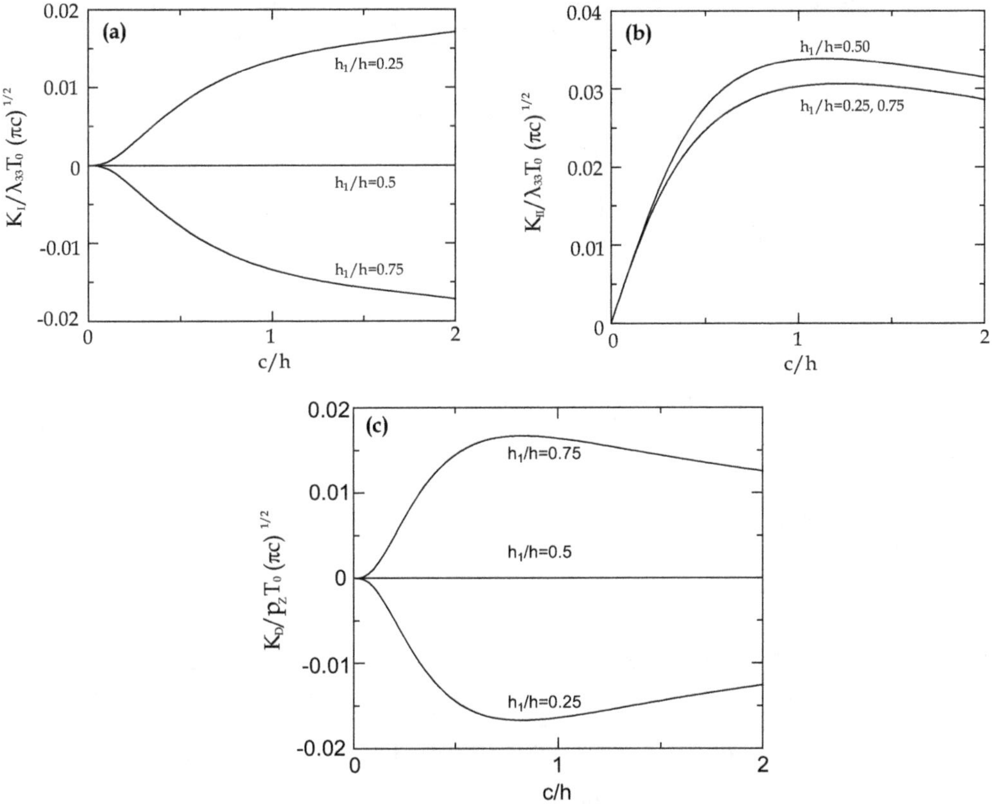

Fig. 4. (a) The effect of the crack size on the stress intensity factor K_{I}. (b) The effect of the crack size on the stress intensity factor K_{II}. (c) The effect of the crack size on the electric displacement intensity factor K_{D}

for $h_1/h=0.5$ are zero, and $[K_\mathrm{I}]_{h_1/h=0.25}=-[K_\mathrm{I}]_{h_1/h=0.75}$, $[K_\mathrm{II}]_{h_1/h=0.25}=[K_\mathrm{II}]_{h_1/h=0.75}$ $[K_\mathrm{D}]_{h_1/h=0.25}=-[K_\mathrm{D}]_{h_1/h=0.75}$. The absolute value of $K_\mathrm{I}/\lambda_{33}T_0(\pi c)^{1/2}$ for $h_1/h=0.25$ and 0.75 monotonically increases with increasing c/h, but the value of $K_\mathrm{II}/\lambda_{33}T_0(\pi c)^{1/2}$ and the absolute value of $K_\mathrm{D}/p_zT_0(\pi c)^{1/2}$ increase at first, reach maximum values and then decrease with increasing c/h. The value of K_I for $h_1/h=0.75$ becomes negative so that the contact of the crack faces would occur. The results presented here without considering this effect may not be exact but would be more conservative. Since the contact of the crack faces will increase the friction between the faces and make thermo-electrical transfer across the crack faces easier, the stress and electric displacement intensity factors would be lowered by these two factors.

In the final set of calculations, we investigate the influence of the crack location on the intensity factors. Figure 5 indicates the effect of the crack location on K_I, K_II and K_D for $c/h=0.5$. As h_1/h increases, the values of K_I and K_D tend to decrease or increase monotonically. The value of $K_\mathrm{II}/\lambda_{33}T_0(\pi c)^{1/2}$ decreases if the crack approaches the free boundaries $(h_1/h \to 0.0 \text{ or } 1.0)$, and the peak value of $K_\mathrm{II}/\lambda_{33}T_0(\pi c)^{1/2}$ =0.0277 occurs at $h_1/h=0.5$.

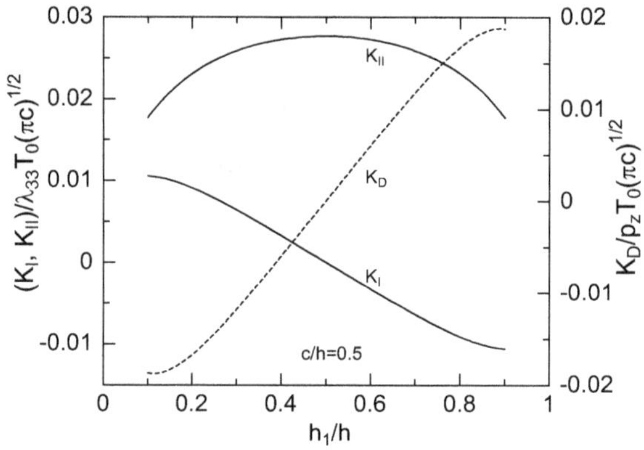

Fig. 5 The effect of the crack location on the stress intensity factors K_I, K_II and the electric displacement intensity factor K_D

Conclusion

An example of the application of Hankel transform for solving a mixed-mode thermo-electro-elastic fracture problem of a piezoelectric material strip with a parallel penny-shaped crack is explained. The effects of the crack size (c/h) and the crack location (h_1/h) on the fracture behavior are analyzed. The following facts can be found from the numerical results.

1. The large shear stress occurs in the strip without crack due to the disturbed temperature field.
2. The normalized intensity factors are under the great influence of the geometric parameters h_1/h and c/h.

3. For the case of $h_1 / h > 0.5$, mode I stress intensity factor becomes negative so that the contact of the crack faces would occur.
4. The intensity factors of crack near the free surfaces due to the thermal load are not so large.

Appendix A

The constants $p_{kij}^{(1)}$ $(i, j = 1,2, k = 1,2,...,6)$ are

$$\left.\begin{aligned} p_{1ij}^{(1)} &= \left(c_{13} - c_{33}k_{ij}\tau_{ij}{}^2\right)C_{ij} - e_{33}\tau_{ij}N_{ij} - \lambda_{33}, \\ p_{2ij}^{(1)} &= c_{44}\left(1 + k_{ij}\right)\tau_{ij}C_{ij} + e_{15}N_{ij}, \\ p_{3ij}^{(1)} &= \left(e_{31} - e_{33}k_{ij}\tau_{ij}{}^2\right)C_{ij} + \varepsilon_{33}\tau_{ij}N_{ij} + p_z, \\ p_{4ij}^{(1)} &= -k_{ij}\tau_{ij}C_{ij},\ p_{5ij}^{(1)} = C_{ij},\ p_{6ij}^{(1)} = N_{ij} \end{aligned}\right\} \quad (i, j = 1,2) \tag{A.1}$$

where

$$\left.\begin{aligned} C_{ij} &= \frac{b'_{21} - b'_{22}\tau_{ij}^2}{m_{ij}^{(1)} + k_{ij}m_{ij}^{(2)}}, \\ N_{ij} &= \frac{n_{ij}^{(11)} + k_{ij}n_{ij}^{(12)}}{m_{ij}^{(1)} + k_{ij}m_{ij}^{(2)}}, \\ k_{ij} &= -\frac{\tau_{ij}^2 n_{ij}^{(11)} + n_{ij}^{(21)}}{\tau_{ij}^2 n_{ij}^{(12)} + n_{ij}^{(22)}} \end{aligned}\right\} \quad (i, j = 1,2) \tag{A.2}$$

with

$$\left.\begin{aligned} m_{ij}^{(1)} &= a'_{41} - a'_{42}\tau_{ij}^2, \\ m_{ij}^{(2)} &= \left(a'_{44}\tau_{ij}^2 - a'_{43}\right)\tau_{ij}^2, \\ n_{ij}^{(11)} &= \left\{(H_{1r} + H_{3r})\left(b'_{21} - b'_{22}\tau_{ij}^2\right) + H_{4r}m_{ij}^{(1)}\right\}\tau_{ij}, \\ n_{ij}^{(21)} &= \left\{(H_{1z} + H_{3z})\left(b'_{21} - b'_{22}\tau_{ij}^2\right) + H_{4z}m_{ij}^{(1)}\right\}\tau_{ij}, \\ n_{ij}^{(12)} &= \left\{\left(H_{1r} - H_{2r}\tau_{ij}^2\right)\left(b'_{21} - b'_{22}\tau_{ij}^2\right) + H_{4r}m_{ij}^{(2)}\right\}\tau_{ij}, \\ n_{ij}^{(22)} &= \left\{\left(H_{1z} - H_{2z}\tau_{ij}^2\right)\left(b'_{21} - b'_{22}\tau_{ij}^2\right) + H_{4z}m_{ij}^{(2)}\right\}\tau_{ij} \end{aligned}\right\} \quad (i, j = 1,2) \tag{A.3}$$

$$\left.\begin{aligned} &H_{1r} = \frac{c_{44}\varepsilon_{33} + e_{15}e_{33}}{e_{33}\varepsilon_{11} - e_{15}\varepsilon_{33}},\ H_{1z} = \frac{c_{44}\varepsilon_{11} + e_{15}^2}{e_{15}\varepsilon_{33} - e_{33}\varepsilon_{11}},\ H_{2r} = \frac{c_{33}\varepsilon_{33} + e_{33}^2}{e_{33}\varepsilon_{11} - e_{15}\varepsilon_{33}},\ H_{2z} = \frac{c_{33}\varepsilon_{11} + e_{15}e_{33}}{e_{15}\varepsilon_{33} - e_{33}\varepsilon_{11}}, \\ &H_{3r} = \frac{c_{13}\varepsilon_{33} + e_{31}e_{33}}{e_{33}\varepsilon_{11} - e_{15}\varepsilon_{33}},\ H_{3z} = \frac{c_{13}\varepsilon_{11} + e_{15}e_{31}}{e_{15}\varepsilon_{33} - e_{33}\varepsilon_{11}},\ H_{4r} = \frac{p_z e_{33} - \lambda_{33}\varepsilon_{33}}{e_{33}\varepsilon_{11} - e_{15}\varepsilon_{33}},\ H_{4z} = \frac{p_z e_{15} - \lambda_{33}\varepsilon_{11}}{e_{15}\varepsilon_{33} - e_{33}\varepsilon_{11}} \end{aligned}\right\} \tag{A.4}$$

$$\left.\begin{array}{ll} a'_{41}=c_{11}, & a'_{42}=c_{44}+(e_{15}+e_{31})(H_{1r}+H_{3r}), \\ a'_{43}=c_{13}+c_{44}+(e_{15}+e_{31})H_{1r}, & a'_{44}=(e_{15}+e_{31})H_{2r}, \\ b'_{21}=\lambda_{11}, & b'_{22}=-(e_{15}+e_{31})H_{4r} \end{array}\right\} \tag{A.5}$$

Appendix B

The constants γ_{ij} $(i=1,2, j=1,2,...,6)$ are the roots of the following characteristic equations:

$$\begin{array}{l} (f_4g'_2+g_4f'_2)\gamma_{ij}^6+(f_4g'_0+f_2g'_2+g_4f'_0+g_2f'_2)\gamma_{ij}^4+ \\ +(f_2g'_0+f_0g'_2+g_2f'_0+g_0f'_2)\gamma_j^2+(f_0g'_0+g_0f'_0)=0 \end{array} \quad (i=1,2,\ j=1,2,...,6) \tag{B.1}$$

where $\Re[\gamma_{1j}]<\Re[\gamma_{1j+1}]$, $\Re[\gamma_{2j}]>\Re[\gamma_{2j+1}](j=1,2,...,5)$ and

$$\left.\begin{array}{l} f_4=c_{44}e_{33}, \\ f_2=(c_{13}+c_{44})(e_{15}+e_{31})-c_{11}e_{33}-c_{44}e_{15}, \\ f_0=c_{11}e_{15}, \\ f'_2=c_{33}(e_{15}+e_{31})-e_{33}(c_{13}+c_{44}), \\ f'_0=-c_{44}(e_{15}+e_{31})+e_{15}(c_{13}+c_{44}) \end{array}\right\} \tag{B.2}$$

$$\left.\begin{array}{l} g_4=c_{44}\varepsilon_{33}, \\ g_2=-(e_{15}+e_{31})^2-c_{11}\varepsilon_{33}-c_{44}\varepsilon_{11}, \\ g_0=c_{11}\varepsilon_{11}, \\ g'_2=e_{33}(e_{15}+e_{31})+\varepsilon_{33}(c_{13}+c_{44}), \\ g'_0=-e_{15}(e_{15}+e_{31})-\varepsilon_{11}(c_{13}+c_{44}) \end{array}\right\} \tag{B.3}$$

The functions $p_{kij}^{(2)}(s)$ $(i=1,2, j,k=1,2,...,6)$ are

$$\left.\begin{array}{r} p_{1ij}^{(2)}=c_{13}a_{ij}+\gamma_{ij}\left(c_{33}-e_{33}b_{ij}\right), \\ p_{2ij}^{(2)}=c_{44}\left(\gamma_{ij}a_{ij}-1\right)+e_{15}b_{ij}, \\ p_{3ij}^{(2)}=e_{31}a_{ij}+\gamma_{ij}\left(e_{310}+\varepsilon_{330}b_{ij}\right), \\ p_{4ij}^{(2)}=1, \\ p_{5ij}^{(2)}=a_{ij}, \\ p_{6ij}^{(2)}=b_{ij} \end{array}\right\} \quad (i=1,2, j=1,2,...,6) \tag{B.4}$$

where a_{ij} and b_{ij} $(i=1,2, j=1,2,...,6)$ are given by

$$\left.\begin{array}{l} a_{ij}=\dfrac{g'_2\gamma_{ij}^2+g'_0}{g_4\gamma_{ij}^4+g_2\gamma_{ij}^2+g_0}, \\ b_{ij}=-\dfrac{(c_{44}\gamma_{ij}^2-c_{11})a_{ij}-c_{13}-c_{44}}{e_{15}+e_{31}} \end{array}\right\} \quad (i=1,2,\ j=1,2,...,6) \tag{B.5}$$

Appendix C

The functions $d_{ijk}(s)\ (i=1,2, j=1,2,...,6, k=1,2,3)$ are given by

$$\left.\begin{aligned} d_{1jk}(s) &= q_{j,k+9}(s), \\ d_{2jk}(s) &= q_{j+6,k+9}(s) \end{aligned}\right\} \quad (j=1,2,...,6, k=1,2,3) \tag{C.1}$$

where the functions $q_{j,k}(s)\ (j,k=1,2,...,12)$ are the elements of a square matrix $Q=\mathbf{\Delta}^{-1}$ of order 12. The elements $\delta_{j,k}(s)\ (j,k=1,2,...,12)$ of the square matrix Δ are given by

$$\left.\begin{aligned} \delta_{j,k}(s) &= p_{j1k}^{(2)} \exp(s\gamma_{1k}h_1) && (j=1,2,3), \\ \delta_{j+3,k+6}(s) &= p_{j2k}^{(2)} \exp(-s\gamma_{2k}h_2) && (j=1,2,3), \\ \delta_{j+6,k}(s) &= p_{j1k}^{(2)} && (j=1,2,...,6), \\ \delta_{j+6,k+6}(s) &= -p_{j2k}^{(2)} && (j=1,2,...,6) \end{aligned}\right\} (k=1,2,...,6) \tag{C.2}$$

The functions $d_{ij}^{T}(s)\ (i=1,2, j=1,2,...,6)$ are

$$\left.\begin{aligned} d_{1j}^{T}(s) &= \sum_{k=1}^{12} q_{j,k}(s)u_k(s), \\ d_{2j}^{T}(s) &= \sum_{k=1}^{12} q_{j+6,k}(s)u_k(s) \end{aligned}\right\} (j=1,2,...,6) \tag{C.3}$$

where

$$\left.\begin{aligned} u_k(s) &= -\frac{R_0(s)}{s}\sum_{j=1}^{2} p_{k1j}^{(1)} R_{1j}(s)\exp(s\tau_{1j}h_1) && (k=1,2,3), \\ u_{k+3}(s) &= -\frac{R_0(s)}{s}\sum_{j=1}^{2} p_{k2j}^{(1)} R_{2j}(s)\exp(-s\tau_{2j}h_2) && (k=1,2,3), \\ u_{k+6}(s) &= -\frac{R_0(s)}{s}\sum_{j=1}^{2} \left\{ p_{k1j}^{(1)} R_{1j}(s) - p_{k2j}^{(1)} R_{2j}(s) \right\} && (k=1,2,...,6) \end{aligned}\right\} \tag{C.4}$$

References

Ashida, F. & Tauchert, T.R. (1998). Transient Response of a Piezothermoelastic Circular Disk under Axisymmetric Heating. *Acta Mechanica*, Vol. 128, pp. 1-14, 0001-5970

Dag, S., Ilhan, K.A. & Erdogan, F. (2006), Mixed-Mode Stress Intensity Factors for an Embedded Crack in an Orthotropic FGM Coating, Proceedings of the International Conference FGM IX, 978-0-7354-0492-2, Oahu Island, Hawaii, October 2006

Erdogan, F. & Wu, B.H. (1996). Crack Problems in FGM Layers under Thermal Stresses. *Journal of Thermal Stresses*, Vol. 19, pp. 237-265, 0149-5739

Rao, S.S. & Sunar, M. (1994). Piezoelectricity and Its Use in Disturbance Sensing and Control of Flexible Structures: a Survey. *Applied Mechanics Review,* Vol. 47, pp. 113-123, 0003-6900

Sih, G.C. (Ed.). (1972). *Methods of Analysis and Solution of Crack Problems,* Noordhoff, International Publishing, 978-9048182466, Leyden

Sneddon, I.N. & Lowengrub, M. (1969). *Crack Problems in the Classical Theory of Elasticity,* John Wiley & Sons, Inc., 978-0471808459, New York

Tauchert, T.R. (1992). Piezothermoelastic Behavior of a Laminated Plate. *Journal of Thermal Stresses,* Vol. 15, pp. 25-37, 0149-5739

Ueda, S. (2006a). The Crack Problem in Piezoelectric Strip under Thermoelectric Loading. *Journal of Thermal Stresses,* Vol. 29, pp. 295-316, 0149-5739

Ueda, S. (2006b). Thermal Stress Intensity Factors for a Normal Crack in a Piezoelectric Strip. *Journal of Thermal Stresses,* Vol. 29, pp. 1107-1126, 0149-5739

3

Fourier Series and Fourier Transform with Applications in Nanomaterials Structure

Florica Matei[1] and Nicolae Aldea[2]
[1]*University of Agricultural Science and Veterinary Medicine, Cluj-Napoca,*
[2]*National Institute for Research and Development of Isotopic and Molecular Technologies, Cluj-Napoca,*
Romania

Introduction

One of the most important problems in the physics and chemistry of the nanostructured materials consists in the local and the global structure determination by X-ray diffraction and X-ray absorption spectroscopy methods. This contribution is dedicated to the applications of the Fourier series and Fourier transform as important tools in the determination of the nanomaterials structure. The structure investigation of the nanostructured materials require the understanding of the mathematical concepts regarding the Fourier series and Fourier transform presented here without theirs proofs. The Fourier series is the traditional tool dedicated to the composition of the periodical signals and its decomposition in discreet harmonics as well as for the solving of the differential equations. Whereas the Fourier transform is more appropriate tool in the study of the non periodical signals and for the solving of the first kind integral equations. From physical point of view the Fourier series are used to describe the model of the global structure of nanostructured materials that consist in: average crystallite size, microstrain of the lattice and distribution functions of the crystallites and microstrain versus size. Whereas the model for the local structure of the nanomaterials involves the direct and inverse Fourier transform. The information obtained consist in the number of atoms from each coordination shell and their radial distances.

Fourier series and theirs applications

One of the most often model studied in physics is the one of oscillatory movement of a material point. The oscillation of the electrical charge into an electrical field, the vibration of a tuning fork that generated sound waves or the electronic vibration into atoms that generate light waves are studied in the same mode (Richard et al., 2005). The motion equations related to the above phenomena have similar form; therefore the phenomena treated are analogous. From mathematical point of view these are modeled by the ordinary differential equations, most of them with constant coefficients. Due to the particular form of the equation any linear combination of the solution it is also a solution and the mathematical substantiation is given by the superposition principle. It consists in, if u_1, u_2, ..., u_k are solutions for the homogenous linear equations $L[u]=0$, then the linear combination (or the superposition) is a solution of $L[u]=0$ for any choice of the constants. The previous statement shows that the general solution of a linear equation is a superposition of its linearly independent particularly solution that compose a base in the finite dimension space of the solution. The superposition is true for any algebraic equations as well as any homogenous linear ordinary differential equations.

Physical concept and mathematical background

The analysis of the linear harmonic oscillatory motion for a material point of mass m round about equilibrium position due to an elastically force F=-Kx it is given by the harmonic equation that is a differential equations which appear very frequently in the analysis of physical phenomena (Tang, 2007)

$$m\frac{d^2x}{dt^2}+Kx=0 \tag{1}$$

with the solution $x(t)=A\cos(2\pi f_0 t-\varphi_0)$ where A, K, f_0, $\omega=\sqrt{\frac{K}{m}}$, $\omega=2\pi f_0$ and φ_0 represent the motion's amplitude, elastic constant, fundamental frequency, angular speed and phase shift, respectively. Generalizing let consider the physical signal given by

$$x(t)=\underbrace{a_1\sin(2\pi f_0 t)}_{f_0\ line}+\underbrace{a_2\sin(4\pi f_0 t)}_{2f_0\ line}+\underbrace{a_3\sin(6\pi f_0 t)}_{3f_0\ line}+...+\underbrace{a_n\sin(2\pi n f_0 t)}_{nf_0\ line} \tag{2}$$

that is periodic but non harmonic process, the physical signal being a synthesis of n spectral lines with the frequencies f_0, $2f_0$, $3f_0$, ..., nf_0 and the amplitudes $a_1, a_2, \ldots, a_n$, respectively. The practical problem that had lead to Fourier series was to solve the heat equation which is a parabolic partial differential equation. Before the Fourier contribution no solution for the general form of the heat equation was known. The Fourier idea was to consider the solution as a linear combination of sine or cosine waves in according with the superposition principle. The solution space for of the partial differential equation are infinite dimensional spaces thus there are needed an infinite number of independent solutions. Therefore is not possible to find all independent particular solutions of a linear partial differential equations. The key found by Joseph Fourier in his article "Théorie analitique de la chaleur", published in 1811, was to form a series with the basic solutions. The ortonormality is the key concept of the Fourier analysis. The general representation of the Fourier series with coefficients a, b_n and c_n is given by:

$$x(t)=\frac{a}{2}+\sum_{n=1}^{\infty}\left[b_n\cos(nt)+c_n\sin(nt)\right] \tag{3}$$

The Fourier series are used in the study of periodical movements, acoustics, electrodynamics, optics, thermodynamics and especially in physical spectroscopy as well as in fingerprints recognition and many other technical domains. It was proved (Walker, 1996) that any physical signal of the period T=1/f_0 can be represented as an harmonic function with the frequencies f_0, $2f_0$, $3f_0$, ...

$$x(t)=\frac{a}{2}+\sum_{n=1}^{\infty}\left[b_n\cos(2\pi n f_0 t)+c_n\sin(2\pi n f_0 t)\right] \tag{4}$$

The Fourier coefficients are obtained in the following way:

i. by integration of the previous relation between $[-T/2,T/2]$

$$a=\frac{2}{T}\int_{-T/2}^{T/2}x(t)dt \tag{5}$$

ii. by multiplication of (4) with $\cos(2\pi n f_0 t)$ and integration

$$b_n=\frac{2}{T}\int_{-T/2}^{T/2}x(t)\cos(2\pi n f_0 t)dt \tag{6}$$

iii. by multiplication of (4) with $\sin(2\pi n f_0 t)$ and integration

$$c_n=\frac{2}{T}\int_{-T/2}^{T/2}x(t)\sin(2\pi n f_0 t)dt \tag{7}$$

Some observations about physical signal modeled by Fourier series are given below.

i. Value $a/2$ represents the mean value for the physical signal on $[-T/2,T/2]$.
ii. If $x(t+T/2)=-x(t)$ then Fourier series of x has only amplitudes with odd index (Tang, 2007) all the other terms will vanish:

$$x(t)=\sum_{n=0}^{\infty}\left[b_{2n+1}\cos(2\pi n f_0 t)+c_n\sin(2\pi n f_0 t)\right] \tag{8}$$

iii. In practice the argument of the function x can be a scalar as time, frequency, length, angle, and so on, thus (4) is defined on period [0, *T*], spectral interval [0,*f*], spatial interval $[0,L]$, wave length $[0,\lambda]$ or the whole trigonometric circle, respectively.
iv. Using the Fourier coefficients the Parseval's equality is given by

$$\int_{-T/2}^{T/2}x^2(t)\,dt=\int_{-T/2}^{T/2}x(t)\left\{\frac{a}{2}+\sum_{n=1}^{\infty}\left[b_n\cos(2\pi n f_0 t)+c_n\sin(2\pi n f_0 t)\right]\right\}dt= \\ =\frac{T}{2}\left[\frac{a^2}{2}+\sum_{n=1}^{\infty}\left(b_n^2+c_n^2\right)\right] \tag{9}$$

The right term of the above equality represents the energy density of the signal x. Thus the Parseval equality shows that the whole density energy is contained in the squares of the all amplitudes of harmonic terms defined on the interval $[-T/2,T/2]$. The Parseval equality holds for any function whose square is integrable. The next problem is to analyze the convergence of the series. Because the aim of this chapter are the application of the Fourier series it will be only mentioned the basic principles of the Fourier analysis. If the interval $[-T/2,T/2]$ can be decomposed in a finite number of intervals on that the function f is continuous and monotonic, then the function f has a Fourier series representation. The next consideration is connected with the $L^2[A,B]$ space defined below

$$L^2[A,B]=\left\{x:[A,B]\to R\left|\int_A^B |x(t)|^2\,dt<\infty\right.\right\} \tag{10}$$

The complete system for the $L^2[A,B]$ is given by

$$\begin{aligned}&\sqrt{\frac{2}{B-A}},\ \sqrt{\frac{2}{B-A}}\cos\frac{2\pi t}{B-A},\ \sqrt{\frac{2}{B-A}}\sin\frac{2\pi t}{B-A},\dots,\\ &\sqrt{\frac{2}{B-A}}\cos\frac{2\pi kt}{B-A},\ \sqrt{\frac{2}{B-A}}\sin\frac{2\pi kt}{B-A},\dots\end{aligned} \tag{11}$$

then any function from $L^2[A,B]$ can be written as a linear combination of its complete system and the Fourier coefficients are

$$\begin{aligned}&a=\sqrt{\frac{2}{B-A}}\int_A^B x(t)dt,\\ &\beta_k=\sqrt{\frac{2}{B-A}}\int_A^B x(t)\cos\frac{2\pi kt}{B-A}dt,\quad \gamma_k=\sqrt{\frac{2}{B-A}}\int_A^B x(t)\sin\frac{2\pi kt}{B-A}dt.\end{aligned} \tag{12}$$

Trigonometric polynomials and Fourier coefficients determination

One of the useful mathematical tools in order to apply the Fourier series in data analysis is the trigonometric polynomial. From physical point of view the trigonometric polynomials are used to characterize the periodic signals. The general form of the real trigonometric polynomial of degree M is given by:

$$P(t)=\frac{\alpha}{2}+\sum_{k=1}^{M}\left(\beta_k\cos\left(\frac{2\pi kt}{T}\right)+\gamma_k\sin\left(\frac{2\pi kt}{T}\right)\right) \tag{13}$$

where α, β_k, γ_k, T being real constants. Let denote by S the square deviation of the function x from trigonometric polynomial P defined on interval of length T

$$S(\alpha,\beta_1,\dots,\beta_M,\gamma_1,\dots,\gamma_M)=\int_{-T/2}^{T/2}(x(t)-P(t))^2\,dt \tag{14}$$

Let enumerate some properties of trigonometric polynomials that will be used in this chapter (Bachmann et al., 2002).

i. Let x be a function in $L^2[-T/2,T/2]$, then from all polynomials of degree M the minimum of the square deviation is obtained for the trigonometric polynomial with the coefficient equal with the Fourier coefficients of function x;
ii. If the physical signal is defined on an arbitrary interval $[A,B]$ with period B-A then the trigonometric polynomial associated is given by

$$P(t)=\frac{\alpha}{2}+\sum_{k=1}^{M}\left(\beta_k\cos\left(\frac{2\pi kt}{B-A}\right)+\gamma_k\sin\left(\frac{2\pi kt}{B-A}\right)\right); \tag{15}$$

iii. The Fourier coefficients associated to polynomial P are obtained by the least square method and the linear system is

$$\begin{cases}\frac{\alpha}{2}\int_A^B dt+\sum_{K=1}^{M}\left\{\int_A^B[\beta_k\cos C_k-\gamma_k\sin C_k]dt\right\}=\int_A^B Y(t)dt\\ \frac{\alpha}{2}\int_A^B\cos C_m dt+\sum_{K=1}^{M}\left\{\int_A^B[\beta_k\cos C_k-\gamma_k\sin C_k]\cos C_m dt\right\}=\int_A^B Y(t)\cos C_m dt\\ \frac{\alpha}{2}\int_A^B\sin C_m dt+\sum_{K=1}^{M}\left\{\int_A^B[\beta_k\cos C_k-\gamma_k\sin C_k]\sin C_m dt\right\}=\int_A^B Y(t)\sin C_m dt\end{cases} \tag{16}$$

where $C_k=Ckt$, $C_m=Cmt$, $C=2\pi/(B-A)$, Y is the approximated function by the trigonometric polynomial of degree M and $m=1, \ldots, M$. The system (16) has $2M+1$ equations and due to the orthogonality properties of the functions $\cos C_k$ and $\sin C_k$ the solution of system (16) is

$$\alpha=\frac{2}{B-A}\int_A^B Y(t)dt,\ \beta_k=\frac{2}{B-A}\int_A^B Y(t)\cos\frac{2\pi kt}{B-A}dt \text{ and } \gamma_k=\frac{2}{B-A}\int_A^B Y(t)\sin\frac{2\pi kt}{B-A}dt; \tag{17}$$

iv. Previous considerations are often used in the process of global approximation of the discreet physical signals. Let consider the sequence of experimental values $(y_k,P_k)_{k=\overline{1,N}}$, with discretization step defined by $\Delta t=(B-A)/(N-1)$ thus $t_k=A+(k-1)\Delta t$ and then the approximate values of Fourier coefficients are given by

$$\begin{aligned}&\alpha\approx\frac{2}{N-1}\sum_{k=1}^{N}P_k\quad \beta_j\approx\frac{2}{N-1}\sum_{k=1}^{N}\left[P_k\cos\left(\frac{2\pi jA}{B-A}+\frac{2\pi j(k-1)}{N-1}\right)\right]\\ &\gamma_j\approx\frac{2}{N-1}\sum_{k=1}^{N}\left[P_k\sin\left(\frac{2\pi jA}{B-A}+\frac{2\pi j(k-1)}{N-1}\right)\right],\quad j=\overline{1,M};\end{aligned} \tag{18}$$

v. If the trigonometric polynomial pass through all the experimental points, that is

$$\begin{aligned}&P(t_i)=Y(t_i)\quad\Leftrightarrow\\ &\frac{\alpha}{2}+\sum_{k=1}^{M}(\beta_k\cos Cky_i+\gamma_k\sin Cky_i)=Y_i,\quad i=\overline{1..N}\end{aligned} \tag{19}$$

The system (16) is equivalent with $\sum_{i=1}^{N}(P(t_i)-Y_i)^2=0$ and represents the same condition find in the least squares method for the discreet case. The coefficients are given by (17) relations. In this case the degree of the trigonometric polynomial, M, has to satisfy the relation $2M+1=N$ where N represents the number of experimental points;

vi. The degree of approximation is given by the residual index defined by

$$R = 100\sum_{i=1}^{N} \frac{\left| y_i^{\exp} - P_i \right|}{y_i^{\exp}} \tag{20}$$

where $y_i^{\exp}$ represents the sequence of experimental values;

vii. The first derivative of the physical signal approximated by the trigonometric polynomial is given by

$$\frac{dP(t)}{dt} \approx -\frac{2\pi}{B-A}\sum_{i=1}^{M}\left(k\beta_k \sin\frac{2\pi kt}{B-A} - k\gamma_k \cos\frac{2\pi kt}{B-A} \right) \tag{21}$$

Previous relation can be useful only when the physical signal is less affected by the noise;

viii. Other application of the trigonometric polynomials is the determination of the integral intensity, *I*, of the physical signal

$$I \approx \frac{\alpha}{2}(B-A). \tag{22}$$

Application of the Fourier series in X-ray diffraction

The spectrum for X-ray diffraction for the nickel foil is represented in Fig.1. It has been registered with the Huber goniometer that used an incident fascicle with synchrotron radiation with the wavelength 1.8276 Å. The experimental data was recorded with constant step $\Delta\theta = 0.033^0$ and its number of pairs is n=766.

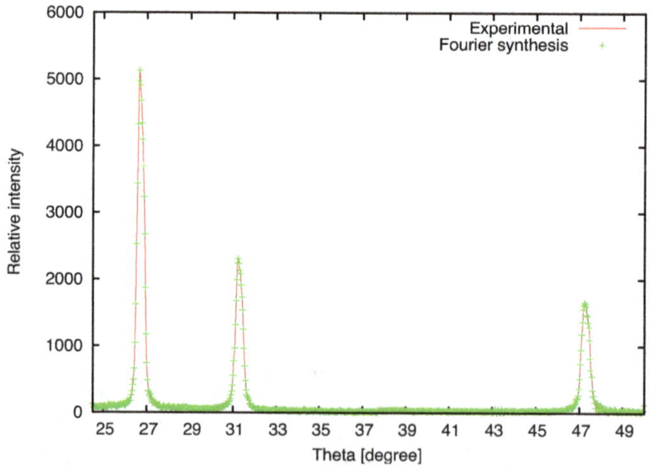

Fig. 1. Experimental spectrum of the nickel foil

The Miller indexes of the X-ray line profiles measurements are (111), (200), respectively (220). Data analysis of the experimental spectrum was realized by our package program (Aldea & Indrea, 1990). The Fig. 2 shows the square magnitude of Fourier coefficients versus the harmonic indexes. They are used in the Warren - Averbach model (Warren, 1990) for the average crystallite size, microstrains of the lattice, total probability of the defaults and the distribution functions of the crystallites and the microstrains determination.

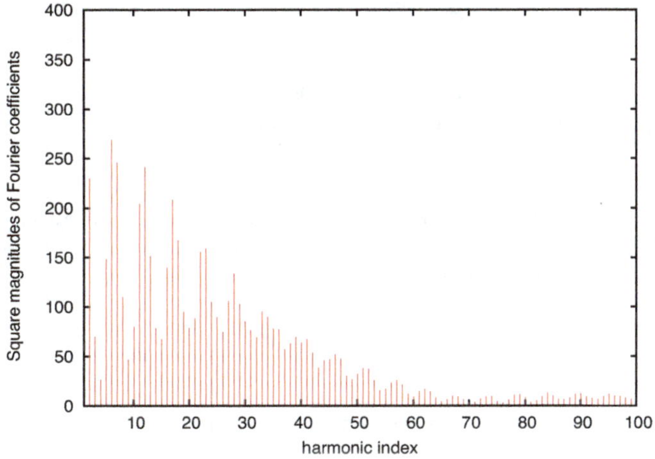

Fig. 2. The square magnitudes of the Fourier coefficients

The broadening of X-ray line profiles can be determined from the first derivative of the experimental spectrum analysis. The first derivative of the nickel foil spectrum in Fig. 3 is given.

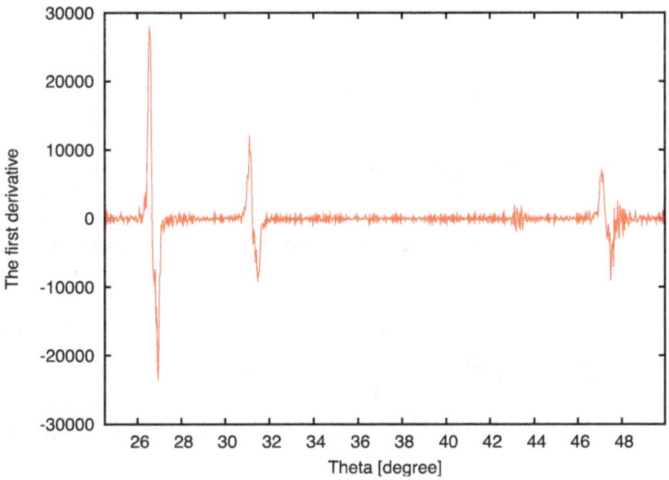

Fig. 3. The first derivative for the computed signal obtained through Fourier synthesis

Application of the trigonometric polynomial in the integration of partial differential equations

One of the most important roles of the trigonometric polynomial is that played in the solving of the partial differential equations. The first example of this procedure is applied to the heat equation in one dimensional space in the conditions of the following problem:

$$\begin{cases} \dfrac{\partial u}{\partial t} = a^2 \dfrac{\partial^2 u}{\partial x^2} \\ u(0,t) = u(L,t) = 0 \text{ (boundary condition)} \\ u(x,t) = g(x) \text{ (intial condition)} \end{cases} \tag{23}$$

where a^2 is the diffusion coefficient, it represents the heat conductibility of the material expressed in cm^2/s. The solution of (23) is obtained using the separation of variables and the Fourier series technique. The solution has the form

$$u(t,x) = \sum_{n=1}^{\infty} B_n(t) \sin\frac{n\pi x}{L} \tag{24}$$

where B_n are determined using the Fourier series technique.

$$B_n(t) = \frac{2}{L}\int_0^L u(x,t)\sin\frac{n\pi x}{L}dx \tag{25}$$

Replacing (24) in the heat equation is obtained

$$B_n^{'}(t) + \left(\frac{n\pi a}{L}\right)^2 B_n(t) = 0, \quad n = 0,1,2,\cdots \tag{26}$$

From the initial condition is obtained

$$b_n = B_n(0) = \frac{2}{L}\int_0^L u(x,0)\sin\frac{n\pi x}{L}dx \quad n = 0,1,2,\cdots \tag{27}$$

The relation (27) shows that the right member represents the n Fourier coefficient for the function that gives initial temperature g. By solving the ordinary differential equation (26) is obtained

$$B_n(t) = b_n \exp\left((-n\pi a / L)^2 t\right) \quad n = 0,1,2,\cdots \tag{28}$$

Replacing (28) in (24), the general solution of the problem (23) for the heat equation is

$$u(t,x) = \sum_{n=1}^{\infty} b_n \exp\left(-\left(\frac{n\pi a}{L}\right)^2 t\right)\sin\frac{n\pi x}{L} \tag{29}$$

where the gaussian part plays the role of damping factor.

Let consider, as the second example, a vibrating string of length L fixed on both ended in the absence of any external force, $0 \le x \le L$ and t>0, its motion is describes by the wave equation

$$\begin{cases} \frac{\partial^2 y}{\partial t^2} = c^2 \frac{\partial^2 y}{\partial x^2} \\ y(0,t) = y(L,t) = 0 \quad \text{(boundary condition)} \\ y(x,0) = f(x) \quad \text{(initial conditions)} \\ \frac{\partial y}{\partial t}(x,0) = g(t) \end{cases} \tag{30}$$

The function y represents the position of each oscillating point versus Ox axis. The function f describes the initial string position and g describes the initial speed of the string. The constant c^2 represents the ratio between straining force and the linear density of the string; the force has the same direction as the movement of the string element. The solution for the problem (30) is obtained using the same technique as in the case of the heat equation and it has the form

$$y(x,t) = \sum_{n=1}^{\infty} \left[a_n \cos(nc\pi t / L) + b_n t \text{sinc}(nct / L) \right] \sin \frac{n\pi x}{L} \tag{31}$$

where "sinc" represents the normalized sinc function, $a_n = B_n(0) = \frac{2}{L}\int_0^L f(x) \sin\frac{n\pi x}{L} dx$ and $b_n = B_n^{'}(0) = \frac{2}{L}\int_0^L g(x) \sin\frac{n\pi x}{L} dx$ for all $n = 0,1,2,\ldots$

Schrödinger equation is the other example presented that describes the quantum behavior in time and space of a particle with m mass inside the potential V and it is given by

$$i\hbar \frac{\partial \Psi}{\partial t} = -\frac{\hbar^2}{2m}\frac{\partial^2 \Psi}{\partial x^2} + V(x)\Psi \tag{32}$$

where $\hbar$ is Planck constant divided by 2π. If the potential energy is vanished the previous equation becomes the free particle equation and this case will be analyzed forward.

$$\begin{cases} i\hbar \frac{\partial \Psi}{\partial t} = -\frac{\hbar^2}{2m}\frac{\partial^2 \Psi}{\partial x^2} \\ \Psi(0,t) = \Psi(L,t) = 0 \\ \Psi(x,0) = f(x) \end{cases} \tag{33}$$

From physical point of view Ψ represents a probability density generator and $|\Psi|^2 = \Psi\Psi^*$ describes the existence probability of a particle of mass m at position x and time t, thus

$$\|\Psi\|^2 = \int_0^L |\Psi(x,t)|^2 dx = 1 \tag{34}$$

Analogous with the previous two examples the solution of the problem (33) is given by

$$\Psi(x,t)=\sum_{n=1}^{\infty}\exp\left[-\frac{i\hbar}{2m}\left(\frac{n\pi}{L}\right)^{2}t\right]c_n\sin\frac{nx\pi}{L}. \tag{35}$$

where $C_n(0)=c_n$, $n=0,\pm 1,\pm 2,\cdots$ and

$$C_n(t)=\frac{2}{L}\int_0^L\Psi(x,t)\sin\frac{nx\pi}{L}\mathrm{dx} \tag{36}$$

Generalization of the Fourier series for the function of infinite period

In the physical and chemical signals analyzing it is often find a non periodical signals defined on the whole real axis. There are many examples in the physics spectroscopy where the signals damp in time due in principal by the absorption process thus there can not be modeled by the periodical functions. The nuclear magnetic resonance (NMR), Fourier transform infrared spectroscopy (FTIR) as well as X-ray absorption spectroscopy (XAS) dedicated to K or L near and extended edges are based on non periodical signals analysis.

Mathematical background of the discreet and inverse Fourier transform

Let consider the complex form of the Fourier series for a signal h defined on the interval $[-T/2,T/2]$ it will be introduced the Fourier transform of h based on the concept of infinite period ($T\to\infty$)

$$h(t)=\frac{1}{T}\sum_{n=-\infty}^{\infty}\underbrace{\left\{\int_{-T/2}^{T/2}h(s)\exp\left(-2\pi i\frac{n}{T}s\right)ds\right\}}_{\text{Fourier coefficients}}\exp\left(2\pi i\frac{n}{T}t\right) \tag{37}$$

where the fundamental frequencies f_0 is expressed by $f_0=1/T$. If it be denoted by $f_n=n/T$ and $\Delta f=f_{n+1}-f_n=1/T$ then the relation (37) becomes

$$h(t)=\sum_{n=-\infty}^{\infty}\left\{\int_{-T/2}^{T/2}h(s)\exp(-2\pi i f_n s)ds\right\}\exp(2\pi i f_n t)\Delta f \tag{38}$$

If $\Delta f\to 0$ then period T goes to infinity and

$$h(t)=\int_{-\infty}^{\infty}\underbrace{\left\{\int_{-\infty}^{\infty}h(s)\exp(-2\pi i f s)\,ds\right\}}_{H(f)}\exp(2\pi i ft)df \tag{39}$$

The function h will be found in the scientific literature named as the Fourier integral (Brigham, 1988) and the expression

$$H(f)=\int_{-\infty}^{\infty}h(s)\exp(-2\pi i f s)\,ds \tag{40}$$

represents the Fourier transform of the function h. From the relation (40) is it possible to obtain the function h by inverse Fourier transform given by

$$h(t) = \int_{-\infty}^{\infty} H(f)\exp(2\pi i f t)\, df \tag{41}$$

The argument of the exponential function from relation (40) is dimensionless. From physical point of view this is very important to emphasize it. For instance, if the argument represents time [s] or distance [m] thus the argument of Fourier transform has dimension [1/s] and [1/m], therefore the product of dimensions for the arguments of Fourier transform is dimensionless.

From physical point of view the difference between the Fourier series and the Fourier transform is illustrated considering two signals one periodic, g, and the other non periodic, h. The non periodical signal, h, is often find in NMR spectroscopy.

$$g(t) = \begin{cases} 0, & \text{if } t<0 \\ a\exp(-\gamma t/2)\exp(2\pi i f_0 t), & \text{if } (n-1)\,\mathrm{T} \le t \le nT, \quad n=1,2,\cdots;\gamma>0 \end{cases} \tag{42}$$

$$h(t) = \begin{cases} 0, & \text{if } t<0 \\ a\exp(-\gamma t/2)\exp(2\pi i f_0 t), & \text{if } t\ge 0,\ \gamma>0 \end{cases} \tag{43}$$

The Fourier coefficients A_n and theirs magnitudes associated with periodical signal (42) are given by relation

$$A_n = a\frac{\exp\left(-\frac{\gamma}{2f_0}\right)-1}{2\pi i f_0(1-n)-\frac{\gamma}{2}} \text{ and } |A_n|^2 = a^2\frac{\left(\exp\left(-\frac{\gamma}{2f_0}\right)-1\right)^2}{4\pi^2 f_0^2(1-n)^2+\frac{\gamma^2}{4}} \tag{44}$$

Fig. 4. shows the graphic for the real part of the periodical g signal for a=10, γ=5 ms^{-1} and f_0=2 KHz.

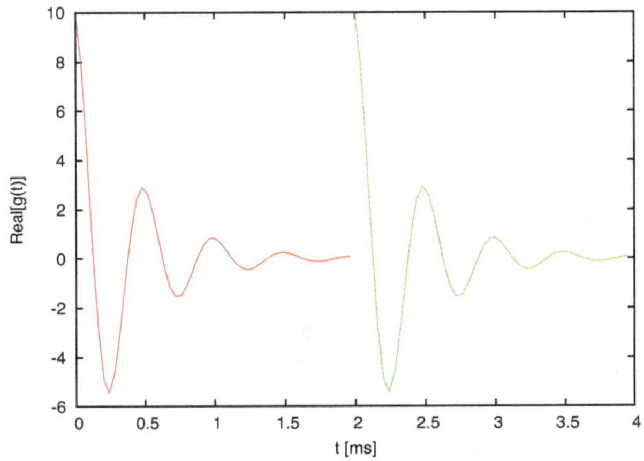

Fig. 4. The periodical physical signal g

Fig. 5 represents the spectral distribution of g signal that it is defined by the square magnitude of the Fourier coefficients.

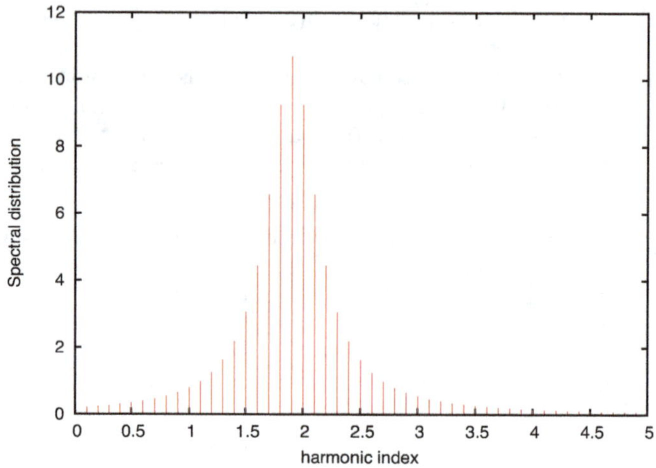

Fig. 5. The spectral distribution of the signal g

The real component of the function (43) is represented in Fig. 6 and the square magnitude of the Fourier transform is given by relation (45) and it is represented in Fig. 7.

$$|H(f)|^2 = \frac{a^2}{4\pi^2(f_0 - f)^2 + \frac{\gamma^2}{4}} \tag{45}$$

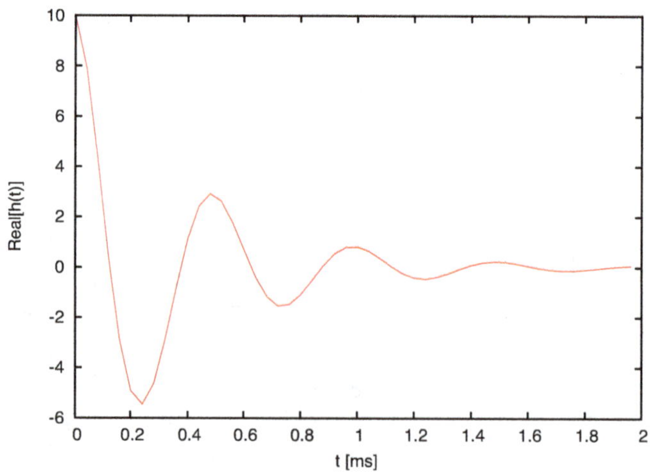

Fig. 6. The real part of non periodical signal h

The maximum value for the spectral distribution take place when $f = f_0$, thus $|H_{\max}(f)|^2 = 4a^2/\gamma^2$. The full width at half maxim (FWHM) is denoted by $\Delta f_{1/2}$ and $\Delta f_{1/2} = \gamma/2\pi$

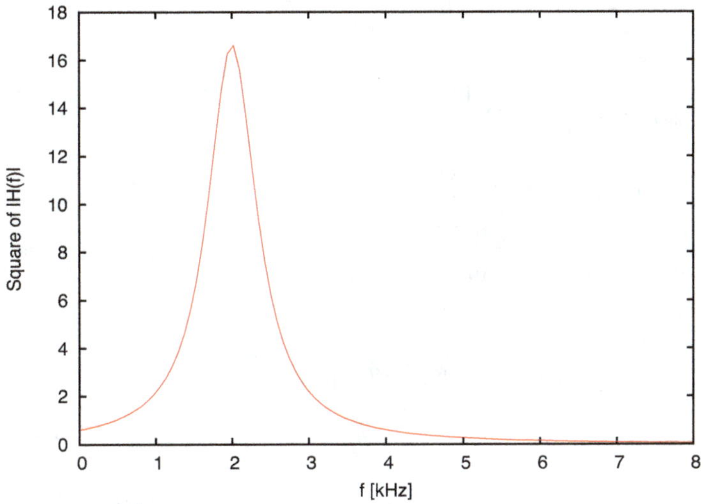

Fig. 7. The square magnitude of Fourier transform of the signal *h*

Some times in signals analysis theory, the inverse of the damping parameter γ is named the relaxation time denoted by τ. Between the relaxation time and FWHM there is the relation $\tau \Delta f_{1/2} = \dfrac{1}{2\pi}$.

From physical reason it is easier to analyze the resonant answer of a physical signal than free damping oscillations. In NMR spectroscoply this signal is known as Free Impulse Decay (FID). Studying the resonant answer it is possible to obtain the relaxation time parameter of the free oscillations. The relaxation time derives from the width of the spectrum obtained from the Fourier transform of the FID. Therefore it is easier to determine the relaxation time from Fourier transform of *h* signal instead of fit technique applied to FID.

Above it was shown that the Fourier series of periodical signal is represented as a sum of periodical functions with discreet frequencies f_0, $2f_0$, $3f_0$, ... as shown in Fig. 5. The amplitudes of the signals associated to each frequency are given by the spectral distribution named Fourier analysis. The difference between Fourier series and Fourier transform is that the latter has the frequencies as argument which continuously varies. Whereas Fourier transform of the signal *h* allows spectral decomposition of it with frequencies defined on the whole real axis.

The Fourier transform for discreet signals

In practice the function *h* represents a physical signal resulted from an experiment. The signal can be discretizated on *N* samples with a constant step Δt. From physical reasons, the experimental signals can not be acquisitioned on the entire real axis thus the working interval is $\left[-N\Delta t / 2, (N/2-1)\Delta t\right]$ (Mandal & Asif, 2007). Instead of the function *H* there are a set of pairs $(n\Delta f, H_n)$, $n = \overline{0, N-1}$ where Δf represents the discretization step of data. Let consider the following relation between discretization steps

$$\Delta f = \frac{1}{N\Delta t} \tag{46}$$

the Fourier transform associated to the set $(h(k\Delta t))_{k=\overline{-N/2,N/2-1}}$ is contained in the H vector with the components

$$H(n\Delta f) = \Delta t \sum_{k=-N/2}^{N/2-1} h(k\Delta t)\exp\left(-\frac{2\pi ikn}{N}\right), \quad n = \overline{0, N-1}. \tag{47}$$

It is more convenient, for the computation, if all the indexes are positive, for this it is assumed that $q = k + N$ for k<0. Relation (47) becomes

$$H(n\Delta f) = \Delta t \sum_{k=0}^{N-1} h(k\Delta t)\exp\left(-\frac{2\pi ikn}{N}\right). \tag{48}$$

Using the same consideration, the inverse Fourier transform has the form

$$h(n\Delta f) = \Delta f \sum_{k=0}^{N-1} \underbrace{\left[\operatorname{Re} H(k\Delta f) + i\operatorname{Im} H(k\Delta f)\right]}_{H(k\Delta f)} \exp\left(\frac{2\pi ikn}{N}\right) \tag{49}$$

If the physical signal is recorded on negative arguments then for the numerical computation of the Fourier transform for the components $(k\Delta t, h_k)$, $k = -N/2 \ldots N/2 - 1$ must be arrange in the following order

$$\begin{gathered} (0,1,2,\cdots,N/2-1) \rightarrow (h_0, h_1, h_2, \cdots, h_{N/2-1}) \\ \left(\frac{N}{2}, \frac{N}{2}+1, \frac{N}{2}+2, \cdots, N-1\right) \rightarrow (h_{-N/2}, h_{-N/2+1}, h_{-N/2+2}, \cdots, h_{-1}). \end{gathered} \tag{50}$$

The main algorithms for the Fourier transform. The Filon quadrature and the Cooley-Tukey method

Most of the times the physical-chemical signals can not be expressed analytical, thus it is impossible to use the relation (40). Therefore in computation is used the discreet form of Fourier transform given by the relation (48). Generally speaking the physical signals are recorded around thousands of points; additionally the Fourier transform of the signal is important to compute on the same numbers. By using relation (48) the computation time is too long. This problem was solved by Cooley Tukey algorithm (Brigham, 1988) named in the literature as a Fast Fourier Transform (FFT) method. In the case when the physical signal is registered from pairs in a range of hundreds up to few thousand values can be successfully used the Filon algorithm (Abramowitz & Stegun, 1972). The method assumes that the physical signal is defined on the interval $[t_0, t_{2n}]$ with step Δt then the real component of the Fourier transform is approximated by

$$\begin{aligned} \int_{t_0}^{t_{2n}} h(t)\cos(2\pi tf)dt \approx \Delta t\,\alpha(2\pi f\Delta t)h_{2n}\sin(2\pi t_{2n}f) - \\ - \Delta t\left[h_0 \sin(2\pi t_0 f) - \beta(2\pi f\Delta t)C_{2n} - \gamma(2\pi f\Delta t)C_{2n-1}\right] \end{aligned} \tag{51}$$

where

$$C_{2n} = \sum_{j=0}^{n} h_{2j} \cos(2\pi f t_{2j}) - \frac{1}{2}[h_{2n} \cos(2\pi f t_{2n}) + h_0 \cos(2\pi f t_0)]$$
$$C_{2n-1} = \sum_{j=0}^{n} h_{2j-1} \cos(2\pi f t_{2j-1}), \quad \alpha(\theta) = \frac{1}{\theta} + \frac{\sin 2\theta}{2\theta^2} - \frac{2\sin^2\theta}{2\theta^3} \tag{52}$$
$$\beta(\theta) = 2\left(\frac{1+\cos^2\theta}{\theta^2} - \frac{\sin 2\theta}{\theta^3}\right), \quad \gamma(\theta) = 4\left(\frac{\sin\theta}{\theta^3} - \frac{\cos\theta}{\theta^2}\right)$$

and the imaginary component of the Fourier transform is given by

$$\int_{t_0}^{t_{2n}} h(t)\sin(2\pi t f)dt \approx \Delta t\, \alpha(2\pi f \Delta t) h_0 \cos(2\pi t_0 f) - \tag{53}$$
$$-\Delta t\left[h_{2n}\cos(2\pi t_{2n} f) - \beta(2\pi f \Delta t) S_{2n} - \gamma(2\pi f \Delta t) S_{2n-1}\right]$$

where

$$S_{2n} = \sum_{j=0}^{n} h_{2j} \sin(2\pi f t_{2j}) - \frac{1}{2}[h_{2n} \sin(2\pi f t_{2n}) + h_0 \sin(2\pi f t_0)] \tag{54}$$
$$S_{2n-1} = \sum_{j=0}^{n} h_{2j-1} \sin(2\pi f t_{2j-1}).$$

By taking into account the relation (46) used in the FFT method it is not possible to compute the Fourier transform for all value of the frequency. This disadvantage can lead to poor resolution of the Fourier transform *H*. Meanwhile the Filon algorithm is more time consuming but its application offers a more reliable resolution. A detailed analysis of these algorithms applied in the extended X-ray absorption fine structure (EXAFS) spectroscopy can be found in the paper (Aldea & Pintea, 2009).

Application of the Fourier transform in X-ray absorption spectroscopy and X-ray diffraction

The study of XAS can yield electronic and structural information about the local environment around a specific atomic constituent in the amorphous materials (Kolobov et al., 2005),

Additional, this method provides information about the location and chemical state of any catalytic atom on any support (Miller et al., 2006) as well as the nanoparticle of transition metal oxides (Chen et al., 2002; Turcu et al., 2004)). X-ray absorption near edge structure (XANES) is sensitive to local geometries and electronic structure of atoms that constitute the nanoparticles. The changes of the coordination geometry and the oxidation state upon decreasing the crystallite size and the interaction with molecules absorbed on nanoparticles surface can be extracted from XANES spectrum.

The EXAFS is a specific element of the scattering technique in which a core electron ejected by an X-ray photon probes the local environment of the absorbing atom. The ejected photoelectron backscattered by the neighboring atoms around the absorbing atom interferes constructively with the outgoing electron wave, depending on the energy of the photoelectron. The energy of the photoelectron is equal to the difference between the X-ray energy photon and a threshold energy associated with the ejection of the electron.

X-ray diffraction (XRD) line broadening investigations of nanostructured materials have been limited to find the average crystallite size from the integral breadth or the FWHM of the diffraction profile. In the case of nanostructured materials due to the difficulty of performing satisfactory intensity measurements on the higher order reflections, it is impossible to obtain two orders of (hkl) profile. Consequently, it is not possible to apply the classical method of Warren and Averbach (Warren, 1990). On the other hand we developed a rigorous analysis of the X-ray line profile (XRLP) in terms of Fourier transform where zero strains assumption is not required. The apparatus employed in a measurement generally affects the obtained data and a considerable amount of work has been done to make resolution corrections. In the case of XRLP, the convolution of true data function by the instrumental function produced by a well-annealed sample is described by Fredholm integral equation of the first kind (Aldea et al., 2005; Aldea & Turcu, 2009). A rigorous way for solving this equation is Stokes method based on Fourier transform technique. The local and global structure of nanosized nickel crystallites were determined from EXAFS and XRD analysis.

EXAFS analysis

The interference between the outgoing and the backscattered electron waves has the effect of modulating the X-ray absorption coefficient. The EXAFS function $\chi(k)$ is defined in terms of the atomic absorption coefficient by

$$\chi(k) = \frac{\mu(k) - \mu_0(k)}{\mu_0(k)} \tag{55}$$

where k is the electron wave vector, $\mu(k)$ refers to the absorption by an atom from the material of interest and $\mu_0(k)$ refers to the atom in the free state. The theories of the EXAFS based on the single scattering approximation of the ejected photoelectron by atoms in the immediate vicinity of the absorbing atom gives an expression for $\chi(k)$

$$\chi(k) = \sum_j A_j(k)\sin\left[2kr_j + \delta_j(k)\right] \tag{56}$$

where the summation extends over j coordination shell, r_j is the radial distance from the jth shell and $\delta_j(k)$ is the total phase shift function. The amplitude function $A_j(k)$ is given by

$$A_j(k) = \left(\frac{N_j}{kr_j^2}\right)F(k,r_j,\pi)\exp\left(-2\left(r_j / \lambda_j(k) - k^2\sigma_j^2\right)\right) \tag{57}$$

In this expression N_j is the number of atoms in the jth shell, σ_j is the root mean squares deviation of distance about r_j, $F(k, r_j, \pi)$ is the backscattering amplitude and $\lambda_j(k)$ is the mean free path function for the inelastic scattering. The backscattering factor and the phase shift depend on the kind of atom responsible for scattering and its coordination shell (Aldea et al., 2007). The analysis of EXAFS data for obtaining structural information $[N_j, r_j, \sigma_j, \lambda_j(k)]$

generally proceeds by the use of the Fourier transform. From $\chi(k)$, the radial structure function (RSF) can be derived. The single shell may be isolated by Fourier transform,

$$\Phi(r) = \int_{-\infty}^{\infty} k^n \chi(k) WF(k) \exp(-2ikr) dk. \tag{58}$$

The EXAFS signal is weighted by k^n (n=1, 2, 3) to get the distribution function of atom distances. Different apodization windows WF(k) are available as Kaiser, Hanning or Gauss filters. An inverse Fourier transform of the RSF can be obtained for any coordination shell,

$$\chi_j(k) = \frac{1}{k^n} WF(k) \int_{R_{1j}}^{R_{2j}} \Phi(r) \exp(2ikr) dr. \tag{59}$$

The theoretical equation for $\chi_j(k)$ function is given by:

$$\chi_j(k) = A_j(k) \sin\left[2kr_j + \delta_j(k)\right], \tag{60}$$

where the index j refers to the jth coordination shell. The structural parameters for the first coordination shell are determined by fitting the theoretical function $\chi_j(k)$ given by the relation (60) with the experimental signal $\chi_j(k)$ derived from relation (59). In the empirical EXAFS calculation, $F(k,r,\pi)$ and $\delta_j(k)$ are conveniently parameterized (Aldea et al., 2007). Eight coefficients are introduced for each shell:

$$F_s(k,r,\pi) = c_0\left[\exp(c_1 k + c_2 k^2)\right] / k^{c_3} \tag{61}$$

$$\delta_s(k) = a_{-1} k^{-1} + a_0 + a_1 k + a_2 k^2 \tag{62}$$

The coefficients c_0, c_1, c_2, c_3, a_{-1}, a_0, a_1 and a_2 are derived from the EXAFS spectrum of a compound whose structure is accurately known. The values N_s and r_s for each coordination shell for the standard sample are known. The trial values of the eight coefficients can be calculated by algebraic consideration and then they are varied until the fit between the observed and calculated EXAFS is optimized.

XRD analysis

X-ray diffraction pattern of a crystal can be described in terms of scattering intensity as function of scattering direction defined by the scattering angle 2θ, or by the scattering parameter $s = (2\sin\theta)/\lambda$, where λ is the wavelength of the incident radiation. It is discussed the X-ray diffraction for the mosaic structure model in which the atoms are arranged in blocks, each block itself being an ideal crystal, but with adjacent blocks not accurately fitted together. The experimental XRLP, h, represents the convolution between the true sample f and the instrumental function g

$$h(s) = \int_{-\infty}^{+\infty} g(s - s^*) f(s^*) \, ds^* \tag{63}$$

The equation (63) is equivalent with the following relation

$$H(L) = G(L) F(L), \tag{64}$$

where $F(L)$, $H(L)$ and $G(L)$ are the Fourier transforms of the true sample, experimental XRLP and instrumental function, respectively. The variable L is the perpendicular distance to the (hkl) reflection planes. The generalized Fermi function (GFF) (Aldea et al., 2000) is a simple function with a minimal number of parameters, suitable for the XRLP global approximation based on minimization methods and it is defined by relation:

$$h(s) = \frac{A}{e^{a(s-c)} + e^{-b(s-c)}}, \tag{65}$$

where A, a, b, c are unknown parameters. The values A, c describe the amplitude, the position of the XRLP and a, b control its shape. In the case when X-ray line profiles h and g are approximated by GFF distribution then the solution of Fedholm integral equation of the first kind represents the true sample function and it is given by

$$f(s) = \frac{2A_h\rho_g}{\pi A_g} \frac{\cos\frac{\pi\rho_h}{2\rho_g}\cosh\rho_h s}{\cos\frac{\pi\rho_h}{\rho_g} + \cos 2\rho_h s} \tag{66}$$

where the arguments of trigonometric and hyperbolic functions depend on the shape parameters of the h, g signals, respectively. They are expressed by $\rho_h = (a_h + b_h)/2$ and $\rho_g = (a_g + b_g)/2$.

EXAFS results

The extraction of the EXAFS signal is based on the threshold energy of the nickel K edge determination followed by background removal of pre-edge and after-edge base line fitting with different possible modeling functions where $\mu_0(k)$ and $\mu(k)$ evaluation are presented in Fig. 8.

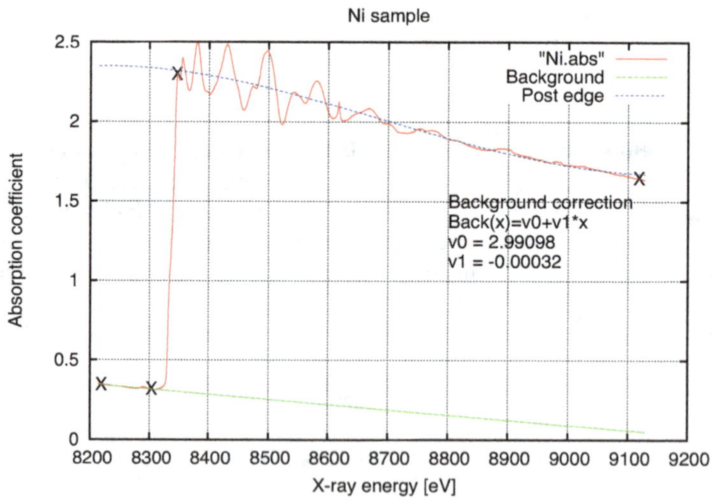

Fig. 8. The absorption coefficient of the nickel K edge

In according with relation (55) EXAFS signals modulated by Hanning and Gauss filters were performed in the range 2 Å -14 Å and they are shown in Fig. 9. In order to obtain the atomic distances distribution it was computed the RSF, using the relation (58) and the Filon algorithm.

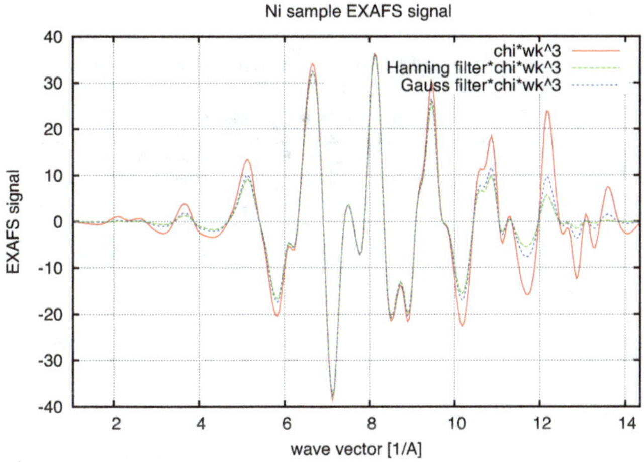

Fig. 9. EXAFS signal for the nickel crystallites

The mean Ni-Ni distances of the first coordination shell for standard sample at room temperature are closed to values of R_1=2.49Å. Based on relation (46) between Δk and Δr steps, the computation of the RSF using the FFT of the EXAFS signal gives a non reliable resolution. To avoid this disadvantage it used the Filon algorithm for Fourier transform procedure. Based on this procedure the Fourier transform of $k^3\chi(k)WF(k)$, performed in the range 0.51 Å and 2.79 Å, are shown in Fig. 10 for the standard Ni foil investigation. In order to minimize the spurious errors in the RSF it was considered Gauss filter as the window function.

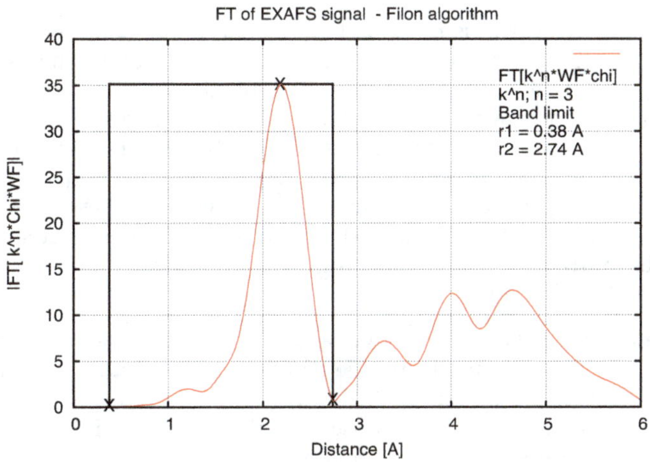

Fig. 10. The Fourier transform of the EXAFS spectrum for the nickel foil

Each peak from $|\Phi(r)|$ is shifted from the true distance due to the phase shift function that is included in the EXAFS signal. We proceed by taking the inverse Fourier transform given by relation (59) of the first neighboring peak, and then extracting the amplitude function $A_j(k)$ and the phase shift function $\delta(k)$ in according with the relations (61) and (62).

By Lavenberg-Marquard fit applied to the relation (60) and from the experimental contribution for each coordination shell, are evaluated the interatomic distances, the number of neighbors

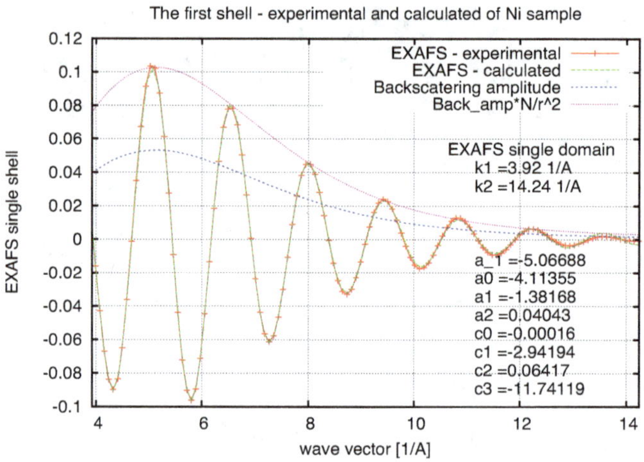

Fig. 11. Experimental and calculated EXAFS signals of the first coordination shell of the nickel foil

and the edge position. Fig. 11 shows the calculated and the experimental EXAFS functions $\chi_1(k)$ of the first shell for the investigated sample.

XRD results

Practically speaking, however, it is not easy to obtain accurate values of the crystallite size and the microstrain without extreme care in the experimental measurements and analysis of XRD data. The Fourier analysis of XRLP validity depends strongly on the magnitude and the nature of the errors propagated in the data analysis. In paper (Aldea et al., 2000) are treated three systematic errors: the uncorrected constant background, the truncation and the effect of the sampling for the observed profile at a finite number of points that appear in discrete Fourier analysis. In order to minimize propagation of these systematic errors, a global approximation of the XRLP is adopted instead of the discrete calculations. Therefore, the analysis of the diffraction line broadening in X-ray powder pattern was analytically calculated using the GFF facilities.

The reason of this choice, as described above, was simplicity and the mathematical elegance of the analytical Fourier transform magnitude and the integral width of the true XRLP. The robustness of the GFF approximation for the XRLP arises from possibility of using the analytical form of the Fourier transform instead of the numerical FFT. The validity of the microstructural parameters are closely related to accuracy of the Fourier transform magnitude of the true XRLP. The experimental relative intensities with respect to θ values and the nickel foil as instrumental broadening effect are shown in Fig. 12. The next steps consist in the background correction of the XRLP by the polynomial procedures and the determination of the best parameters of GFF distributions by nonlinear least squares fit. In order to determine the average crystallite size, the lattice microstrain and the probability of defects were computed the true XRLP by the Fourier transform technique and it is illustrated in Fig. 13, the curve is centred on its mass centre s_0.

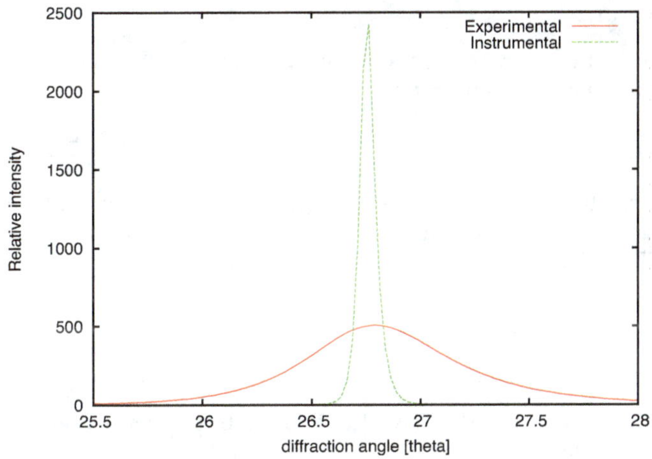

Fig. 12. The experimental XRLP (*h*) and the instrumental signals (*g*)

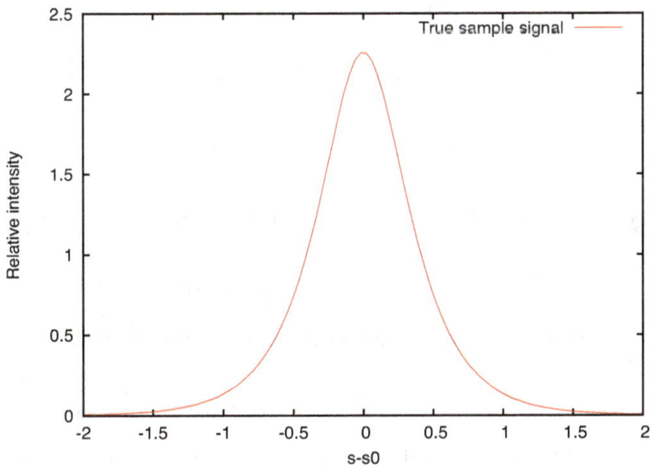

Fig. 13. The true sample signal (*f*)

Conclusions

In this contribution it has presented the mathematical background of Fourier series and Fourier transform used in nanomaterials structure field.

The conclusions that can be drawn from this contribution are:

i. The physical periodical signals are successfully modeled using the trigonometric polynomial such us global approximation of the XRLP and the spectral distribution determination based on the Fourier analysis;
ii. The most important tools applied in EXAFS is based on the direct and inverse Fourier transform methods;
iii. The examples presented are based on the original contributions published in the scientific literature.

The experimental data used in analyses consists in measurements that have done to Beijing Synchrotron Radiation Facilities from High Institute of Physics.

Acknowledgement

The authors are grateful to Beijing Synchrotron Radiation Facilities staff for beam time and for their technical assistance in XAS and XRD measurements. This work was partially supported by UEFISCDI, projects number 32-119/2008 and 22-098/2008.

Appendix

In this appendix are given the main analytical properties of the Fourier transform.

i. *Linearity.* If the signals x and y have the Fourier transform X and Y then the Fourier transform of $\alpha x + \beta y$ is $\alpha X + \beta Y$.
ii. *Symmetry.* If the Fourier transform of the function h is H, then

$$h(-f) = \int_{-\infty}^{\infty} H(t)\exp(-2\pi i f t)dt \tag{67}$$

iii. *Scaling.* If h has the Fourier transform H then

$$\int_{-\infty}^{\infty} h(kt)\exp(-2\pi ift)dt = \frac{1}{|k|}H(f / k), \quad k \in R^{*} \tag{68}$$

iv. *Shifting.* If the Fourier transform of h is H and h is translated with t_0 then

$$\int_{-\infty}^{\infty} h(t - t_0)\exp(-2\pi\, ift)dt = \exp(-2\pi\, if\, t_0)H(f) \tag{69}$$

v. If the signal h_e is even function and the Fourier transform of h_e is H_e then

$$H_e(f) = \int_{-\infty}^{\infty} h_e(t)\cos(2\pi\, ft)dt \tag{70}$$

vi. If the signal h_o is odd function and the Fourier transform of h_o is H_o then

$$H_o(f) = -i\int_{-\infty}^{\infty} h_o(t)\sin(2\pi\, ft)dt \tag{71}$$

vii. Any real function defined on a symmetrical interval has the following decomposition

$$h(t) = h_o(t) + h_e(t)\,, \tag{72}$$

thus

$$H(f) = H_o(f) + H_e(f) \tag{73}$$

For the next properties are assuming the function h is sufficiently smooth so that can be acceptable the differentiation and the integration.

viii. The Fourier transform of the derivative of the function h is given by

$$H_{\frac{dh}{dt}}(f) = (2\pi i\, f)H_h(f) \tag{74}$$

where H_h represents the Fourier Transform of the function h, and for the n times derivative

$$H_{\frac{d^n h}{dt^n}}(f) = (2\pi i\, f)^n H_h(f) \tag{75}$$

ix. The derivative of the Fourier transform is given by

$$\frac{dH_h(f)}{df} = -2\pi i H_{th}(f) \tag{76}$$

and for the n times derivative of the Fourier transform

$$\frac{d^n H_h(f)}{df^n} = (-2\pi i)^n H_{t^n h}(f) \tag{77}$$

References

Abramowitz, M. & Stegun, I. (1972). *Handbook of Mathematical Functions,* Dover Publication, ISBN 0-486-61272-4, NY, USA

Aldea, N. & Indrea, E. (1990). XRLINE, a program to evaluate the crystallite size of supported metal-catalysts by single X-ray profile fourier-analysis. *Computer Physics Communications,* Vol.60, No.1, (August 1990), pp. 155-163, ISSN: 0010-4655

Aldea, N.; Gluhoi, A.; Marginean, P.; Cosma, C. & Yaning X. (2000). Extended X-ray absorption fine structure and X-ray diffraction studies on supported nickel catalysts. *Spectrochimica Acta B,* Vol.55, No.7, (July 2000), pp. 997-1008, ISSN: 0584-8547

Aldea, N.; Barz, B.; Silipas, T. D.; Aldea, F. & Wu, Z. (2005). Mathematical study of metal nanoparticle size determination by single X-ray line profile analysis. *Journal of Optoelectronics and Advanced Materials,* Vol.7, No.6, (December 2005), pp. 3093-3100, ISSN: 1454-4164

Aldea, N.; Marginean, P.; Rednic, V.; Pintea, S.; Barz, B.; Gluhoi, A.; Nieuwenhuys, B. E.; Xie Y.; Aldea, F. & Neumann, M. (2007). Crystalline and electronic structure of gold nanoclusters determined by EXAFS, XRD and XPS methods. *Journal of Optoelectronics and Advanced Materials,* Vol.9, No.5, (May 2007), pp. 1555-1560, ISSN 1454-4164

Aldea, N.; Turcu, R.; Nan, A.; Craciunescu, I; Pana, O; Yaning, X.; Wu, Z.; Bica, D; Vekas, L. & Matei, F. (2009). Investigation of nanostructured Fe_3O_4 polypyrrole core-shell composites by X-ray absorbtion spectroscopy and X-ray diffraction using synchrotron radiation. *Journal of Nanoparticle Research*, Vol.11, No.6, (August 2009), pp. 1429-1439, ISSN: 1388-0764.

Aldea, N.; Pintea, S.; Rednic, V.; Matei, F. & Xie Y. (2009). Comparative study of EXAFS spectra for close-shell systems. *Journal of Optoelectronics and Advanced Materials*, Vol.11, No.12, (December 2009), pp. 2167 – 2171, ISSN: 1454-4164

Bachmann, G.; Narici, L. & Beckenstein, E. (2002). *Fourier and Wavelet Analysis* (2nd edition), Springer, ISBN 978-0-387-98899-3, New York, USA

Brigham, E. (1988). *The Fast Fourier Transform and Its Applications*, Prentice Hall, ISBN 0133075052, New Jersey, USA

Chen, L.X.; Liu, T.; Thurnauer, M.C.; Csencsits, R. & Rajh, T. (2002). Fe_2O_3 nanoparticle structures investigated by X-ray absorption near-edge structure, surface modifications, and model calculations. *Journal of Physical Chemistry B*, Vol.106, No.34, (August 2002), pp. 8539-8546, ISSN 1520-6106.

Kolobov, A. V.; Fons, P.; Tominaga, J.; Frenkel, A. I.; Ankudinov, A. L. & Uruga, T. (2005). Local Structure of Ge-Sb-Te and its modification Upon the Phase Transition. *Journal of Ovonic Research*, Vol.1, No.1, (February 2005), pp. 21 – 24, ISSN 1584 - 9953

Mandal, M & Asif, A. (2007). *Continuous and Discrete Time Signals and Systems*, Cambridge University Press, ISBN 9780521854559, London , UK.

Miller, J.T.; Kropf, A.J.; Zha, Y.; Regalbuto, J.R.; Delannoy, L.; Louis, C.; Bus, E. & Bokhoven, J.A. (2006). The effect of gold particle size on Au{single bond}Au bond length and reactivity toward oxygen in supported catalysts. *Journal of Catalysis*, Vol.240, No.2, (June 2006), pp. 222-234, ISSN: 00219517

Richard, P.; Leighton, R.B. & Sands M. (2005). *The Feynman Lectures on Physics, Vol. 1: Mainly Mechanics, Radiation, and Heat,* (2nd edition), Addison Wesley, ISBN 978-0805390469, Boston, USA

Tang, K-T. (2007). *Mathematical Methods for Engineers and Scientists 3, Fourier Analysis, Partial Differential Equations and Variational Methods* (vol. 3), Springer-Verlag, ISBN 978-3540446958, Berlin-Heidelberg, Germany

Turcu, R.; Peter, I.; Pana, O.; Giurgiu, L; Aldea, N.; Barz, B.; Grecu, M.N. & Coldea, A. (2004). Structural and magnetic properties of polypyrrole nanocomposites. *Molecular Crystals and Liquid Crystals*, Vol. 417, pp. 719-727, ISSN 1058-725X

Walker, J. (1996). *Fast Fourier Transforms*, (2nd edition), CRC Press, ISBN 978-0849371639, Boca Raton, USA

Warren, B. E. (1990). *X-Ray Diffraction*, Dover Publications, ISBN 0486663175, NY, USA

4

Fourier Transform Sound Radiation

F. X. Xin and T. J. Lu
State Key Laboratory for Mechanical Structure Strength and Vibration, School of Aerospace, Xi'an Jiaotong University, Xi'an, China

Introduction

With the increasing use of periodically rib-stiffened composite sandwich structures as the cabin skin of aircrafts, marine ships and express trains etc. [1-5], great efforts have been made in the pursuit of efficient theoretical methods for predicting the vibration and acoustic behaviors of these lightweight structures, so as to design optimized configurations competent for practical low-noise requirements.

Active control algorithms with sensors and actuators have been developed to reduce structural vibration and sound radiation [6], which however inevitably brings the penalty of increasing system complexity and financial costs. Alternatively, passive measures such as inserting fibrous sound absorptive materials in the partitioned cavity of sandwich structures may be a preferable choice to achieve a compromise between noise-reduction efficiency and financial cost. For instance, the fuselages of commercial aircrafts are commonly made of periodically rib-stiffened composite structures filled with fiberglass to enhance thermal and sound insulation [6-9]. This provides strong impetus for the development of effective theoretical models to predict the sound radiation characteristics of periodically rib-stiffened sandwich structures filled with sound absorptive materials.

There exist numerous theoretical models for the vibroacoustic behaviors of periodic rib-stiffened structures, which may be grouped into two main categorizes: one is based on the Fourier transform method [2-4,10-13], which is able to handle both sound radiation and sound transmission problems; the other is built upon the space-harmonic approach [14-18], which are suited particularly for sound transmission problems. Mace [2] employed the Fourier transform method to solve the problem of sound radiation from a fluid-loaded infinite plate reinforced by two sets of parallel stiffeners when excited by a point force; for simplification, only the tensional force of the rib-stiffeners was considered. Subsequently, Mace [3] proposed a theoretical model for the radiation of sound from an infinite fluid loaded plate when the plate is reinforced with two sets of orthogonal line stiffeners; again, only the tensional force of the rib-stiffeners was accounted for. Similarly, by only taking account of the normal force interaction between panel and rib-stiffeners, Yin et al. [4] presented a simplified theoretical model for acoustic radiation from a point-driven, fluid-loaded infinite laminated composite plate reinforced by periodic parallel rib-stiffeners.

As an essentially equivalent method, the space-harmonic approach evolved from progressive wave propagation was initiated by Mead and Pujara [14] when they studied the acoustical response of periodical stiffened beams subjected to a spatial and temporal harmonic pressure. It was demonstrated that as few as three terms of space harmonics could lead to solutions of acceptable accuracy. By combining the space-harmonic approach and virtual energy method, Lee and Kim [15] analyzed the sound transmission characteristics of a thin plate stiffened by equally spaced line stiffeners. By modeling the rib-stiffeners as a combination of translational springs and rotational springs, Wang et al. [16] proposed an analytical model for sound transmission loss across double-leaf partitions stiffened with periodically placed studs. Recently, Xin and Lu [18] developed a comprehensive analytical model for sound transmission through orthogonally rib-stiffened sandwich structures: all possible motions of the rib-stiffeners were accounted for by introducing the tensional forces, bending moments and torsional moments as well as the corresponding inertial terms into the governing equations of the two face panels.

None of the above mentioned investigations dealt with sound radiation and/or sound transmission issues of composite sandwich structures filled with porous sound absorptive materials. As far as the sound radiation/transmission problems of double partitions with cavity absorption is of concern, a number of theoretical [12,13], numerical [19,20] and experimental [21] studies do exist. However, all of these studies *did not* consider the effects of structural rib-connections between two face panels, which may be far away from the factual engineering structures. To address this deficiency, a comprehensive theoretical model is developed here for the radiation of sound from an infinite orthogonally rib-stiffened sandwich structure filled with fibrous sound absorptive material in the partitioned cavity when excited by a time-harmonic point force. The equivalent forces and moments (both bending and torsional) imposed on the two face panels by the rib-stiffeners are accounted for by considering all possible motions of the rib-stiffeners. By employing the well-known equivalent fluid model [12,22], wave propagation in the fibrous sound absorptive material can be accurately described. Both viscous drag forces and thermal exchanges between air and solid fibers are accounted for by introducing frequency dependent dynamic density and bulk modulus. Taking advantage of the periodical property of the composite sandwich structure, the Fourier transform technique is adopted to solve both the structural and acoustical governing equations. In limiting cases the developed model can be favorably degraded to deal with sound radiation issues of sandwich structures with vacuum or air cavities. Therefore, model validation is carried out by comparing the present predictions for simplified sandwich structures with those available in the open literature. To explore the influence of fibrous sound absorptive materials on sound radiation of orthogonally rib-stiffened composite structures, numerical results are presented, with relevant physical features interpreted in detail. Conclusions drawn from the present theoretical study may provide fundamental principles for factual engineering design of rib-stiffened composite structures filled with fibrous sound absorptive materials.

Structural dynamic responses to time-harmonic point force

Analytical formulation of panel vibration

Consider an infinite sandwich structure as shown schematically in Fig. 1, which is reinforced by two periodic sets of orthogonal rib-stiffeners having periodic uniform separations l_x and l_y in the x- and y-directions, respectively. A right-handed Cartesian coordinate system (x, y, z) is established, with its x-axis and y-axis positioned separately along one pair of the orthogonal rib-stiffeners, the positive direction of the z-axis pointing downward (Fig. 1). Three different kinds of sandwich structures will be considered in the proceeding sections, namely, the gap between the two parallel face panels and portioned by the orthogonal lattice cores is in vacuum, air filled, or filled with fibrous sound absorptive material (e.g., fiberglass), respectively. A theoretical model will be formulated for the complex structure (i.e., orthogonally rib-stiffened sandwich structure filled with fibrous sound absorptive material), which can be degraded to deal with the other two sandwiches.

Assume that a time-harmonic point force $q_0 e^{i\omega t}$ acts on the surface of the upper panel at location (x_0, y_0); see Fig. 1. Consequently, a radially outspreading bending wave propagates in the upper panel from the source (x_0, y_0). The vibration of the upper panel is transmitted to the bottom panel via the orthogonal rib-stiffeners and sound-absorbing material (or air cavity). Subsequently, the bottom panel vibrates and radiates sound pressure waves.

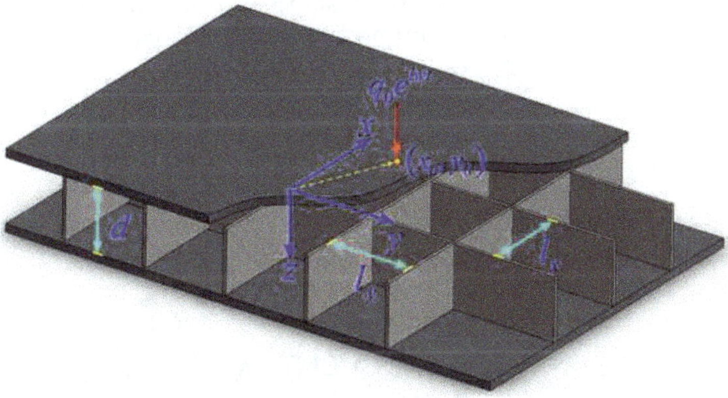

Fig. 1. Schematic illustration of orthogonally rib-stiffened sandwich structure (three different kinds: vacuum cavity, air cavity, and fiberglass filled cavity) excited by time-harmonic point force at (x_0, y_0)

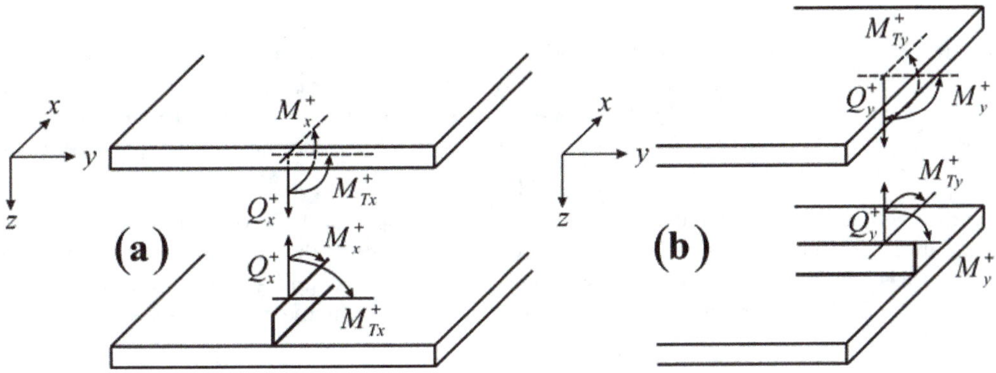

Fig. 2. Conventions for tensional forces, bending moments and torsional moments between upper panel and (a) x-wise and (b) y-wise stiffeners, which also hold at the interface between bottom panel and x- and y-wise stiffeners

Upon point force excitation, the vibration of the upper and bottom panels can be described using two dynamic governing equations, where the influence of the rib-stiffeners exists in the form of tensional forces (general force plus inertial force), bending moments (general bending moment plus inertial bending moment), and torsional moments (general torsional moment plus inertial torsional moment). With the inertial effects of the rib-stiffeners accounted for, the resultant tensional forces, bending and torsional moments acting on the upper and bottom panels per rib-stiffener are not equal, denoted here by (Q^+, M^+, M_T^+) and (Q^-, M^-, M_T^-), respectively. Fig. 2 shows the convention employed for denoting the tensional forces as well as the bending and torsional moments between the upper panel and the x- and y-wise stiffeners. The same apply at the interface between the bottom panel and the x- and y-wise stiffeners.

The dynamic responses of the sandwich structure are time-harmonic as the excitation is in the form of $q_0 e^{i\omega t}$. For simplicity, the harmonic time term $e^{i\omega t}$ is suppressed henceforth. With the equivalent forces and moments of the lattice core and the pressure in the fibrous sound absorptive material (or air cavity) accounted for, the equations governing panel vibrations are given by:

$$\begin{aligned} D_1\nabla^4 w_1 + m_1\frac{\partial^2 w_1}{\partial t^2} = & \sum_{m=-\infty}^{+\infty}\left[Q_y^+\delta(x-ml_x)+\frac{\partial}{\partial y}\left\{M_y^+\delta(x-ml_x)\right\}+\frac{\partial}{\partial x}\left\{M_{Ty}^+\delta(x-ml_x)\right\}\right] \\ & +\sum_{n=-\infty}^{+\infty}\left[Q_x^+\delta(y-nl_y)+\frac{\partial}{\partial x}\left\{M_x^+\delta(y-nl_y)\right\}+\frac{\partial}{\partial y}\left\{M_{Tx}^+\delta(y-nl_y)\right\}\right] \\ & +q_0\delta(x-x_0)\delta(y-y_0)-p_{cav}(x,y,h_1) \end{aligned} \quad (1)$$

$$\begin{aligned} D_2\nabla^4 w_2 + m_2\frac{\partial^2 w_2}{\partial t^2} = & -\sum_{m=-\infty}^{+\infty}\left[Q_y^-\delta(x-ml_x)+\frac{\partial}{\partial y}\left\{M_y^-\delta(x-ml_x)\right\}+\frac{\partial}{\partial x}\left\{M_{Ty}^-\delta(x-ml_x)\right\}\right] \\ & -\sum_{n=-\infty}^{+\infty}\left[Q_x^-\delta(y-nl_y)+\frac{\partial}{\partial x}\left\{M_x^-\delta(y-nl_y)\right\}+\frac{\partial}{\partial y}\left\{M_{Tx}^-\delta(y-nl_y)\right\}\right] \\ & +p_{cav}(x,y,h_1+d) \end{aligned} \quad (2)$$

where $\nabla^4 \equiv \left(\partial^2/\partial x^2+\partial^2/\partial y^2\right)^2$; $\delta(\cdot)$ is the Dirac delta function; (w_1, w_2), (m_1, m_2) and (D_1, D_2) are the displacement, surface mass density and flexural rigidity of the upper panel and bottom panel, respectively. The material loss factor η_j ($j = 1, 2$ for upper panel and bottom panel, respectively) is introduced with the complex Young's modulus $D_j = E_j h_j^3\left(1+i\eta_j\right)/12\left(1-\nu_j^2\right)$ (where j =1, 2).

As the factual forces and moments exerting on the upper and bottom panels are not the same due to the consideration of inertial forces and moments, the terms associated with the two panels are denoted separately by superscripts + (upper) and – (bottom). Subscripts x and y are introduced to represent those terms arising from the x- and y-wise stiffeners, respectively.

Taking into account the inertial effects (due to stiffener mass) and applying both the Hooke's law and Newton's second law, one obtains the tensional forces arising from the rib-stiffeners as [18]:

$$Q_x^+ = -\frac{K_x\left(K_x - m_x\omega^2\right)}{2K_x - m_x\omega^2}w_1 + \frac{K_x^2}{2K_x - m_x\omega^2}w_2 \tag{3}$$

$$Q_x^- = -\frac{K_x^2}{2K_x - m_x\omega^2}w_1 + \frac{K_x\left(K_x - m_x\omega^2\right)}{2K_x - m_x\omega^2}w_2 \tag{4}$$

$$Q_y^+ = -\frac{K_y\left(K_y - m_y\omega^2\right)}{2K_y - m_y\omega^2}w_1 + \frac{K_y^2}{2K_y - m_y\omega^2}w_2 \tag{5}$$

$$Q_y^- = -\frac{K_y^2}{2K_y - m_y\omega^2}w_1 + \frac{K_y\left(K_y - m_y\omega^2\right)}{2K_y - m_y\omega^2}w_2 \tag{6}$$

where ω is the circle frequency and (K_x, K_y) are the tensional stiffness of half the rib-stiffeners per unit length.

Similarly, the bending moments of the rib-stiffeners can be expressed as [18]:

$$M_x^+ = \frac{E_xI_x^*\left(E_xI_x^* - \rho_xI_x\omega^2\right)}{2E_xI_x^* - \rho_xI_x\omega^2}\frac{\partial^2 w_1}{\partial x^2} - \frac{E_x^2I_x^{*2}}{2E_xI_x^* - \rho_xI_x\omega^2}\frac{\partial^2 w_2}{\partial x^2} \tag{7}$$

$$M_x^- = \frac{E_x^2I_x^{*2}}{2E_xI_x^* - \rho_xI_x\omega^2}\frac{\partial^2 w_1}{\partial x^2} - \frac{E_xI_x^*\left(E_xI_x^* - \rho_xI_x\omega^2\right)}{2E_xI_x^* - \rho_xI_x\omega^2}\frac{\partial^2 w_2}{\partial x^2} \tag{8}$$

$$M_y^+ = \frac{E_yI_y^*\left(E_yI_y^* - \rho_yI_y\omega^2\right)}{2E_yI_y^* - \rho_yI_y\omega^2}\frac{\partial^2 w_1}{\partial y^2} - \frac{E_y^2I_y^{*2}}{2E_yI_y^* - \rho_yI_y\omega^2}\frac{\partial^2 w_2}{\partial y^2} \tag{9}$$

$$M_y^- = \frac{E_y^2I_y^{*2}}{2E_yI_y^* - \rho_yI_y\omega^2}\frac{\partial^2 w_1}{\partial y^2} - \frac{E_yI_y^*\left(E_yI_y^* - \rho_yI_y\omega^2\right)}{2E_yI_y^* - \rho_yI_y\omega^2}\frac{\partial^2 w_2}{\partial y^2} \tag{10}$$

where ($E_xI_x^*$, $E_yI_y^*$) are the bending stiffness of half the rib-stiffeners, (ρ_x, ρ_y) and (I_x, I_y) are mass density and polar moment of inertia for the rib-stiffeners, with subscripts x and y indicating the direction of the stiffener.

Following similar procedures for deriving the tensional forces and bending moments, one obtains the torsional moments of the rib-stiffeners as [18]:

$$M_{Tx}^+ = \frac{G_xJ_x^*\left(G_xJ_x^* - \rho_xJ_x\omega^2\right)}{2G_xJ_x^* - \rho_xJ_x\omega^2}\frac{\partial^2 w_1}{\partial x\partial y} - \frac{G_x^2J_x^{*2}}{2G_xJ_x^* - \rho_xJ_x\omega^2}\frac{\partial^2 w_2}{\partial x\partial y} \tag{11}$$

$$M_{Tx}^- = \frac{G_x^2J_x^{*2}}{2G_xJ_x^* - \rho_xJ_x\omega^2}\frac{\partial^2 w_1}{\partial x\partial y} - \frac{G_xJ_x^*\left(G_xJ_x^* - \rho_xJ_x\omega^2\right)}{2G_xJ_x^* - \rho_xJ_x\omega^2}\frac{\partial^2 w_2}{\partial x\partial y} \tag{12}$$

$$M_{Ty}^{+}=\frac{G_yJ_y^*\left(G_yJ_y^*-\rho_yJ_y\omega^2\right)}{2G_yJ_y^*-\rho_yJ_y\omega^2}\frac{\partial^2 w_1}{\partial y\partial x}-\frac{G_y^2J_y^{*2}}{2G_yJ_y^*-\rho_yJ_y\omega^2}\frac{\partial^2 w_2}{\partial y\partial x} \tag{13}$$

$$M_{Ty}^{-}=\frac{G_y^2J_y^{*2}}{2G_yJ_y^*-\rho_yJ_y\omega^2}\frac{\partial^2 w_1}{\partial y\partial x}-\frac{G_yJ_y^*\left(G_yJ_y^*-\rho_yJ_y\omega^2\right)}{2G_yJ_y^*-\rho_yJ_y\omega^2}\frac{\partial^2 w_2}{\partial y\partial x} \tag{14}$$

where ($G_xJ_x^*$, $G_yJ_y^*$) are the torsional stiffness of half the rib-stiffeners and (J_x, J_y) are the torsional moment of inertia for the rib-stiffeners.

In the above expressions for the tensional forces, bending moments and torsional moments of a rib-stiffener, the geometrical properties of its cross-section are given by:

$$K_x=\frac{E_xt_x}{d/2},\ K_y=\frac{E_yt_y}{d/2} \tag{15}$$

$$I_x^*=\frac{t_x(d/2)^3}{12},\ I_y^*=\frac{t_y(d/2)^3}{12},\ I_x=\frac{t_xd^3}{12},\ I_y=\frac{t_yd^3}{12} \tag{16}$$

$$J_x^*=\frac{t_x^3d}{2}\left[\frac{1}{3}-\frac{64}{\pi^5}\frac{2t_x}{d}\sum_{n=1,3,5,\ldots}^{\infty}\frac{\tanh(n\pi d/4t_x)}{n^5}\right] \tag{17}$$

$$J_y^*=\frac{t_y^3d}{2}\left[\frac{1}{3}-\frac{64}{\pi^5}\frac{2t_y}{d}\sum_{n=1,3,5,\ldots}^{\infty}\frac{\tanh(n\pi d/4t_y)}{n^5}\right] \tag{18}$$

$$J_x=t_x^3d\left[\frac{1}{3}-\frac{64}{\pi^5}\frac{t_x}{d}\sum_{n=1,3,5,\ldots}^{\infty}\frac{\tanh(n\pi d/2t_x)}{n^5}\right] \tag{19}$$

$$J_y=t_y^3d\left[\frac{1}{3}-\frac{64}{\pi^5}\frac{t_y}{d}\sum_{n=1,3,5,\ldots}^{\infty}\frac{\tanh(n\pi d/2t_y)}{n^5}\right] \tag{20}$$

where E_x and E_y are separately the Young's modulus of the *x*- and *y*-wise stiffener materials.

To simplify Eqs. (3)-(14), the following set of specified characteristics is introduced to replace the coefficients of general displacements.

1. Replacement of tensional force coefficients:

$$R_{Q1}=\frac{K_x\left(K_x-m_x\omega^2\right)}{2K_x-m_x\omega^2},\ R_{Q2}=\frac{K_x^2}{2K_x-m_x\omega^2} \tag{21}$$

$$R_{Q3}=\frac{K_y\left(K_y-m_y\omega^2\right)}{2K_y-m_y\omega^2},\ R_{Q4}=\frac{K_y^2}{2K_y-m_y\omega^2} \tag{22}$$

2. Replacement of bending moment coefficients:

$$R_{M1}=\frac{E_xI_x^*\left(E_xI_x^*-\rho_xI_x\omega^2\right)}{2E_xI_x^*-\rho_xI_x\omega^2},\ R_{M2}=\frac{E_x^2I_x^{*2}}{2E_xI_x^*-\rho_xI_x\omega^2} \tag{23}$$

$$R_{M3}=\frac{E_yI_y^*\left(E_yI_y^*-\rho_yI_y\omega^2\right)}{2E_yI_y^*-\rho_yI_y\omega^2},\ R_{M4}=\frac{E_y^2I_y^{*2}}{2E_yI_y^*-\rho_yI_y\omega^2}, \tag{24}$$

3. Replacement of torsional moment coefficients:

$$R_{T1}=\frac{G_xJ_x^*\left(G_xJ_x^*-\rho_xJ_x\omega^2\right)}{2G_xJ_x^*-\rho_xJ_x\omega^2},\ R_{T2}=\frac{G_x^2J_x^{*2}}{2G_xJ_x^*-\rho_xJ_x\omega^2}, \tag{25}$$

$$R_{T3}=\frac{G_yJ_y^*\left(G_yJ_y^*-\rho_yJ_y\omega^2\right)}{2G_yJ_y^*-\rho_yJ_y\omega^2},\ R_{T4}=\frac{G_y^2J_y^{*2}}{2G_yJ_y^*-\rho_yJ_y\omega^2}, \tag{26}$$

Using Eqs. (21)-(26), one can simplify the expressions of the tensional forces, bending moments and torsional moments, as:

1. Tensional forces

$$Q_x^+=-R_{Q1}w_1+R_{Q2}w_2,\ Q_x^-=-R_{Q2}w_1+R_{Q1}w_2 \tag{27}$$

$$Q_y^+=-R_{Q3}w_1+R_{Q4}w_2,\ Q_y^-=-R_{Q4}w_1+R_{Q3}w_2 \tag{28}$$

2. Bending moments

$$M_x^+=R_{M1}\frac{\partial^2 w_1}{\partial x^2}-R_{M2}\frac{\partial^2 w_2}{\partial x^2},\ M_x^-=R_{M2}\frac{\partial^2 w_1}{\partial x^2}-R_{M1}\frac{\partial^2 w_2}{\partial x^2}, \tag{29}$$

$$M_y^+=R_{M3}\frac{\partial^2 w_1}{\partial y^2}-R_{M4}\frac{\partial^2 w_2}{\partial y^2},\ M_y^-=R_{M4}\frac{\partial^2 w_1}{\partial y^2}-R_{M3}\frac{\partial^2 w_2}{\partial y^2}, \tag{30}$$

3. Torsional moments

$$M_{Tx}^+=R_{T1}\frac{\partial^2 w_1}{\partial x\partial y}-R_{T2}\frac{\partial^2 w_2}{\partial x\partial y},\ M_{Tx}^-=R_{T2}\frac{\partial^2 w_1}{\partial x\partial y}-R_{T1}\frac{\partial^2 w_2}{\partial x\partial y}, \tag{31}$$

$$M_{Ty}^+=R_{T3}\frac{\partial^2 w_1}{\partial y\partial x}-R_{T4}\frac{\partial^2 w_2}{\partial y\partial x},\ M_{Ty}^-=R_{T4}\frac{\partial^2 w_1}{\partial y\partial x}-R_{T3}\frac{\partial^2 w_2}{\partial y\partial x}, \tag{32}$$

Solutions

Given the 2D (two-dimensional) periodic nature of the sandwich structure as shown in Fig. 1, applying the Poisson summation formula [3,23], the wave components in the structure can be expressed by using space harmonic series, as:

$$\sum_{m=-\infty}^{+\infty} \delta(x - ml_x) = \frac{1}{l_x} \sum_{m=-\infty}^{+\infty} e^{-i(2m\pi/l_x)x}, \quad \sum_{n=-\infty}^{+\infty} \delta(y - nl_y) = \frac{1}{l_y} \sum_{n=-\infty}^{+\infty} e^{-i(2n\pi/l_y)y} \tag{33}$$

The Fourier transform pair of a function with respect to (x, y) and (α, β) can be defined as:

$$w(x,y) = \int_{-\infty}^{+\infty} \int_{-\infty}^{+\infty} \tilde{w}(\alpha,\beta) \qquad e^{i(\alpha x + \beta y)} d\alpha d\beta \tag{34}$$

$$\tilde{w}(\alpha,\beta) = \left(\frac{1}{2\pi}\right)^2 \int_{-\infty}^{+\infty} \int_{-\infty}^{+\infty} w(x,y) \quad e^{-i(\alpha x + \beta y)} dx dy \tag{35}$$

Applying the Poisson summation formula and then taking the Fourier transform of Eqs. (1) and (2), one gets:

$$\begin{gathered}\left[D_1\left(\alpha^2+\beta^2\right)^2 - m_1\omega^2\right]\tilde{w}_1(\alpha,\beta) = \frac{1}{l_x}\sum_m \left[\tilde{Q}_y^+(\alpha_m,\beta) + i\beta\tilde{M}_y^+(\alpha_m,\beta) + i\alpha\tilde{M}_{Ty}^+(\alpha_m,\beta)\right] \\ + \frac{1}{l_y}\sum_n \left[\tilde{Q}_x^+(\alpha,\beta_n) + i\alpha\tilde{M}_x^+(\alpha,\beta_n) + i\beta\tilde{M}_{Tx}^+(\alpha,\beta_n)\right] + \frac{q_0}{(2\pi)^2}e^{-i(\alpha x_0 + \beta y_0)} - \tilde{p}_{cav}(\alpha,\beta,h_1)\end{gathered} \tag{36}$$

$$\begin{gathered}\left[D_2\left(\alpha^2+\beta^2\right)^2 - m_2\omega^2\right]\tilde{w}_2(\alpha,\beta) = -\frac{1}{l_x}\sum_m \left[\tilde{Q}_y^-(\alpha_m,\beta) + i\beta\tilde{M}_y^-(\alpha_m,\beta) + i\alpha\tilde{M}_{Ty}^-(\alpha_m,\beta)\right] \\ - \frac{1}{l_y}\sum_n \left[\tilde{Q}_x^-(\alpha,\beta_n) + i\alpha\tilde{M}_x^-(\alpha,\beta_n) + i\beta\tilde{M}_{Tx}^-(\alpha,\beta_n)\right] + \tilde{p}_{cav}(\alpha,\beta,h_1+d)\end{gathered} \tag{37}$$

where $\alpha_m = \alpha + 2m\pi / l_x$, $\beta_n = \beta + 2n\pi / l_y$, and the dependence of a term on the wavenumbers (α, β) is indicated using the hat sign ~, meaning the corresponding Fourier transform of this term. Expressions for the Fourier transform of the tensional forces, bending and torsional moments are presented below.

1. Fourier transforms of tensional forces

$$\tilde{Q}_x^+(a,\beta_n) = -R_{Q1}\tilde{w}_1(a,\beta_n) + R_{Q2}\tilde{w}_2(a,\beta_n) \tag{38}$$

$$\tilde{Q}_x^-(a,\beta_n) = -R_{Q2}\tilde{w}_1(a,\beta_n) + R_{Q1}\tilde{w}_2(a,\beta_n) \tag{39}$$

$$\tilde{Q}_y^+(a_m,\beta) = -R_{Q3}\tilde{w}_1(a_m,\beta) + R_{Q4}\tilde{w}_2(a_m,\beta) \tag{40}$$

$$\tilde{Q}_y^-(a_m,\beta) = -R_{Q4}\tilde{w}_1(a_m,\beta) + R_{Q3}\tilde{w}_2(a_m,\beta) \tag{41}$$

2. Fourier transforms of bending moments

$$\tilde{M}_x^+(a,\beta_n) = -\alpha^2\left[R_{M1}\tilde{w}_1(a,\beta_n) - R_{M2}\tilde{w}_2(a,\beta_n)\right] \tag{42}$$

$$\tilde{M}_x^-(a,\beta_n) = -\alpha^2\left[R_{M2}\tilde{w}_1(a,\beta_n) - R_{M1}\tilde{w}_2(a,\beta_n)\right] \tag{43}$$

$$\tilde{M}_y^{+}(a_m,\beta)=-\beta^2\left[R_{M3}\tilde{w}_1(a_m,\beta)-R_{M4}\tilde{w}_2(a_m,\beta)\right] \tag{44}$$

$$\tilde{M}_y^{-}(a_m,\beta)=-\beta^2\left[R_{M4}\tilde{w}_1(a_m,\beta)-R_{M3}\tilde{w}_2(a_m,\beta)\right] \tag{45}$$

3. Fourier transforms of torsional moments

$$\tilde{M}_{Tx}^{+}(a,\beta_n)=-\alpha\beta_n\left[R_{T1}\tilde{w}_1(a,\beta_n)-R_{T2}\tilde{w}_2(a,\beta_n)\right] \tag{46}$$

$$\tilde{M}_{Tx}^{-}(a,\beta_n)=-\alpha\beta_n\left[R_{T2}\tilde{w}_1(a,\beta_n)-R_{T1}\tilde{w}_2(a,\beta_n)\right] \tag{47}$$

$$\tilde{M}_{Ty}^{+}(a_m,\beta)=-\alpha_m\beta\left[R_{T3}\tilde{w}_1(a_m,\beta)-R_{T4}\tilde{w}_2(a_m,\beta)\right] \tag{48}$$

$$\tilde{M}_{Ty}^{-}(a_m,\beta)=-\alpha_m\beta\left[R_{T4}\tilde{w}_1(a_m,\beta)-R_{T3}\tilde{w}_2(a_m,\beta)\right] \tag{49}$$

The acoustic pressure and fluid-structure coupling

The absorption of sound absorption by porous materials mainly arises from viscous drag forces and thermal exchange loss when sound penetrates through the material [19,24-26]. There exist numerous theoretical models to address these issues, while different models may be specialized to deal with different types of porous materials. For instance, Lu et al. proposed a model for high porosity cellular metallic foams with open cells [24,25,27,28] and another model for semi-open metal foams [26]. As for fibrous materials considered here, there are two main classes of models [19]. The first one models the fibrous material as an equivalent fluid with effective density and bulk modulus [22,29,30]: under the assumption of the solid fibers being a rigid skeleton, only one compression wave propagates in the air-saturated medium, which thereby is governed by the Helmholtz equation. The other one employs the more rigorous theory of Biot [31,32] with the elasticity of the skeleton taken into account, the solution of which often seeks help from the finite element method (FEM) and suffers from huge computational expenses.

In view of the complexity of the proposed structural vibration model and the primary focus of the present study on sound radiation of the sandwich structure as a whole, the well-developed empirical expressions (i.e., equivalent fluid model) of Allard and Champoux [22] is adopted to model the acoustic pressure in fibrous absorption materials such as glass/rock wools widely used in noise absorption engineering. In terms of scholar description, these may be defined as Newtonian fluid saturated rigid frame fibrous materials, with the frame fibers randomly distributed. Although Allard and Champoux [22] called their empirical equations as the equivalent fluid model, this model is in fact based on Johnson et al.'s two phases theory [33]. It accurately accounted for the viscous forces between fluid and solid and the physical transposition in the process of sound propagation, by adopting two variables - the dynamic density $\rho(\omega)$ and the dynamic bulk modulus $K(\omega)$ - assuming that the fibrous material is isotropic. The equivalent fluid model has been demonstrated to be capable of providing accurate predictions of sound wave propagation across fibrous sound absorptive materials, over a wide frequency range, and hence has been widely acknowledged by the acoustic community [19,24-26]. To be more precise, the equivalent fluid model is valid for most glass/rock wools for f/R smaller than 1.0 $kg^{-1}m^3$, where f is the frequency and R is the flow resistivity of the fibrous material [22]. Generally, the flow resistivity R of typical glass/rock wools is approximately 20000 Nm/s^4, and hence the equivalent fluid model works well for frequencies below 20 kHz.

According to the equivalent fluid model, wave propagation in fibrous sound absorptive material (e.g., fiberglass or mineral wool) is governed by [12,13,22,34]:

$$\left(\partial^2/\partial x^2+\partial^2/\partial y^2+\partial^2/\partial z^2\right)p_{cav}+k_{cav}^2 p_{cav}=0 \tag{50}$$

where p_{cav} is the sound pressure in the fibrous material and k_{cav} is the corresponding complex wavenumber, which is related to the dynamic density $\rho(\omega)$ and dynamic bulk modulus $K(\omega)$ of the fibrous material by:

$$k_{cav}=2\pi f\sqrt{\rho(\omega)/K(\omega)} \tag{51}$$

In accordance with the complex physical phenomena taking place in the fibrous material, such as thermal exchanges between air and fibers showing a significant transition with increasing frequency (i.e., isothermal process at low frequency turning to adiabatic process at high frequency) [22], the equivalent density and bulk modulus are both dynamic. In other words, the dynamic density and dynamic bulk modulus are frequency dependent, given by [22]:

$$\rho(\omega)=\rho_0\left[1+(R/\rho_0 f)G_1(\rho_0 f/R)/i2\pi\right] \tag{52}$$

$$K(\omega)=\gamma_s P_0\left(\gamma_s-\frac{\gamma_s-1}{1+(1/i8\pi N_{pr})(\rho_0 f/R)^{-1}G_2(\rho_0 f/R)}\right)^{-1} \tag{53}$$

where $G_1(\rho_0 f/R)=\sqrt{1+i\pi(\rho_0 f/R)}$, $G_2(\rho_0 f/R)=G_1\left[(\rho_0 f/R)4N_{pr}\right]$, R is the static flow resistivity of the fibrous material, γ_s and ρ_0 are separately the specific heat ratio (i.e., $\gamma_s=c_p/c_v$, c_p and c_v being the specific heat per unit mass of the air at constant pressure and constant volume) and density of air, P_0 is the air equilibrium pressure and N_{pr} is the Prandtl number. As a further understanding of physical meanings, the dynamic density $\rho(\omega)$ contains the inertial and viscous forces per unit volume of air in fibrous material, while the dynamic bulk modulus $K(\omega)$ gives the relationship between the averaged molecular displacement of air and the averaged variation of pressure. As a conclusion of Lu et al.'s model, it is found that the viscous drag forces operating at the fiber surface govern the complex density $\rho(\omega)$ and the thermal forces control the complex bulk modulus $K(\omega)$. As seen in Eqs. (52) and (53), these two quantities is strongly dependent on the term $\rho_0 f/R$, reflecting the inherent dynamic property of sound absorbing process and flow resistance being the fundamental origin of sound absorption.

Generally, in contrast with the facesheet to stiffener interaction and the facesheet to fibrous-material (or air) interaction, the stiffener to fibrous-material (or air) interaction is negligible. It is easy to understand that the direct structural connection between the facesheets and the rib-stiffeners is far stronger than the stiffener to fibrous-material (or air) interaction. As for the facesheet to fibrous-material interaction, note that the stiffener separations l_x and l_y are generally much larger than the stiffener height d, implying that the contact surface area between the facesheet and fibrous material is much larger than that between the stiffener and fibrous material. Therefore, whilst the vibration of the facesheet is affected significantly by the fibrous material in contact, it has negligible influence on the motion of the short stiffeners. As a result, the fluid-structure coupling here only needs to consider the facesheet to fibrous-material interaction. To ensure the equality of panel velocity and fluid velocity on the panel surface, the momentum equation (i.e., continuity condition of fluid-structure coupling [7,10,35]) is applied:

$$\left[\partial p_{cav}/\partial z\right]_{z=h_1} = \rho_{cav}\omega^2 w_1 \, , \; \left[\partial p_{cav}/\partial z\right]_{z=h_1+d} = \rho_{cav}\omega^2 w_2 \tag{54}$$

where the complex density ρ_{cav} of the fibrous material is related to the complex wavenumber k_{cav} and porosity σ as [12]:

$$\frac{k_{cav}^2}{k_0^2} = \frac{\gamma_s \sigma \rho_{cav}}{\rho_0} \tag{55}$$

Applying the Fourier transform to Eqs. (50) and (54), one obtains:

$$\tilde{p}_{cav}(\alpha,\beta,z) = -\frac{\rho_{cav}\omega^2}{\gamma(\alpha,\beta)\left(e^{\gamma d} - e^{-\gamma d}\right)} \left\{ \begin{array}{l} \left[\tilde{w}_1(\alpha,\beta)e^{\gamma(h_1+d)} - \tilde{w}_2(\alpha,\beta)e^{\gamma h_1}\right]e^{-\gamma z} \\ +\left[\tilde{w}_1(\alpha,\beta)e^{-\gamma(h_1+d)} - \tilde{w}_2(\alpha,\beta)e^{-\gamma h_1}\right]e^{\gamma z} \end{array} \right\} \tag{56}$$

where $\gamma^2 = \alpha^2 + \beta^2 - k_{cav}^2$. More specifically, the pressures acting on the upper and bottom panels are given by:

$$\tilde{p}_{cav}(\alpha,\beta,h_1) = -\frac{\rho_{cav}\omega^2\left[\tilde{w}_1(\alpha,\beta)\cosh(\gamma d) - \tilde{w}_2(\alpha,\beta)\right]}{\gamma(\alpha,\beta)\sinh(\gamma d)} \tag{57}$$

$$\tilde{p}_{cav}(\alpha,\beta,h_1+d) = -\frac{\rho_{cav}\omega^2\left[\tilde{w}_1(\alpha,\beta) - \tilde{w}_2(\alpha,\beta)\cosh(\gamma d)\right]}{\gamma(\alpha,\beta)\sinh(\gamma d)} \tag{58}$$

Substituting Eqs. (57) and (58) into Eqs. (36) and (37), respectively, one can rewrite the governing equations as:

$$\begin{aligned} &\left[D_1\left(\alpha^2+\beta^2\right)^2 - m_1\omega^2 - \frac{\rho_{cav}\omega^2\cosh(\gamma d)}{\gamma(\alpha,\beta)\sinh(\gamma d)}\right]\tilde{w}_1(\alpha,\beta) + \frac{\rho_{cav}\omega^2}{\gamma(\alpha,\beta)\sinh(\gamma d)}\tilde{w}_2(\alpha,\beta) \\ &= \frac{1}{l_x}\sum_m\left[-R_{Q3} - i\beta^3 R_{M3} - i\alpha\alpha_m\beta R_{T3}\right]\tilde{w}_1(\alpha_m,\beta) + \frac{1}{l_y}\sum_n\left[-R_{Q1} - i\alpha^3 R_{M1} - i\alpha\beta\beta_n R_{T1}\right]\tilde{w}_1(\alpha,\beta_n) \\ &+\frac{1}{l_x}\sum_m\left[R_{Q4} + i\beta^3 R_{M4} + i\alpha\alpha_m\beta R_{T4}\right]\tilde{w}_2(\alpha_m,\beta) + \frac{1}{l_y}\sum_n\left[R_{Q2} + i\alpha^3 R_{M2} + i\alpha\beta\beta_n R_{T2}\right]\tilde{w}_2(\alpha,\beta_n) \\ &+\frac{q_0}{(2\pi)^2}e^{-i(\alpha x_0+\beta y_0)} \end{aligned} \tag{59}$$

$$
\begin{aligned}
&\left[D_2\left(\alpha^2+\beta^2\right)^2-m_2\omega^2-\frac{\rho_{cav}\omega^2\cosh(\gamma d)}{\gamma(\alpha,\beta)\sinh(\gamma d)}\right]\tilde{w}_2(\alpha,\beta)+\frac{\rho_{cav}\omega^2}{\gamma(\alpha,\beta)\sinh(\gamma d)}\tilde{w}_1(\alpha,\beta)\\
&=-\frac{1}{l_x}\sum_m\left[-R_{Q4}-i\beta^3R_{M4}-i\alpha\alpha_m\beta R_{T4}\right]\tilde{w}_1(\alpha_m,\beta)\\
&-\frac{1}{l_y}\sum_n\left[-R_{Q2}-i\alpha^3R_{M2}-i\alpha\beta\beta_nR_{T2}\right]\tilde{w}_1(\alpha,\beta_n)\\
&-\frac{1}{l_x}\sum_m\left[R_{Q3}+i\beta^3R_{M3}+i\alpha\alpha_m\beta R_{T3}\right]\tilde{w}_2(\alpha_m,\beta)\\
&-\frac{1}{l_y}\sum_n\left[R_{Q1}+i\alpha^3R_{M1}+i\alpha\beta\beta_nR_{T1}\right]\tilde{w}_2(\alpha,\beta_n)
\end{aligned}
\tag{60}
$$

To simplify the derivation procedures, the following definitions are introduced:

$$
f_1(\alpha,\beta)=\left(\alpha^2+\beta^2\right)^2-\frac{m_1\omega^2}{D_1}-\frac{\rho_{cav}\omega^2\cosh(\gamma d)}{D_1\gamma(\alpha,\beta)\sinh(\gamma d)}
\tag{61}
$$

$$
f_2(\alpha,\beta)=\left(\alpha^2+\beta^2\right)^2-\frac{m_2\omega^2}{D_2}-\frac{\rho_{cav}\omega^2\cosh(\gamma d)}{D_2\gamma(\alpha,\beta)\sinh(\gamma d)}
\tag{62}
$$

As seen in Eqs. (59) and (60), the panel displacements $\tilde{w}_1(\alpha,\beta)$ and $\tilde{w}_2(\alpha,\beta)$ to be solved are not independent but have coupling terms $\tilde{w}_1(\alpha_m,\beta)$, $\tilde{w}_1(\alpha,\beta_n)$, $\tilde{w}_2(\alpha_m,\beta)$ and $\tilde{w}_2(\alpha,\beta_n)$ in the corresponding sum formula. To solve these unknowns, one needs to replace (α , β) by (α'_m, β'_n), leading to two sets of simultaneous algebraic equations, as:

$$
\begin{aligned}
&D_1f_1(\alpha'_m,\beta'_n)\tilde{w}_1(\alpha'_m,\beta'_n)+\frac{\rho_{cav}\omega^2}{\gamma(\alpha'_m,\beta'_n)\sinh\left[\gamma(\alpha'_m,\beta'_n)\cdot d\right]}\tilde{w}_2(\alpha'_m,\beta'_n)\\
&+\frac{1}{l_x}\left[\left(R_{Q3}+i\beta_n'^3R_{M3}\right)\sum_m\tilde{w}_1(\alpha_m,\beta'_n)+i\alpha'_m\beta'_nR_{T3}\sum_m\alpha_m\tilde{w}_1(\alpha_m,\beta'_n)\right]\\
&+\frac{1}{l_y}\left[\left(R_{Q1}+i\alpha_m'^3R_{M1}\right)\sum_n\tilde{w}_1(\alpha'_m,\beta_n)+i\alpha'_m\beta'_nR_{T1}\sum_n\beta_n\tilde{w}_1(\alpha'_m,\beta_n)\right]\\
&-\frac{1}{l_x}\left[\left(R_{Q4}+i\beta_n'^3R_{M4}\right)\sum_m\tilde{w}_2(\alpha_m,\beta'_n)+i\alpha'_m\beta'_nR_{T4}\sum_m\alpha_m\tilde{w}_2(\alpha_m,\beta'_n)\right]\\
&-\frac{1}{l_y}\left[\left(R_{Q2}+i\alpha_m'^3R_{M2}\right)\sum_n\tilde{w}_2(\alpha'_m,\beta_n)+i\alpha'_m\beta'_nR_{T2}\sum_n\beta_n\tilde{w}_2(\alpha'_m,\beta_n)\right]\\
&=\frac{q_0}{(2\pi)^2}e^{-i(\alpha'_mx_0+\beta'_ny_0)}
\end{aligned}
\tag{63}
$$

$$\begin{aligned}
&\frac{\rho_{cav}\omega^2}{\gamma(\alpha'_m,\beta'_n)\sinh\left[\gamma(\alpha'_m,\beta'_n)\cdot d\right]}\tilde{w}_1(\alpha'_m,\beta'_n)+D_2f_2(\alpha'_m,\beta'_n)\tilde{w}_2(\alpha'_m,\beta'_n)\\
&-\frac{1}{l_x}\left[\left(R_{Q4}+i\beta_n'^3R_{M4}\right)\sum_m\tilde{w}_1(\alpha_m,\beta'_n)+i\alpha'_m\beta'_nR_{T4}\sum_m\alpha_m\tilde{w}_1(\alpha_m,\beta'_n)\right]\\
&-\frac{1}{l_y}\left[\left(R_{Q2}+i\alpha_m'^3R_{M2}\right)\sum_n\tilde{w}_1(\alpha'_m,\beta_n)+i\alpha'_m\beta'_nR_{T2}\sum_n\beta_n\tilde{w}_1(\alpha'_m,\beta_n)\right]\\
&+\frac{1}{l_x}\left[\left(R_{Q3}+i\beta_n'^3R_{M3}\right)\sum_m\tilde{w}_2(\alpha_m,\beta'_n)+i\alpha'_m\beta'_nR_{T3}\sum_m\alpha_m\tilde{w}_2(\alpha_m,\beta'_n)\right]\\
&+\frac{1}{l_y}\left[\left(R_{Q1}+i\alpha_m'^3R_{M1}\right)\sum_n\tilde{w}_2(\alpha'_m,\beta_n)+i\alpha'_m\beta'_nR_{T1}\sum_n\beta_n\tilde{w}_2(\alpha'_m,\beta_n)\right]\\
&=0
\end{aligned} \tag{64}$$

which contain two sets of infinite unknowns: $\tilde{w}_1(\alpha'_m,\beta'_n)$ and $\tilde{w}_2(\alpha'_m,\beta'_n)$, with $m=-\infty$ to $+\infty$ and $n=-\infty$ to $+\infty$. Insofar as the solution converges, these equations can be truncated to retain one set of finite unknowns $\tilde{w}_1(\alpha'_m,\beta'_n)$ and $\tilde{w}_2(\alpha'_m,\beta'_n)$, with $m=-\hat{m}$ to $\hat{m}$ and $n=-\hat{n}$ to $\hat{n}$ (both $\hat{m}$ and $\hat{n}$ are positive integers), and hence can be numerically solved.

Far field sound radiated pressure

Owing to the fluid-structure interaction of the vibrating panel (bottom panel in the present case) and its surrounding fluid, sound pressure will be radiated from the fluid-structure interface into the far field. Therefore, once the response of the bottom panel $\tilde{w}_2(\alpha,\beta)$ is numerically solved, the radiated sound pressure at the far field can be obtained by employing the established sound radiation theory.

With the origin of the spherical coordinates (r,θ,φ) located at the excitation point (x_0, y_0), the far field sound pressure $p(r,\theta,\varphi)$ radiated from a vibrating surface with displacement $w(x,y)$ is given by [30]:

$$p(r,\theta,\varphi)=-\rho_0\omega^2\frac{e^{ik_0r}}{2\pi r}e^{i(\alpha x_0+\beta y_0)}\int_{-\infty}^{+\infty}\int_{-\infty}^{+\infty}w(x,y)e^{-i(\alpha x+\beta y)}dxdy \tag{65}$$

where $k_0=\omega/c_0$, c_0 and ρ_0 are separately the sound speed and air density, and the wavenumbers α and β are:

$$\alpha=k_0\cos\varphi\sin\theta,\ \beta=k_0\sin\varphi\sin\theta \tag{66}$$

By adopting the Fourier transform of Eq. (35), Eq. (65) becomes:

$$p(r,\theta,\varphi)=-2\pi\rho_0\omega^2\left(e^{ik_0r}/r\right)e^{i(\alpha x_0+\beta y_0)}\tilde{w}(\alpha,\beta) \tag{67}$$

For reference, the high frequency asymptote of far field sound pressure radiated by an *unstiffened* plate [3] is introduced, as:

$$p_{asy} = \rho_0 q_0 e^{ik_0 r} / 2\pi m r \tag{68}$$

The far field sound pressure radiated by the present orthogonally rib-stiffened sandwich structure with cavity absorption is then given in the form of sound pressure level (SPL) in decibel scales (dB) with respect to p_{asy}, as:

$$SPL = 10 \cdot \log_{10} \left(p / p_{asy} \right)^2 \tag{69}$$

Numerical results and discussions

In this section, representative examples for the on-axis (i.e., on the axis $\theta = \varphi = 0$) far field pressure is numerically calculated to explore the sound radiation characteristics of infinite orthogonally rib-stiffened sandwich structures with fibrous material filled cavity. Note that, on the selected axis (i.e., $\theta = \varphi = 0$), the stationary phase wavenumbers $\bar{\alpha}$ and $\bar{\beta}$ are both equal to zero.

The material properties and structural dimensions of the sandwich structure are taken as follows. Both the face panels and rib-stiffeners are made of aluminum, with Young's modulus $E = 70$ GPa, density $\rho = 2700$ kg/m³, Poisson ratio $\nu = 0.33$, and loss factor $\eta_1 = \eta_2 = 0.01$. The thickness of the two face panel are $h_1 = h_2 = 2$ mm, and that of rib-stiffeners are $t_x = t_y = 1$ mm. The depth of air cavity (i.e., height of rib connections; see Fig. 1) is $d = 0.025$ m. For air at normal temperature and atmospheric pressure, it is assumed that $\rho_0 = 1.21$ kg/m³, $N_{pr} = 0.702$, $\gamma_s = 1.4$, $P_0 = 101320$ N/m² and $c_0 = 343$ m/s. Fiberglass is selected as the cavity filling material, with porosity $\sigma = 0.95$ and flow resistivity $R = 24000$ Nm/s⁴. The time-harmonic point force with unit amplitude acts on the upper panel at location $(l_x/2, l_y/2)$. With these system parameters, the present theoretical model is used to examine the influence of fiberglass material in the cavity (partitioned by the lattice core) on the sound radiation characteristics of the sandwich structure. For comparison, the sound radiation behaviors of sandwich structures with pure air cavity (i.e., air-structure coupling effect included) and vacuum cavity (i.e., fluid-structure coupling effect ignored) are also considered.

Convergence check for numerical solution

As previously mentioned, the infinite simultaneous algebraic equations are truncated so that one only needs to solve a finite system of equations containing a finite number of unknowns. More specifically, only $M = 2\hat{m} + 1$ and $N = 2\hat{n} + 1$ unknowns are retained, associated separately with subscripts m and n, leading to the same number of harmonic wave components in the x - and y -directions. Insofar as a sufficiently large number of terms are retained, the finite system is capable of ensuring the convergence and accuracy of the solution. The well acknowledged criterion [1,15] is employed, which assumes that once the solution converges at a given frequency, it converges for all lower frequencies. Therefore, the required number of unknowns is determined by the highest frequency of interest (10 kHz in the present study). To check the convergence of the solution, a numerical test is carried out by calculating the SPL at 10 kHz, with increasingly more terms used in Eqs. (63) and (64), as shown in Fig. 3. It can be seen from Fig. 3 that, when $\hat{m}$ and $\hat{n}$ both have a value of 10, solution convergence is ensured at 10 kHz. Consequently, following the above mentioned criterion, the values of $\hat{m}$ and $\hat{n}$ are both taken as 10 (i.e., retaining 441 unknowns in the finite system) for all frequencies below 10kHz, which is sufficient to ensure the convergence and accuracy of the solution.

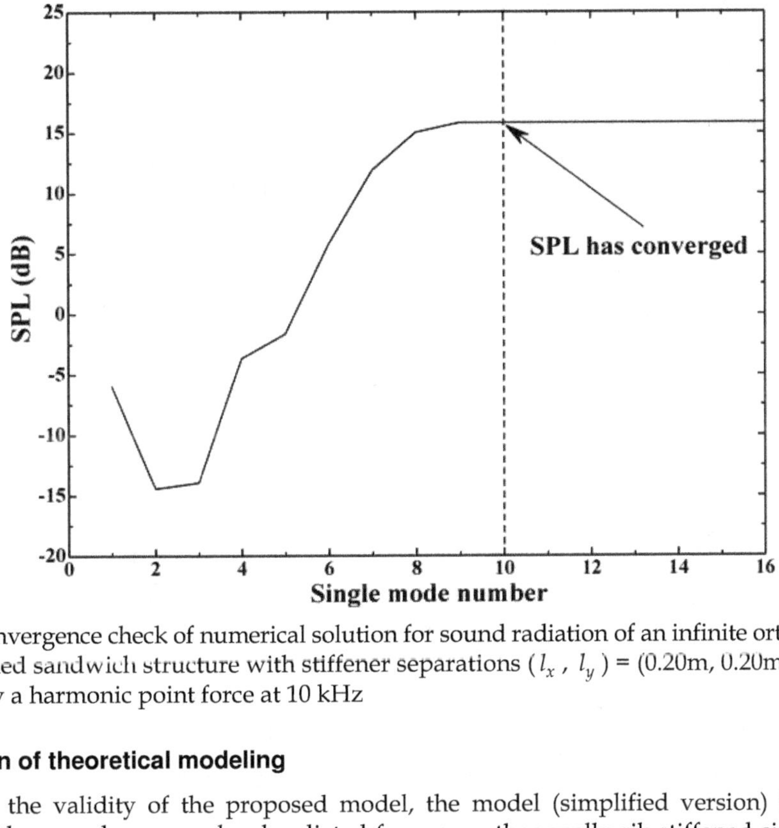

Fig. 3. Convergence check of numerical solution for sound radiation of an infinite orthogonally rib-stiffened sandwich structure with stiffener separations (l_x , l_y) = (0.20m, 0.20m) when excited by a harmonic point force at 10 kHz

Validation of theoretical modeling

To check the validity of the proposed model, the model (simplified version) is used to calculate the sound pressure level radiated from an orthogonally rib-stiffened single panel and the predictions are compared in Fig. 4 with those obtained by Mace [3]. To degrade the present model for sandwich structures to cover rib-stiffened single panels, negligibly small values are assigned to the prime parameters (i.e., Yong's modulus E, density ρ and thickness h) of one face panel of the sandwich whilst the remaining system parameters are identical to those used by Mace [3].

It can be seen from Fig. 4 that overall the present predictions agree excellently well with those of Mace: only slight deviations exist beyond 5000 Hz. These discrepancies in the high frequency range are expected, which can be attributed to the difference in vibration modeling of the rib-stiffeners between the present model and Mace's theory. The rib-stiffeners were modeled as Euler beams in Mace's theory [3], meaning that only the bending moments and the inertial effect of the tensional forces of the rib-stiffeners are considered. In contrast, the present model accounts for all possible motions of the rib-stiffeners, including tensional forces, bending and torsional moments as well as their inertial effects. Therefore, insofar as the dynamic responses and sound radiation of rib-stiffened plates are of concern, the present model provides a more precise theoretical tool than the beam-based theory of Mace. The discrepancies between the two theories in the high frequency range of Fig. 4 just demonstrate the necessity of accurately modeling the motion of the rib-stiffeners.

To further check the accuracy of the present model for the double-panel case, the model is degraded to reproduce Takahashi's results [11] for rib-stiffened double-panel structures, as shown in Fig. 5. The relevant geometrical dimensions and material property parameters are identical as those of Takahashi. Again, the model predictions fit well with Takahashi's theoretical results, with only slight divergences appearing at relatively high frequencies. These divergences are attributed to the additional consideration of inertial effects corresponding to the bending moments and torsional moments in the present model, which Takahashi did not take into account.

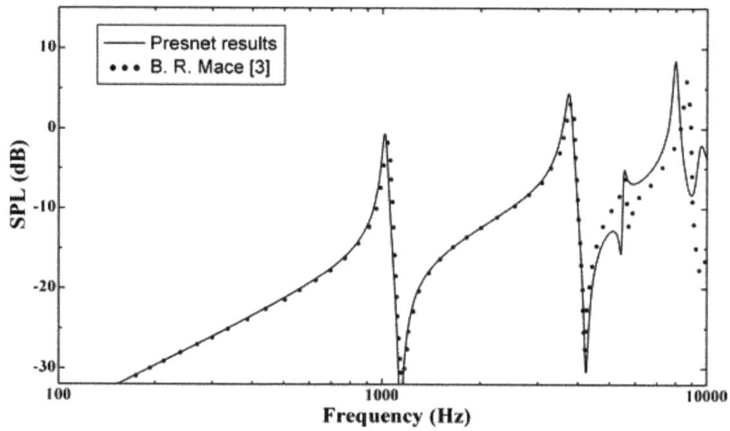

Fig. 4. Comparison between present model predictions and those by Mace [3] for orthogonally rib-stiffened single panel excited by time-harmonic point force at location (0, 0)

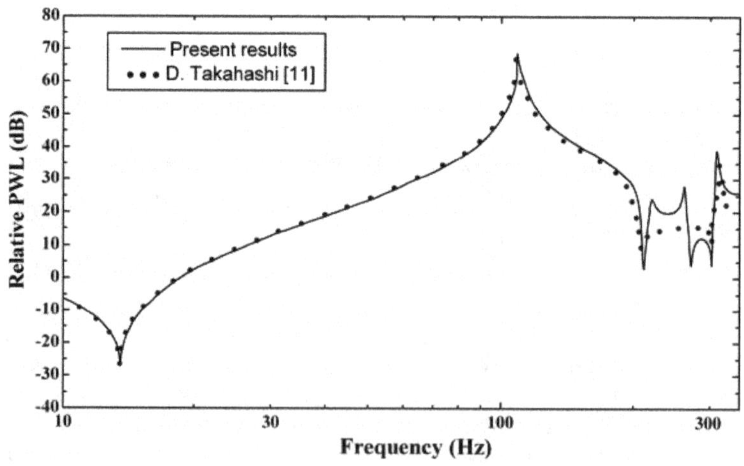

Fig. 5. Comparison between model predictions and theoretical results of Takahashi [11] for rib-stiffened double-panel structure excited by time-harmonic point force at location $\left(l_x/2,\ l_y/2\right)$

To a large extent, the comparisons made above may be regarded as acceptable validations for the proposed theoretical model, because all the theoretical formulations have been involved in the numerical calculation. In particular, if a theoretical model can be degraded to obtain the same results for simplified cases, its accuracy and feasibility would be better than the case when it can only give results similar to those obtained with its counterpart models.

Influence of air-structure coupling effect

Together with the equivalent fluid model for fibrous sound absorptive materials, the present model is able to characterize the sound radiation characteristics of lightweight lattice-cored sandwich structures filled with fibrous materials, such as fiberglass considered here. Note also that the model can be degraded to describe sandwich structures with either air cavity (i.e., air-structure coupling effect included) or vacuum cavity (i.e., fluid-structure coupling effect ignored). Therefore, comparisons amongst the three different kinds of sandwiches under time-harmonic point force excitation can be performed to assess the influences of air-structure coupling effect and fibrous filling material on sound radiation.

To better evaluate the influences of air-structure coupling effect and fibrous material, the location of point force acting on the face panel is selected at the center of one lattice cell, i.e., $(l_x/2, l_y/2)$, away from the conjunction between the face panel and rib-stiffeners. The predicted sound pressure level (SPL) radiated by the three different sandwich structures is plotted in Figs. 6, 7, and 8 as a function of frequency for (l_x, l_y) = (0.20m, 0.20m), (0.35m, 0.35m) and (0.50m, 0.50m), respectively. For each pair of stiffener spacing selected, three kinds of sandwich configurations are compared: (i) vacuum cavity; (ii) air cavity; (iii) cavity filled with fiberglass.

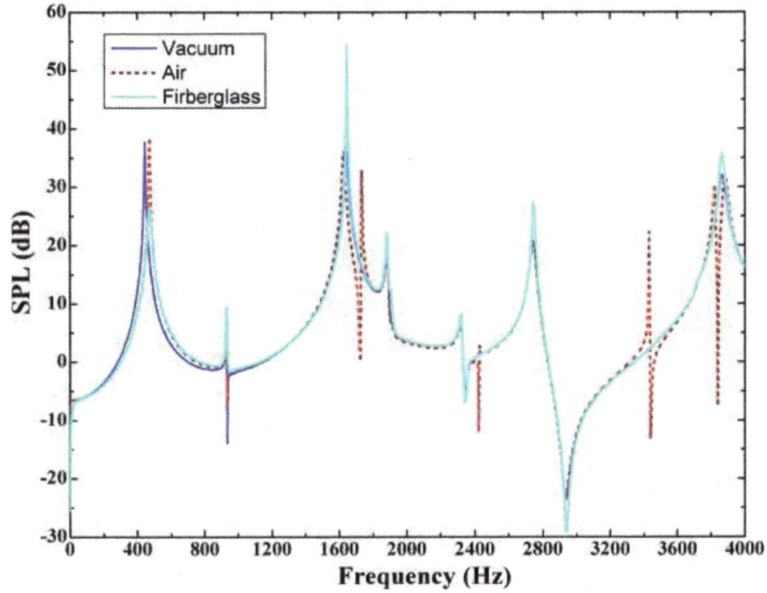

Fig. 6. Sound pressure levels radiated by different orthogonally rib-stiffened sandwich structures plotted as functions of frequency for stiffener separations (l_x, l_y) = (0.20m, 0.20m)

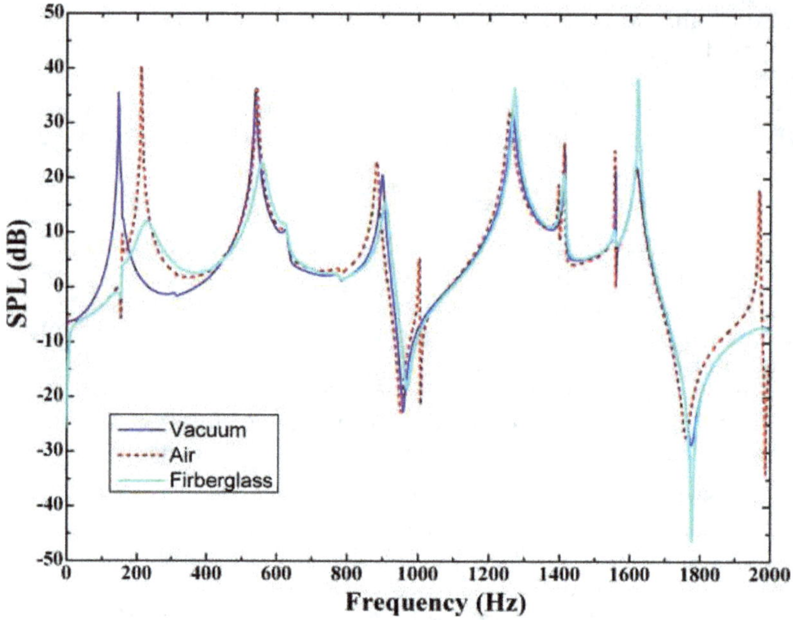

Fig. 7. Sound pressure levels radiated by different orthogonally rib-stiffened sandwich structures plotted as functions of frequency for stiffener separations (l_x =0.35m, l_y =0.35m)

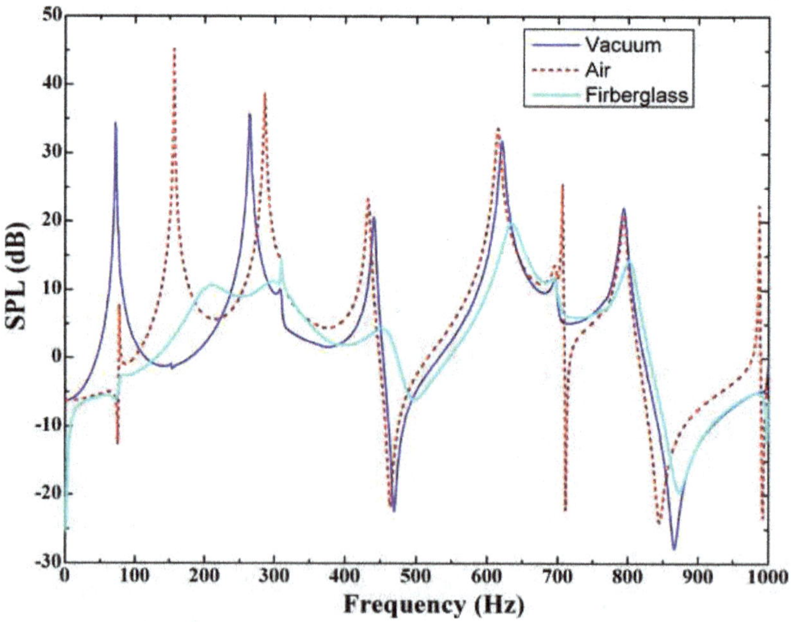

Fig. 8. Sound pressure levels radiated by different orthogonally rib-stiffened sandwich structures plotted as functions of frequency for stiffener separations (l_x =0.50m, l_y =0.50m)

At first glance, it can be seen from Figs. 6-8 that the air cavity case shows several additional peaks and dips on the SPL versus frequency curve. This is caused by air cavity interacting with the face panels through air-structure coupling. Besides these additional peaks and dips, it is also observed that the air-structure coupling effect plays an increasingly significant role in structure sound radiation with increasing rib-stiffener separation. This is reflected by the enlarged deviations between the two curves associated separately with the vacuum case and the air cavity case as the rib-stiffener separation is increased. In particular, when the separation is relatively large, air-structure coupling exerts a visible effect on the location of maximum sound radiation especially in low frequency range. The air-structure coupling is in effect by means of pumping effect, that is, the air cavity partitioned by the face panels and rib-stiffeners has timely changing pressure as its volume alters with the dynamic displacements of these two face panels, often imposing a converse force on the panels. In the case of rib-stiffener separation being relatively large, a considerable area of the panels is exposed to the impinging of air cavity pressure. It is thence understandable that the air-structure coupling effect may not be ignored when the rib-stiffeners are sparsely distributed.

Influence of fibrous sound absorptive filling material

In contrast to the air cavity case, the fiberglass case exhibits almost the same trends as the vacuum one, especially when the stiffener separation is relatively small, although the discrepancies between the two cases increase as the stiffener separation is increased. Note that the air-structure coupling effect is not present in the vacuum case whilst it is eliminated in the fiberglass case (the presence of fiberglass in the cavity significantly changes the behavior of the cavity). This is also the reason why the fiberglass case exhibits almost the same trend as the vacuum one: the discrepancies between the two cases enlarging with increasing stiffener separation actually reflect the combined effect of fiberglass stiffness and damping on structure responses.

It is understandable that the stiffness of the cavity-filling fiberglass reinforces the structural connection between the two face panels, enabling more vibration energies transmitted from the upper panel to the bottom one and thus causing larger sound radiation pressure levels. Conversely, fiberglass can dissipate acoustic energy via viscous drag forces and thermal exchange between the air and fibers, and hence decreases sound radiation. In addition, both the stiffness and damping of the fiberglass material are frequency dependent [17,19,22]. Consequently, the fact that the discrepancies between the vacuum and fiberglass cases increase with increasing stiffener separation can be well explained.

The periodically distributed rib-stiffeners with relatively narrow separations restrict the deformation of fiberglass in-between, offering therefore the fiberglass a larger stiffness than that inserted between those stiffeners having wider separations. That the fiberglass case exhibits the same trend as the vacuum one when the separation is small (e.g., $l_x = 0.20$ m and $l_y = 0.20$ m, as shown in Fig. 6) may be attributed to the balance of the converse effects of fiberglass stiffness and damping on sound radiation. More specifically, whilst damping is dominant at low frequencies, causing decreased sound radiation in the first peak, stiffness dominates at high frequencies, resulting in increased sound radiation in the following peaks (Figs. 7 and 8). As mentioned above, the stiffness of fiberglass decreases with increasing stiffener separation. Therefore, as the separation is increased, the frequency range dominated by stiffness (i.e., stiffness-controlled region) is shifted to higher frequencies and that dominated by damping (i.e., damping-controlled region) is widened. Correspondingly, in Fig. 7 the first three sound radiation peaks of the fiberglass case are lower than those of the vacuum one, and all the sound radiation peaks of the fiberglass case are significantly lower than those of the vacuum one in Fig. 8.

It may thence be deduced that the fiberglass-filled cavity affects structural radiation through the combined effects of fiberglass stiffness and damping (both being frequency dependent), the balance of which is significantly influenced by stiffener separation. It is therefore possible to optimize the stiffener separation and fiberglass porosity (both indirectly related to the stiffness and damping of fiberglass) to reduce structure sound radiation to an acceptable level required in specific cases.

Conclusions

The sound radiation characteristics of an infinite orthogonally rib-stiffened sandwich structure having cavity-filling fibrous sound absorptive material have been formulated by a comprehensive theoretical model when the structure is excited by a time-harmonic point force. The novelty of this work is to provide a general theoretical framework to address sound radiation issues of sandwich structures filled with fibrous sound absorptive materials, which can be degraded to deal with relatively simple structures. In the theoretical model, the vibration behaviors of the rib-stiffeners are accounted for by including all possible forces and moments exerted on the face panels by the rib-stiffeners in the governing equations. The propagation of sound in the fibrous material is modeled by adopting an equivalent fluid model with frequency dependent dynamic density and bulk modulus, with viscous drag force and thermal exchanges between air and fibers taken into account. The technique of Fourier transform is applied to solve the governing equations, resulting in an infinite set of simultaneous algebraic equations, which can be truncated and numerically solved.

Numerical calculations are subsequently carried out to explore the influences of air-structure coupling effect and fibrous sound absorptive materials on structure sound radiation. The model is validated by comparing the present model predictions with previously published data, with excellent agreement achieved especially at low frequencies. Nevertheless, slight deviations emerge at high frequencies, which just demonstrate the superiority of the present model.

Special attention is then focused on the effects of air-structure coupling and fibrous sound absorptive materials on sound penetration. This is explored by comparing three different sandwich structures: partitioned cavity in vacuum, filled with air, and filled with fiberglass. Interesting physical features emerging from the comparison are well interpreted by considering the combined effects of fiberglass stiffness and damping as well as the influence of rib-stiffener separation. It is found that the air-structure coupling effect induces additional peaks and dips in the SPL versus frequency curves, which plays an increasingly significant effect on structure sound radiation as the stiffener separation is increased. In particular, it is concluded that the fiberglass-filled cavity exerts its impact on wave penetration (finally on structural radiation) through the combined effects of fiberglass stiffness and damping (both frequency dependent), the balance of which is significantly affected by stiffener separation. This may provide a convenient and efficient tool to optimize the porosity, cell size and other topological parameters of fiberglass (indirectly altering its stiffness and damping) in conjunction with stiffener separation to reduce the structure vibration and sound radiation to an acceptable level required in specific situations.

As a future research forecast, the theoretical model for sandwich composite structures considered here (i.e., square lattice-cored sandwich structures filled with fibrous materials) can be further extended to study the acoustical performance of sandwich composite structures having laminated composites as skins, since these structures have been increasingly applied in aerospace and astronautic fields.

Acknowledgements

This work is supported by the National Basic Research Program of China (2011CB6103005), the National Natural Science Foundation of China (11072188, 11102148, 10825210 and 11021202) and the Fundamental Research Funds for the Central Universities.

References

[1] Xin FX, Lu TJ, Chen CQ. Sound transmission through simply supported finite double-panel partitions with enclosed air cavity. ASME J Vib Acoust 2010;132(1):011008: 1-11.

[2] Mace BR. Sound radiation from a plate reinforced by two sets of parallel stiffeners. J Sound Vib 1980;71(3):435-441.

[3] Mace BR. Sound radiation from fluid loaded orthogonally stiffened plates. J Sound Vib 1981;79(3):439-452.

[4] Yin XW, Gu XJ, Cui HF, Shen RY. Acoustic radiation from a laminated composite plate reinforced by doubly periodic parallel stiffeners. J Sound Vib 2007;306(3-5):877-889.

[5] Xin FX, Lu TJ, Chen CQ. Dynamic response and acoustic radiation of double-leaf metallic panel partition under sound excitation. Comput Mater Sci 2009;46(3):728–732.

[6] Maury C, Gardonio P, Elliott SJ. Active control of the flow-induced noise transmitted through a panel. AIAA J. 2001;39(10):1860-1867.

[7] Xin FX, Lu TJ, Chen CQ. External mean flow influence on noise transmission through double-leaf aeroelastic plates. AIAA J. 2009;47(8):1939-1951.

[8] Xin FX, Lu TJ. Analytical and experimental investigation on transmission loss of clamped double panels: Implication of boundary effects. J Acoust Soc Am 2009;125(3):1506-1517.

[9] Xin FX, Lu TJ, Chen CQ. Vibroacoustic behavior of clamp mounted double-panel partition with enclosure air cavity. J Acoust Soc Am 2008;124(6):3604-3612.

[10] Lin G-F, Garrelick JM. Sound transmission through periodically framed parallel plates. J Acoust Soc Am 1977;61(4):1014-1018.

[11] Takahashi D. Sound radiation from periodically connected double-plate structures. J Sound Vib 1983;90(4):541-557.

[12] Trochidis A, Kalaroutis A. Sound transmission through double partitions with cavity absorption. J Sound Vib 1986;107(2):321-327.

[13] Alba J, Ramis J, Sanchez-Morcillo VJ. Improvement of the prediction of transmission loss of double partitions with cavity absorption by minimization techniques. J Sound Vib 2004;273(4-5):793-804.

[14] Mead DJ, Pujara KK. Space-harmonic analysis of periodically supported beams: response to convected random loading. J Sound Vib 1971;14(4):525-532.

[15] Lee JH, Kim J. Analysis of sound transmission through periodically stiffened panels by space-harmonic expansion method. J Sound Vib 2002;251(2):349-366.

[16] Wang J, Lu TJ, Woodhouse J, Langley RS et al. Sound transmission through lightweight double-leaf partitions: Theoretical modelling. J Sound Vib 2005;286(4-5):817-847.

[17] Legault J, Atalla N. Numerical and experimental investigation of the effect of structural links on the sound transmission of a lightweight double panel structure. J Sound Vib 2009;324(3-5):712-732.

[18] Xin FX, Lu TJ. Analytical modeling of fluid loaded orthogonally rib-stiffened sandwich structures: Sound transmission. J Mech Phys Solids 2010;58(9):1374-1396.

[19] Panneton R, Atalla N. Numerical prediction of sound transmission through finite multilayer systems with poroelastic materials. J Acoust Soc Am 1996;100(1):346-354.

[20] Sgard FC, Atalla N, Nicolas J. A numerical model for the low frequency diffuse field sound transmission loss of double-wall sound barriers with elastic porous linings. J Acoust Soc Am 2000;108(6):2865-2872.

[21] Brown SM, Niedzielski J, Spalding GR. Effect of sound-absorptive facings on partition airborne-sound transmission loss. J Acoust Soc Am 1978;63(6):1851-1856.

[22] Allard J-F, Champoux Y. New empirical equations for sound propagation in rigid frame fibrous materials. J Acoust Soc Am 1992;91(6):3346-3353.

[23] Rumerman ML. Vibration and wave propagation in ribbed plates. J Acoust Soc Am 1975;57(2):370-373.

[24] Lu TJ, Hess A, Ashby MF. Sound absorption in metallic foams. J Appl Phys 1999;85(11):7528-7539.

[25] Wang XL, Lu TJ. Optimized acoustic properties of cellular solids. J Acoust Soc Am 1999;106(2):756-765.

[26] Lu TJ, Chen F, He D. Sound absorption of cellular metals with semiopen cells. J Acoust Soc Am 2000;108(4):1697-1709.

[27] Lu TJ, Kepets M, Dowling A. Acoustic properties of sintered FeCrAlY foams with open cells (I): Static flow resistance. Sci China Ser E: Tech Sci 2008;51(11):1803-1811.

[28] Lu TJ, Kepets M, Dowling A. Acoustic properties of sintered FeCrAlY foams with open cells (II): Sound attenuation. Sci China Ser E: Tech Sci 2008;51(11):1812-1837.

[29] Zwikker C, Kosten CW, 1949, *Sound absorbing materials,* Elsevier, New York.

[30] Morse PM, Ingard KU, 1968, *Theoretical acoustics,* McGraw-Hill, New York.

[31] Biot MA. Theory of propagation of elastic waves in a fluid-saturated porous solid. I. Low-frequency range. J Acoust Soc Am 1956;28(2):168-178.

[32] Biot MA. Theory of propagation of elastic waves in a fluid-saturated porous solid. II. Higher frequency range. J Acoust Soc Am 1956;28(2):179-191.

[33] Johnson DL, Koplik J, Dashen R. Theory of Dynamic Permeability and Tortuosity in Fluid-Saturated Porous-Media. J. Fluid Mech. 1987;176(379-402.

[34] Xin FX, Lu TJ. Sound radiation of orthogonally rib-stiffened sandwich structures with cavity absorption. Compos Sci Technol 2010;70(15):2198-2206.

[35] Xin FX, Lu TJ. Analytical modeling of sound transmission across finite aeroelastic panels in convected fluids. J Acoust Soc Am 2010;128(3):1097-1107.

5

Eliminating the Undamaging Fatigue Cycles using the Frequency Spectrum Filtering Techniques

S. Abdullah[1], T. E. Putra[2] and M. Z. Nuawi[1]
[1]*Universiti Kebangsaan Malaysia*
[2]*Universitas Syiah Kuala*
[1]*Malaysia*
[2]*Indonesia*

Introduction

Fatigue mechanism in structural failures

In 1982, the Battelle group has been found that between 80 – 90 % of all structural failures occur through a fatigue mechanism. Based on the finding, considerable effort has been applied in order to address this fact (Halfpenny, n.d.). The fatigue behaviour of mechanical components under service loading and its evaluation are usually affected by numerous uncertainties and characterized by several random variables such as material and structural properties, and load variation. The repeated fluctuating loads lead to microscopic physical damage of materials (Nizwan, et al., 2007).

Many vehicle components, like engine, steering, and suspension parts, are frequently subjected to a variety of driving conditions. When an automobile is driven on any road surface and hit a pothole, bump or curb, the lower suspension arm is effected by a significant shock amount compared to other components. The load is then transmitted through the control arm while it serves to maintain the contact between the wheel and the road. Furthermore, this component plays a vital role in the failure of automotive parts and it is submitted to multi axial fatigue loading under service conditions with generally nonproportional and variable amplitude loading (Nadot & Denier, 2004).

Signal processing of fatigue

With the advances in the digital signal processing research, there has been an increasingly strong interest in the application of the fatigue signal analysis for life prediction in automotive components. Several methods for analysis of signals have been proposed and performed for achieving this task. Among those is the Fast Fourier Transform (FFT) which has been widely applied to engineering problems. In the Fourier transform, the frequency components of an entire signal are analysed for producing the frequency information only. Since time information is lost, it is impossible to know when a particular event took place. For stationary signals, for which the frequency information is not changed by time, this drawback is not very important. However, most interesting signals contain numerous nonstationary or transitory characteristics which often occur in the most important part of an experimental measured signal with variable amplitude pattern. This means that this Fourier transform method was found to be an unsuitable method to investigate the behaviour of nonstationary patterns in a signal (Misiti et al., 2008; Valens, 1999).

Therefore, the time-frequency localisation technique was identified to overcome this drawback of analysis of a nonstationary signal. During the last decade, a new mathematical technique, called the wavelet transform (WT), has been frequently used in the field of vibrational diagnostics and also in fault detection. In addition, the wavelet coefficient analysis has also have been applied to detect fatigue transverse cracks in rotors. Its peak absolute value is highly sensitive to the depth of a crack and even a very shallow crack can be detected. The rotor is not required to stop and the detection process can be applied to a rotating shaft making the methodology more versatile, convenient and unambiguous (Darpe, 2007).

Fatigue data editing

For many automotive components, the primary mode of failure can be attributed to fatigue damage resulted from the application of variable amplitude loading (VAL). It contains large percentage of small amplitude cycles and the fatigue damage for these cycles can be small. For this reason, in many cases, the signal was edited by removing these cycles in order to produce representative and meaningful yet economical testing (Stephens et al., 1997). Therefore, it seems appropriate to see a method to summarise a fatigue strain signal. Without editing the service load, the testing time and cost become prohibitive (Abdullah, 2007).

The effectiveness of the edited signals depends on the retention of the fatigue damage and signal statistical parameter values. Those values are measured in order to observe the signal behaviour. They are compared with the original signal to ensure the edited signal obtained can retain the original signal features. A 5 % difference of the fatigue damage and ± 10 % difference of the root-mean-square (r.m.s) and the kurtosis are required in order to produce an edited signal which has equivalent fatigue damage and signal statistics to the original signal.

In the previous study, Nizwan et al., (2007) developed a fatigue data editing using the FFT. This algorithm was performed by eliminating unwanted cycles in a fatigue strain signal. Using Power Spectral Density (PSD), energy distribution in the frequency domain displayed strain characteristic of the input signal. In order to trace higher amplitude cycles, low pass filter was applied to filter out higher frequency containing small amplitude which not causes the fatigue damage. Thus, a smoothen signal was obtained. Despite using the frequency spectrum filtering technique could not shorten the signal, but it still could simplify the signal by reducing the numbers of cycle counting for the signal. The optimum Cut Off Frequency (COF) for lower pass filter application was determined to observe the effect of fatigue data editing technique in frequency domain.

Since the technique could not shorten the original signal, it gave a motivation to the authors for developing a similar data extraction approach in the WT. Therefore, a new algorithm for fatigue feature extraction using the Morlet wavelet was developed. The WT is probably the most recent solution to overcome the nonstationary signals. This time-frequency technique is applied by cutting time domain signal into various frequency components through the compromise between time and frequency-based views of the signal. It presents information in both time and frequency domain in a more useful form (Valens, 1999; Percival & Walden, 2000; Addison, 2002).

This paper discussed on the study of fatigue data editing by using the frequency spectrum filtering techniques. Both the techniques were used to eliminate undamaging fatigue cycles in order to simplify raw signal for the simulation testing purposes The WT results were compared to the findings using the FFT extraction approach in order to see the suitability approach in fatigue history editing.

Literature background

Fatigue signal

Signals are a form of information, as a function of time, coming from many sources obtained by measuring and using some recording method. Most data samples are normally measured using an analogue-to-digital converter, so as to produce an experimental signal at a series of regularly spaced times, known as a discrete time. The analysis of time series aims to determine the statistical characteristics of the original function by manipulating the series of discrete numbers.

Signals can be divided into two main categories, i.e. deterministic and nondeterministic. A deterministic signal can be described by a mathematical relationship between the value of the function and the value of time. Many signals in nature exhibit nondeterministic or random characteristics, which provide a challenge to analysis using signal processing techniques (Tacer & Loughlin, 1998). A signal representing a random physical phenomenon can not be described in a point by point manner by means of a deterministic mathematical equation.

A signal representing a random phenomenon can be characterised as either stationary or nonstationary. A stationary signal is characterised by values of the global signal statistical parameters, such as the mean, variance and root-mean-square (r.m.s.), which are unchanged across the signal length. In the case of nonstationary signals, the values are dependent on the time of measurement. Nonstationary signals can be divided into two categories: mildly nonstationary and heavily nonstationary. A mildly nonstationary signal is defined as a random process with a stable mean, variance and r.m.s. values for most of the recorded data, but with short periods of differing signal statistics due to the presence of transient behaviour. A heavily nonstationary signal is defined as being similar to a mildly nonstationary signal, but with the presence of transient events over a large interval of the time history. Since nonstationary loadings are common in the case of fatigue and vibrational analysis (Giancomin et al., 1999), signal modelling has often been used in the time domain due to its simplicity and efficiency for the purpose of loading simulations.

Global signal statistical parameters

In the case of fatigue researches, the signals consist of a measurement of cyclic loads, i.e. force, strain, and stress against time. A time series typically consists of a set of observations of a variable were taken at equally spaced intervals of time. Global signal statistical parameters are frequently used to classify random signals and monitor the pattern of analyzed signals. For a signal with a numbers of data point n in a sampled sequence, the mean $\overline{x}$ is given by:

$$\overline{x} = \frac{1}{n}\sum_{j=1}^{n} x_j \qquad (1)$$

In the fatigue signals, the calculation of the r.m.s. and the kurtosis are very important in order to retain a certain amount of the signal amplitude range characteristics (Nuawi et al., 2009). The r.m.s. is the signal 2nd statistical moment used to quantify the overall energy content of the oscillatory signal. The r.m.s. relationship is defined as:

$$r.m.s. = \left\{ \frac{1}{n}\sum_{j=1}^{n} x_j^{\ 2} \right\}^{1/2} \qquad (2)$$

The kurtosis is the signal 4th statistical moment. In an engineering field, it is used as a measure of nongaussianity for detection of fault symptoms since it is highly sensitive to spikiness or outlier signal among the instantaneous values. Mathematically, the kurtosis expression is defined as:

$$K = \frac{1}{n(r.m.s)^4}\sum_{j=1}^{n}\left(x_j - \overline{x}\right)^4 \qquad (3)$$

where x_j is the amplitude of signal. In some definitions of the kurtosis, a deduction of 3.0 is added to the definition in order to maintain the kurtosis of a Gaussian distribution to be equal to zero. For clarity and convenience, in this study the original definition of the kurtosis, where the Gaussian distribution has a kurtosis value of approximately 3.0 was used for the analysis. Therefore, a kurtosis value of higher than 3.0 indicates the presence of more extreme values than one that should be found in a Gaussian distribution (Abdullah et al., 2006). This situation indicates that the fatigue damage is higher than Gaussian stresses due to higher amplitude fatigue cycles (Braccesi et al., 2009).

Fatigue life assessment

In material science, fatigue is defined as a process of progressive and localised structural damage duo to cyclic loads, which depend on the stresses and the strains at critical regions of a component (Prawoto, 2002). The repeated or otherwise varying loads never reach a level sufficient for causing failure in a single application. The loads cause plastic deformation (slip) locally which results in a crack beginning, crack propagation, and then a breakage. This plastic deformation might arise through the presence of a small crack or pre-existing defect on the surface of a component. In the case of the fatigue research, the signals consist of a measurement of cyclic loads, i.e. force, strain, and stress against time (Nuawi et al., 2009).

Three major approaches to predicting fatigue life namely stress-life, strain-life, and fracture mechanics. The strain-life fatigue approach relates the plastic deformation that occurs in the localized region where fatigue cracks begin to the durability of a structure under the influence of a mean stress. It is often used for ductile materials at relatively short fatigue life. It also can be used where there is little plasticity at long fatigue life. Therefore, this is a comprehensive approach that can be used in place of the related stress - based approach (Halfpenny, n.d.).

The total strain amplitude ε_a produced by the combination of elastic and plastic amplitude is defined as:

$$\varepsilon_a = \varepsilon_{ea} + \varepsilon_{pa} \tag{4}$$

where the elastic strain amplitude ε_{ea} and the plastic strain amplitude ε_{pa} are defined by:

$$\varepsilon_{ea} = \frac{\sigma_a}{E} = \frac{\sigma'_f}{E}\left(2N_f\right)^b \tag{5}$$

$$\varepsilon_{pa} = \varepsilon'_f\left(2N_f\right)^c \tag{6}$$

where σ_a is the stress amplitude, σ'_f is the fatigue strength coefficient, E is the material modulus of elasticity, N_f is the numbers of cycle to failure for a particular stress range and mean, b is the fatigue strength exponent, ε'_f is the fatigue ductility coefficient, and c is the fatigue ductility exponent.

Combining Eq. (5) and (6) gives the Coffin-Manson relationship, which is mathematically defined as (nCode, n.d.):

$$\varepsilon_a = \frac{\sigma'_f}{E}\left(2N_f\right)^b + \varepsilon'_f\left(2N_f\right)^c \tag{7}$$

where ε_a is the true strain amplitude, σ'_f is the fatigue strength coefficient, E is the material modulus of elasticity, N_f is the numbers of cycle to failure for a particular stress range and mean, b is the fatigue strength exponent, ε'_f is the fatigue ductility coefficient, and c is the fatigue ductility exponent.

Fatigue damage is derived from the number of cycles to failure. The fatigue damage caused by each cycle of repeated loading is calculated by reference to material life curves, such as *S-N* or *ε-N* curves. The fatigue damage D for one cycle is calculated as:

$$D = \frac{1}{N_f} \tag{8}$$

and the total fatigue damage ΣD caused by cycles is expressed as (Abdullah, 2005):

$$\Sigma D = \Sigma\left(\frac{N_i}{N_f}\right) \tag{9}$$

where N_i is the numbers of cycle within a particular stress range and mean. Therefore, fatigue damaging values have the range (0 - 1) where zero denotes no damage (extremely high or infinite numbers of cycle to failure) and 1 means total failure (one cycle to failure). For strain - based fatigue life prediction, current industrial practice uses the Palmgren-Miner linear cumulative damaging rule normally associated with the established strain-life fatigue damaging models, such as the Coffin-Manson.

The FFT

Frequency domain is a term used to describe the analysis of mathematical functions or signals with respect to frequency. A frequency domain graph shows how much of the signal lies within each given frequency band over a range of frequency. A frequency domain representation can also include information on the phase shift that must be applied to each sinusoid in order to be able to recombine the frequency components to recover the original time signal. The frequency domain relates to the Fourier transform or Fourier series by decomposing a function into an infinite or finite number of frequency. This is based on the concept of Fourier series that any waveform can be expressed as a sum of sinusoids (Nizwan et al., 2007).

The Fourier transform has been most commonly used to denoise signals for a frequency - based editing method, which cannot provide any information regarding the time localization of the spectral components. This Fourier representation has been found inadequate in analyzing nonstationary signals (Oh, 2001). Frequency analysis data is typically presented in graphical form as PSD. Essentially, a PSD displays the amplitude of each sinusoidal wave of a particular frequency which is given on the x-axis. The mean squered amplitude of a sinusoidal wave at any frequency can be determined by finding the area under the PSD over that frequency range (nCode, 2005).

A filter is used to remove undesirable frequency information from a dynamic signal. Filter can be broadly classified as being low pass, high pas, bandpass, and notch. Low pass filter permits frequency below a prescribed COF to pass while blocking the passage of frequency information above the COF. Similarly, a high-pass filter permits only frequency above the COF to pass. A bandpass filter combines features of both the low pass and high pass filter. It describes a lower and higher COF to define a band of frequency that is permitted to pass through the filter. A notch filter permits the passage of all frequency except those within narrow frequency band (Nizwan et al., 2007).

The Morlet wavelet

The WT analysis is started with a basic function (called the mother wavelet) scaled and translated to represent the signal being analyzed (Berry, 1999). The transform shifts a window along the signal and calculates the spectrum for every position. The process is repeated many times with a slightly shorter (or longer) window for every new cycle. In the end, the result will be a collection of time-frequency representations of the signal with different resolutions. The WT provides information on when and at what frequency the change in signal behaviour occurs (Valens, 1999). The major advantage is the ability to analyze a localized area of larger signal (local analysis) (Misiti et al., 2008).

Obviously, the WT represents a windowing technique with variable-sized regions. This technique allows the use of long time intervals (more precise low frequency information) and shorter regions (high frequency information). It means the wavelet method solves the resolution problem because the window length is long for low frequency and short for high frequency. Therefore, the frequency resolution is good for low frequency (at high scales) and the time resolution is good at high frequency (at low scales). The major advantage is the ability to analyse a localised area of a larger signal, also known as the local analysis (Misiti et al., 2008).

The Morlet wavelet is one of functions that are generally used in the Continuous Wavelet Transform (CWT) analyses (Gao et al., 2001). The wavelet decomposition calculates a resemblance index between signal being analyzed and the wavelet, called coefficient. It is a result of a regression of an original signal produced at different scales and different sections on the wavelet. It represents correlation between the wavelet and a section of the signal. If the index is large, the resemblance is strong, otherwise it is slight (Misiti et al., 2008).

The WT of any time-varying signal $f(t)$ is defined as the sum of all of the signals time multiplied by a scaled and shifted version of the wavelet function $\psi(t)$ (Kim et al., 2007). The CWT is expressed by the following integral:

$$CWT_{(a,b)} = \int_{-\infty}^{+\infty} f(t)\psi_{a,b}(t)dt \tag{10}$$

The parameter a represents the scale factor which is a reciprocal of frequency, the parameter b indicates the time shifting or translation factor, and t is the time. $\Psi_{a,b}(t)$ denotes the mother wavelet, i.e (Purushotham et al., 2005):

$$\psi_{a,b}(t) = \frac{1}{\sqrt{|a|}}\psi\left(\frac{(t-b)}{a}\right) a,b \in R; a \neq 0 \tag{11}$$

$$CWT_{(a,b)} = \int_{-\infty}^{+\infty} f(t)\frac{1}{\sqrt{a}}\psi\left(\frac{t-b}{a}\right)dt \tag{12}$$

In addition, the wavelet coefficient indicates how energy in the signal is distributed in the time-frequency plane (Darpe, 2007). The energy spectrum (the energy density over frequency) is plotted in order to observe the signal behaviour and its content gives significant information about the random signal pattern.

Fatigue data editing

In a fatigue life assessment, fatigue signal extraction is described as a method for fatigue data editing which lead to summarise a fatigue signal. The method is performed by segment identification and extraction that contribute to the more fatigue damaging events to a metallic material. On the other hand, segments containing lower amplitude cycles are omitted, since these data type theoretically gave minimal or no fatigue damage. The goal of the removal of those parts from the original signal is to generate a new shortened mission signal, for which this signal type can be used to reduce the testing time and costs for fatigue testing (Abdullah, 2005).

The magnitude of time domain spectrum level is used as a parameter to set gate value for the eliminating process. The value is used to slice the original signal. The extracted segment identification is performed by searching two inversion points (one on either side of the peak value) which define the temporal extent of the extracted segment. The identification is based on energy loss concept, i.e. selected segments are at the start and finish points.

The example of the segment identification is described in Fig. 1. In the figure, the selected segment is at gate value of 400 $\mu\varepsilon^2$/Hz. Start point is a valley point if the peak before is higher than the peak after the point. While the finish point is selected if peak after is higher than peak before the point. This concept is performed by Abdullah (2005) based on transient vibration where start and finish points are selected based on transient form. The points are determined based on the signal where the shortening in signal background occurs.

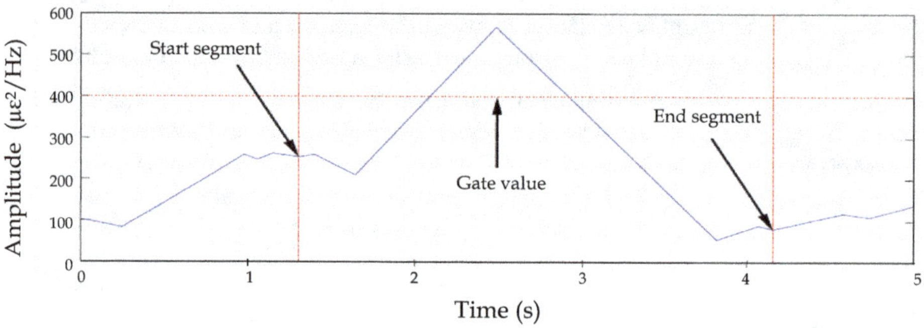

Fig. 1. The extracted segment identification

After all the segments are identified, the time history fatigue signal is then sliced to remove the lower amplitude (less than the gate value) contained in the original time history range with 100 % fatigue damaging retains. For this reason, the majority of the original fatigue damage is retained in the edited signal. All extracted segments (the complete section between the start and the end of the segments) selected based on time location of the wavelet coefficient amplitude are then combined together to produce a new mission time history. The mission signal replicates the signal statistical parameter and total fatigue damaging characteristics of the original time history. The optimum gate value is accordingly determined and it is based on the effectiveness of retaining the characteristics of the original signal in the mission signal. Ideally, the signal has shorter time length but is equivalent in the characteristic values.

Materials and methods

Finite element analysis

Every structure will experiences a transformation if it accepts a load, and an internal force, called stress, will stops the transformation. The load causes the structure to experience direct stress, not only tensile stress but also compressive stress. Structure accepting the tensile stress will become smaller, while structure accepting the compressive stress will cause buckling or permanent damage (Balia & Putra, 2006). If A is the cross-sectional area and F is the applied load, the true stress σ at fracture can be defined mathematically as the following expression (Draper, 2007):

$$\sigma = \frac{P}{A} \tag{13}$$

Data acquisition

For the data acquisition purpose, strain gauge was placed at the highest stress location of the front lower suspension arm of a passenger car. The car was travelled on pavé (cobblestone) road surface with velocity was 20 - 30 km/h. Fig. 2 shows the strain gauge position, the test track and the data acquisition set-up.

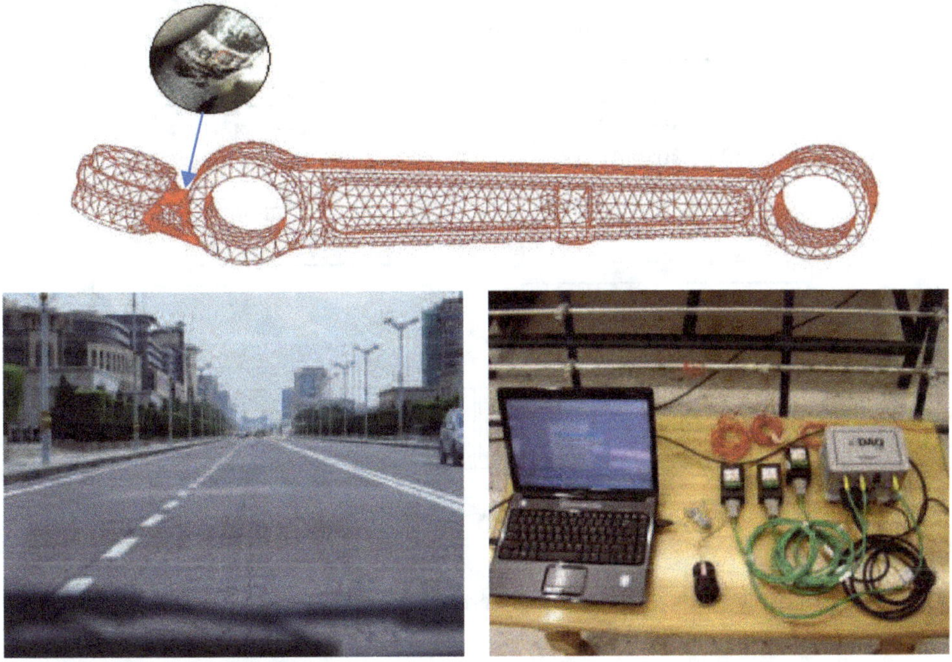

Fig. 2. The strain gauge position, a section of the pavé test track and the data acquisition set-up

The input signal was the variable amplitude loading sampled at 200 Hz since the fatigue damage occurs at lower frequency. From the data collection, it gave the total signal record length of 160 seconds. The collected signal recorded using a fatigue data acquisition system contained many small amplitude and high frequency in occurrence signal background. By

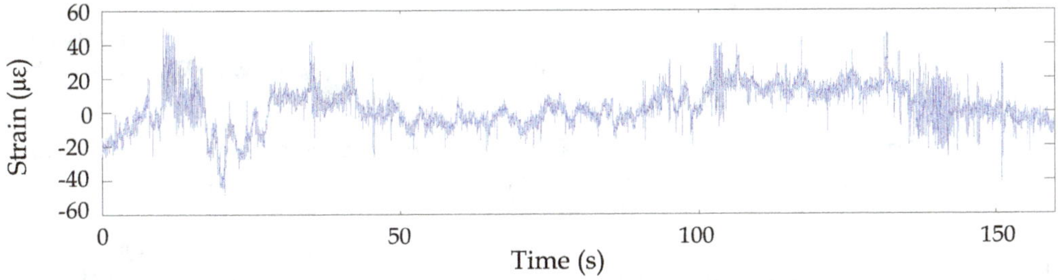

Fig. 3. The time history plot of the original test signal

multiplying the frequency and the signal length, it gave 32,000 data points. Fig. 3 shows the 160 second time history plotted for the original signal. The flowchart of the study is shown in Fig. 4.

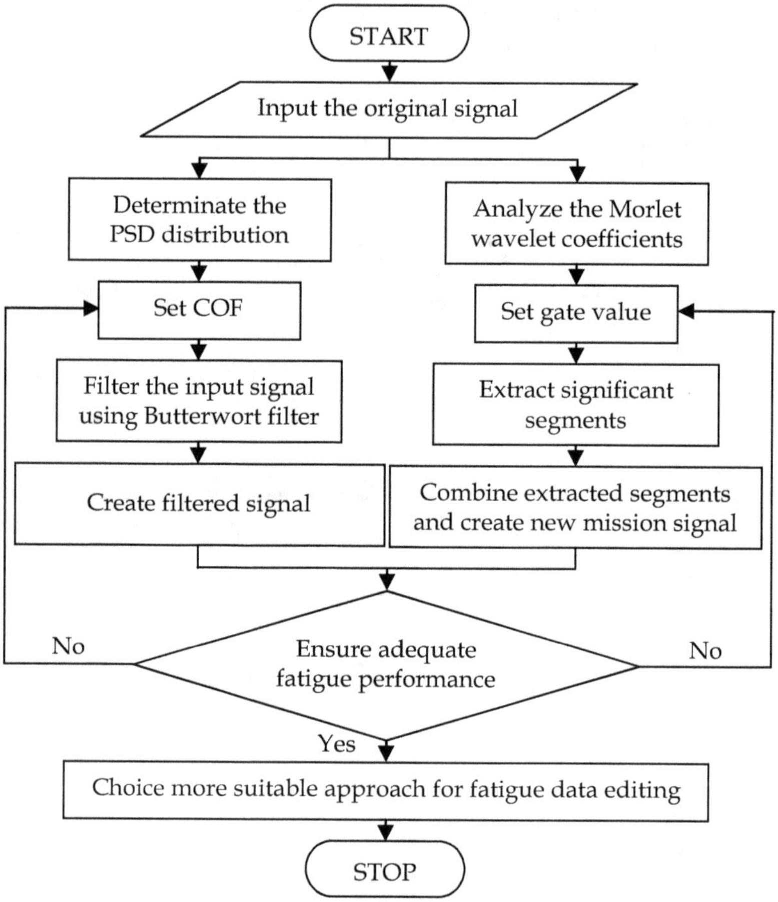

Fig. 4. Simplified flowchart of the undamaging cycle eliminating process

In this study, the selected material for the simulation purpose was the SAE1045 carbon steel shaft. This material was chosen because it was commonly used in automotive industries for fabricating a vehicle lower suspension arm structure (Khalil & Topper, 2003). The material properties and their definitions are given in Table 1 (nCode, 2005).

The fatigue damage and signal statistical parameter values for the each edited signal which was produced from the extraction process with differences COF and gate values would then be determined in order to observe the effectiveness of the methods for the fatigue data editing technique. The fatigue damage was estimated by utilizing the Coffin-Manson strain-life model. The cumulative fatigue damage was then determined from the Palmgren-Miner linear cumulative damaging rule. The edited signal required lower than 5 % difference of fatigue damage to the original signal in order to produce a edited signal which retaining the original fatigue damage.

Properties	Value
Ultimate tensile strength, S_u (MPa)	621
Modulus of elasticity, E (GPa)	204
Fatigue strength coefficient, σ'_f(MPa)	948
Fatigue strength exponent, b	-0.092
Fatigue ductility exponent, c	-0.445
Fatigue ductility coefficient, ε'_f	0.26

Table 1. The mechanical properties of the SAE1045 carbon steel shaft

Results and discussion

Stress distribution

The finite element modeling and analysis have been performed utilizing CATIA and MSC finite element analysis codes respectively. By using Eq. 13, this analysis determinates the maximum stress locations before doing the fatigue test. Based on the analysis performed by Al-Asady et al. (2008), it was obtained the stress distribution at the front lower suspension arm, as shown in Fig. 5.

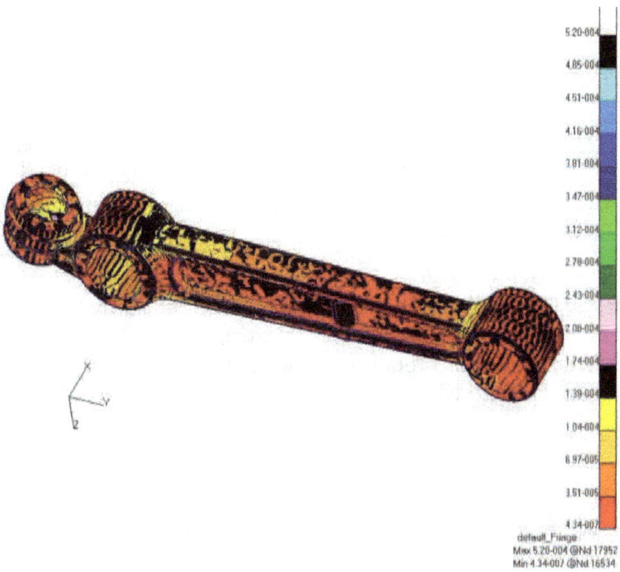

Fig. 5. Stress distribution and strain gauge positions

In this simulation, the load was assumed of 300 kg. This value came from the car weight and the passengers. The stress level is presented by a colour contour, where the white colour shows the highest stress content and followed by black, blue, and so on.

The FFT analysis

In this frequency spectrum filtering analysis, data in time domain was then transformed into the frequency domain using the FFT algorithm in order to obtain its PSD distribution. The FFT formed a complex vector of values which each value represents the amplitude and phase of the particular sinusoidal wave at a particular frequency. The energy distribution showed on the PSD was applied to see the frequency region with lower energy in order to find the COF value for filtering purpose. The PSD display for original data is shown in Fig. 6.

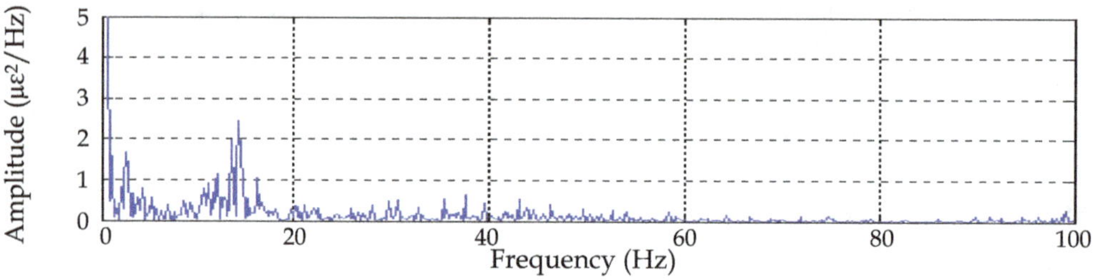

Fig. 6. The PSD distribution of the original signal

For fatigue data, lower amplitude region commonly located at higher frequency spectrum. Therefore, lower pass filter could be applied to remove higher frequency data which contain lower amplitude cycle in time history. Lower amplitude cycle could be removed because it gave minimal fatigue damage.

The filtering analysis with 4th order of Butterworth low-pass filter has been used for this data set. The 30 Hz COF was selected as the initial COF value, where getting from the PSD plotted. With 30 Hz COF, filtering process would filter out the frequency information higher than that value and obtain the new signal which contains only the allowed frequency. The same procedure was then repeated by using other COF values i.e. 35 Hz, 40 Hz, 45 Hz, 50 Hz, 55 Hz, 60 Hz, 65 Hz, 70 Hz, 75 Hz, 80 Hz, 85 Hz, 90 Hz, 92 Hz, 94 Hz, 96 Hz and 98 Hz. The difference in COF values were used in order to observe the effect of filtering behaviour related to fatigue life assessment for each filtered signal.

From the analysis of each filtered signal, the optimum value of COF was accordingly determined and it was based on the effectiveness of retaining the value of the fatigue damage and the signal statistical parameters. Unfortunately, the filtered signal with lower COF gave an obvious deviation in retaining fatigue damage. Nevertheless, by increasing the COF value the fatigue damage was ascending and almost reaches the fatigue damage of the

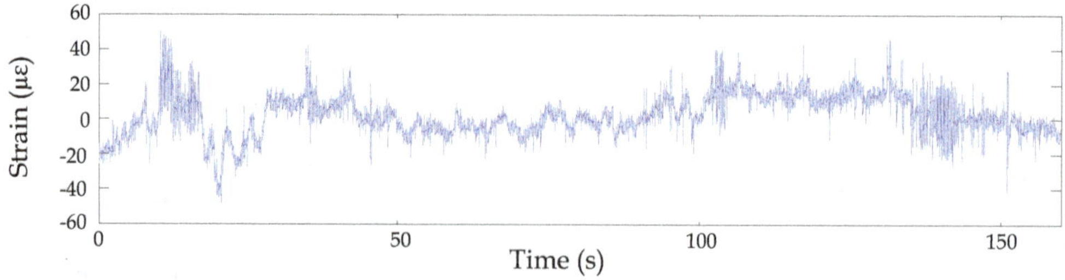

Fig. 7. Signal display after the filtering process

original signal. The filtered signal with COF 96 Hz was selected as the optimum values that giving lower than 5 % difference of fatigue damage. For the statistical parameter analysis, the values of the r.m.s. and the kurtosis for filtered signals were almost equivalent with the original signal, where those values were in the required range for any filtered signal with the differences were below than 10 %. Fig. 7 displays signal after the filtering process.

The Morlet wavelet analysis

For the Morlet wavelet based edited signal, it was started by analyzing the wavelet coefficients, as shown in Fig. 8. In the scalogram, the x-axis denoted the time parameter, the y-axis represented the scale that has an inversely related to the frequency value, and the colour intensity at each x-y point was proportional to the absolute value of the wavelet coefficients as a function of the dilation and translation parameters. It provided the signal energy distribution display with respect to the particular time and frequency information.

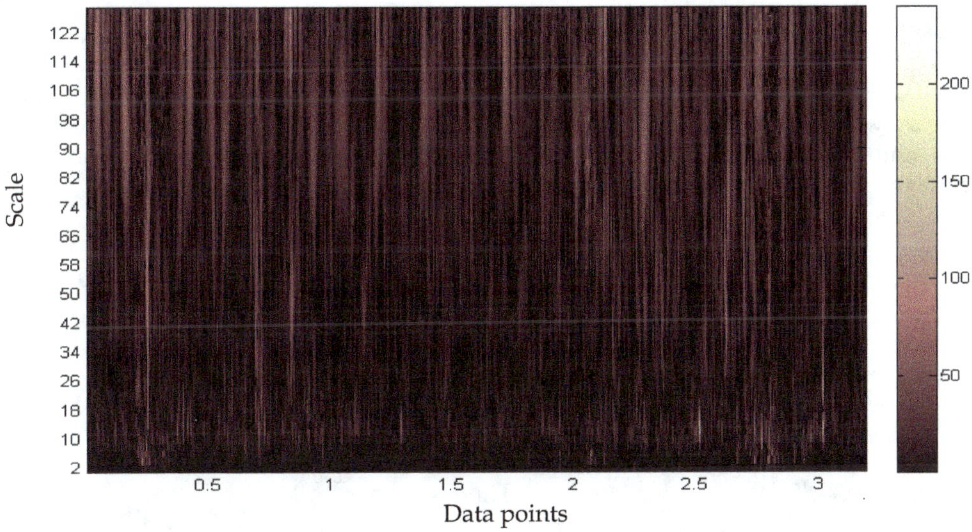

Fig. 8. The Morlet wavelet coefficients in time-frequency representation

Accordingly, a lower scale indicated higher frequency and had small amplitude that means these cycles had lower energy. They gave minimal or no fatigue damaging potential. A large scale was indicative of lower frequency and higher amplitude that indicates these cycles had higher energy causing the fatigue damage. Obviously, the lower frequency indicated higher magnitude distribution, and the lower magnitude distribution was presented at higher frequency event. Using the newly Morlet wavelet - based developed computational algorithm, the wavelet coefficient magnitude segments were transposed into time domain signal, as shown in Fig. 9. The representation showed a two dimensional view of the energy distribution, as observed in time-frequency plane.

This extraction process involved 150 με, 160 με, 170 με, 180 με, 190 με, and 200 με gate values. From the total fatigue damaging calculation results, it was found that 170 με was

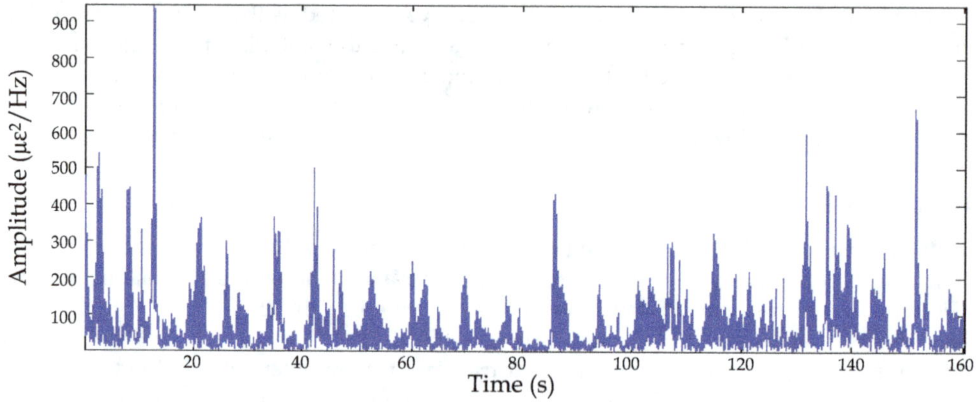

Fig. 9. The Morlet wavelet coefficients in time representation

selected to be the optimum gate value giving lower than 5 % difference of the fatigue damage. In this gate value, the numbers of cycle counting were only 4,561 cycles, which was 58 % less than the original signal. Furthermore, the new edited signal was produced of 68 seconds, which was 92 seconds shorter than the original signal length. The values of the r.m.s. and the kurtosis for the signal were in the required range with the differences were below than 10 %. Fig. 10 shows the plot of the edited signal.

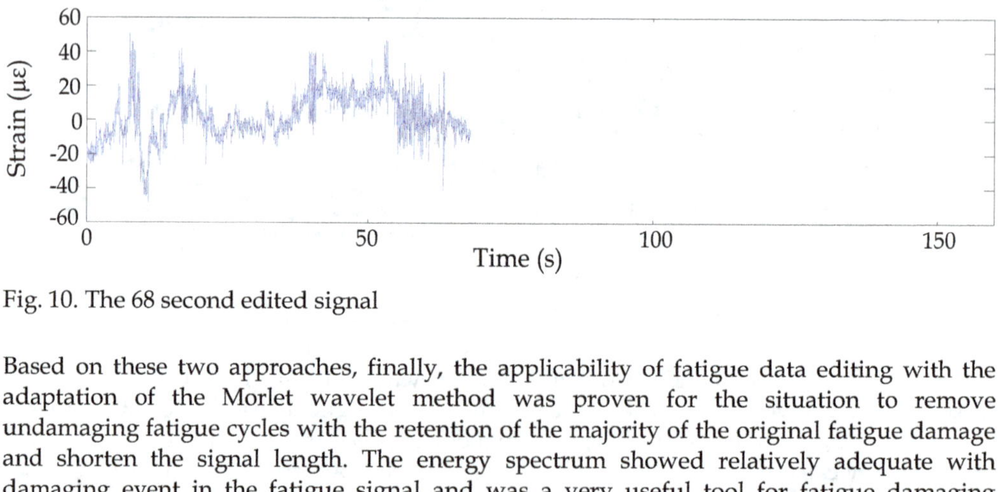

Fig. 10. The 68 second edited signal

Based on these two approaches, finally, the applicability of fatigue data editing with the adaptation of the Morlet wavelet method was proven for the situation to remove undamaging fatigue cycles with the retention of the majority of the original fatigue damage and shorten the signal length. The energy spectrum showed relatively adequate with damaging event in the fatigue signal and was a very useful tool for fatigue damaging detection in the fatigue signal.

Conclusions

This paper discussed on the study of fatigue data editing by using the frequency spectrum filtering techniques. Both the methods have been used to analyze random signal that can be applied to extract fatigue damaging events in the fatigue strain loading. The techniques were used to eliminate undamaging fatigue cycles in order to simplify raw signal for the simulation testing purposes. Based on the simulation analysis, it was found that the Morlet wavelet was more suitable approach for eliminating undamaging cycles. The method removed 58 % of the original cycles in the 68 second edited signal with the retention of at least 95.5 % of the original fatigue damage. It removed 28 % more than the edited signal that was found using the FFT technique. The extraction of fatigue damaging events successfully removed the lower energy cycles in the time history. It created a new edited signal which retains higher fatigue damaging segments containing the majority of the fatigue damage. Hence, this fatigue data summarising computational algorithm can be used in laboratories in order to study the durability characteristics of automotive components. In overall, the analysis findings of this paper suggested that the Morlet wavelet was more suitable for the fatigue data editing.

Acknowledgement

The authors would like to express their gratitude to Universiti Kebangsaan Malaysia and Universitas Syiah Kuala for supporting the research.

References

Abdullah, S. (2005). *Wavelet Bump Extraction (WBE) for Editing Variable Amplitude Fatigue Loadings*, Ph.D. Thesis, The University of Sheffield

Abdullah, S.; Choi, J.C.; Giacomin, J.A. & Yates, J.R. (2006). Bump Extraction Algorithm for Variable Amplitude Fatigue Loading. *International Journal of Fatigue*, Vol.28, pp. 675-691

Abdullah, S. (2007). The Wavelet Transform for Fatigue History Editing: is It Applicable for Automotive Application? *Journal of Engineering and Applied Sciences*, Vol.2, No.2, pp. 342-349

Addison, P.S. (2002). *The Illustrated Wavelet Transform Handbook*, Institute of Physics Publishing, ISBN 0 7503 0692 0, Bristol, UK

Al-Asady, N.A.; Abdullah, S.; Arifin, A.K. & Rahman, M.M. (2008). Effects of Surface Finish on the Fatigue Life Assessment Using Finite Element Analysis: a Case Study of a Lower Suspension Arm, *Proceeding of 2nd Regional Conference on Vehicle Engineering and Technology*

Balia, F.N. & Putra, T.E. (2006). Strength Analysis Due To Static Load on Main Tribune Truss of Harapan Bangsa Stadium, *Proceedings of Conference on Computational Mechanics & Numerical Analysis (CMNA)*, pp. 297-300, Banda Aceh, Indonesia

Berry, S. (1999). Practical Wavelet Signal Processing for Automated Testing, *IEEE*, pp. 653-659, 0-7803-5432-X

Braccesi, C.; Cianetti, F.; Lori, G. & Pioli, D. (2009). The Frequency Domain Approach in Virtual Fatigue Estimation of Non-linear Systems: the Problem of Non-Gaussian States of Stress. *International Journal of Fatigue*, Vol.31, pp. 766-775

Darpe, A.K. (2007). A Novel Way to Detect Transverse Surface Crack in a Rotating Shaft. *Journal of Sound and Vibration*, Vol.305, pp. 151-171

Draper, J. (2007). *Modern Metal Fatigue Analysis*, EMAS Publishing Ltd., Warrington, UK

Gao, J.H.; Wu, R.S. & Wang, B.J. (2001). A New Type of Analyzing Wavelet and Its Applications for Extraction of Instantaneous Spectrum Bandwidth, *Proceedings of*

SEG International Exposition and Annual Meeting, San Antonio, Texas, USA, September 9-14, 2001

Giancomin, J.; Steinwolf, A. & Staszewski, W.J. (1999). A Vibration Mission Synthesis Algorithm for Mildly Nonstationary Road Data, *Proceedings of ATA 6th International Conference on the New Role of Experimentation in the Modern Automotive Product Development Process*, Florence, Italy, November 17-19, 1999

Halfpenny, A. (n.d.). *A Practical Introduction to Fatigue*, nCode International Ltd., Sheffield, UK

Kim, B.S.; Lee, S.H.; Lee, M.G.; Ni, J.; Song, J.Y. & Lee, C.W. (2007). A Comparative Study on Damage Detection in Speed-Up and Coast-Down Process of Grinding Spindle-Typed Rotor-Bearing System. *Journal of Materials Processing Technology*, Vol.187-188, pp. 30-36

Khalil, M. & Topper, T.H. (2003). Prediction of Crack-Opening Stress Levels for 1045 As-Received Steel Under Service Loading Spectra. *International Journal of Fatigue*, Vol.25, pp. 149-157

Misiti, M.; Misiti, Y.; Oppenheim, G. & Poggi, J.M. (2008). *Matlab User's Guide: Wavelet ToolboxTM 4*, The Math Works Inc, MA, USA

Nadot, Y. & Denier, V. (2004). Fatigue Failure of Suspension Arm: Experimental Analysis and Multiaxial Criterion. *Engineering Failure Analysis*, Vol.11, pp. 485-499

nCode. (2005). *ICE-flow: GlyphWorks 4.0 Tutorials*, nCode International Ltd, Sheffield, UK

nCode. (n.d.). *The nCode Book of Fatigue Theory*, nCode International Ltd, Sheffield, UK

Nizwan, C.K.E.; Abdullah, S.; Nuawi, M.Z. & Lamin, F. (2007). A Study of Fatigue Data Editing Using Frequency Spectrum Filtering Technique, *Proceedings of World Engineering Congress*, pp. 372-378, ISBN 978–983–43571–1–5, Penang, Malaysia, August 5-9, 2007

Nuawi, M.Z.; Abdullah, S.; Abdullah, S.; Haris, S.M. & Arifin, A. (2009). *Matlab: A Comprehensive Reference for Engineers*, McGraw-Hill Sdn. Bhd, ISBN 978-983-3850-52-5, Kuala Lumpur, Malaysia

Oh, C.S. (2001). Application of Wavelet Transform in Fatigue History Editing. *International Journal of Fatigue*, Vol.23, pp. 241-250

Percival, D.B. & Walden, A.T. (2000). *Wavelet Methods for Time Series Analysis*, Cambridge University Press, UK

Prawoto, Y. (2002). Linear Elastic Fracture Mechanics (LEFM) Analysis of the Effect of Residual Stress on Fatigue Crack Propagation Rate. *PFANF8*, Vol.2, No.5, pp. 75-83

Purushotham, V.; Narayanan, S. & Prasad, S.A.N. (2005). Multi-fault Diagnosis of Rolling Bearing Elements Using Wavelet Analysis and Hidden Markov Model Based Fault Recognition. *NDT&E International*, Vol.38, pp. 654-664

Stephens, R.I.; Dindinger, P.M. & Gunger, J.E. (1997). Fatigue Damage Editing for Accelerated Durability Testing Using Strain Range and SWT Parameter Criteria. *International Journal of Fatigue*, Vol.19, No.8-9, pp. 599-606

Tacer, B. & Loughlin, P.J. (1998). Non-stationary Signal Classification Using the Joint Moments of Time-Frequency Distributions. *Pattern Recognition*, Vol.31, No. 11, pp. 1635-1641

Valens, C. (1999). A Really Friendly Guide to Wavelets, 07.05.2008, Available from http://pagesperso-orange.fr/polyvalens/clemens/wavelets/wavelets.html

6

Charaterization of Pore Structure and Surface Chemistry of Activated Carbons

Bingzheng Li
Taiyuan University of Science & Technology
P. R. China

Introduction

There is a rapid development in industries since industrial revolution began in the latter half of the 18th century, which not only significantly promotes development of economy all over the world and prosperity in societies, but also extremely improves our lives. However, a large amount of wastewaters including some toxic and hazardous materials such as some metals ions and some non-biodegraded organic compounds are usually produced in many factories during production process. These pollutants such as simple/complex aromatic compounds in wastewaters discharged from factories such as textiles, rubber, paper, plastic, explosive, pharmaceuticals, petroleum refining and cosmetics factories into ditches/rivers are stable to light and oxidation as resistant to aerobic digestion, causing damage to the aquatic life and food web (Kagalkar, et al., 2010, Saratale, et al., 2011, Tabrez&Ahmad, 2009). The basic properties of simple aromatic compounds are shown in Table 1. The compounds can also cause allergic dermatitis and skin irritation. Furthermore, Most of them have been reported to be carcinogenic and mutagenic for aquatic organisms (Lorenc-Grabowska&Gryglewicz, 2007), so discharging of the pollutants is becoming the most challenging threat to environment safety and even human beings' lives. The introduction of toxic and/or non-biodegradable compounds in the natural environment has been a world wide serious problem that has been paid more attention to by many environmentalist and research institutions from different countries.

Simple aromatic compounds	Boiling point (°C)	pKa at 25°C	Aqueous solubility at 25°C(g/L)
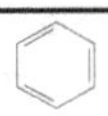	80.4	–	1.79
	115.3	5.17	∞
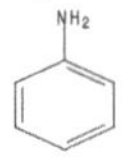	184.4	4.60	34
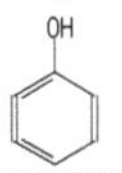	182	9.89	93

Table 1. Basic properties of some simple aromatic compounds (Gokel&Dean, 2004)

To decrease or even eliminate the concentration of the compounds in wastewaters before discharging into environment, the various treatment methods are used for the removal of the pollutants mentioned above from wastewaters, which usually consist of cloud point extraction (Silva, et al., 2009), coagulation using alum (Ghafari, et al., 2009), lime (Foley, et al., 2010), ferric sulphate (Matilainen, et al., 2010) and ferric chloride (Liang, Z., et al., 2009), chemical oxidation (Lee&Von Gunten, 2010) using Fenton reagents (ferrous ion and H_2O_2) (Padoley, et al., 2011), ozone (O_3) (Bundschuh, et al., 2011), chlorine and air (Wang, X., et al., 2011), membrane separation process including nanofiltration (Zahrim, et al., 2011), adsorption (Dos Santos, et al., 2011, Lu, et al., 2011) and so on. Among these, adsorption appears to be one of the best potential and effective methods for removal of pollutants from wastewaters.

The adsorbents can be classified into two types of adsorption materials (natural materials and synthetic ones). Natural adsorbents, usually non-conventional low cost adsorbents used for removal of compounds, include peat/sphagnum moss peat (Naumova, et al., 2011), red mud (oruh, et al., 2011), coir pith (Parab, et al., 2010), leaves (Chakravarty, et al., 2010), activated sludge (Bassin, et al., 2011, Zhang, C., et al., 2011a), waste organic peel (Feng, et al., 2011), tree fern (Liu, X. L., et al., 2011b), lignite (Al-Asheh, et al., 2003), sawdust (Mane&Babu, 2011), banana pith (Namasivayam, et al., 1998), peanut hull (Tanyildizi, 2011), modified chitosan beads (Liu, B., et al., 2011a), natural biopolymers (Vinod&Sashidhar, 2011), biosorption materials (Montazer-Rahmati, et al., 2011), and minerals such as activated ash/clay (Andersson, et al., 2011) and natural sepiolite (Sevim, et al., 2011). However, the adsorption capacity of the above adsorbents is relatively low.

Usually, there are many types of synthetic or artificial adsorbents (convential adsorbents) such as activated carbon (AC) (Li, et al., 2010b), resin (Vinodh, et al., 2011), zeolite (Seifi, et al., 2011), and so on used in adsorption operation, but AC is the most widely used adsorbents for removal of inorganic and organic compounds from wastewaters or waste gases because they have excellent capacities for adsorption of compounds derived from their huge surface area, developed pore texture, as well as easy availability (Khan, et al., 1997a, Khan, et al., 1997b, Okolo, et al., 2000, Viraraghavan&De Maria Alfaro, 1998).

According to its various appearance, AC is divided into cylindrical activated carbon (CAC) (Wang, J. C., et al., 2008), granular activated carbon (GAC) (Huang, L., et al., 2011, Saoudi&Hamdaoui, 2011), powder activated carbon (PAC) (Li, et al., 2009a, Li, et al., 2009b, Li, et al., 2010b) and acivated carbon fiber (ACF) (Li, et al., 2010a). In terms of extent of its pore development, AC is classified as microporous activated carbon/activated carbon fiber (Li, et al., 2009a, Li, et al., 2010a) and mesoporous activated carbon (Zhang, Y., et al., 2011b). In accordance with range of its use, AC is divided into gas activated carbon and water activated carbon, which are used in treatment of feed/waste gases and waters/wastewaters, respectively.

Usually, AC is prepared from different raw materials such as coal, wood, and so on. It is well known that a type of AC is coal-derived AC. Another one is prepared from woods such as pine wood (Hadi, et al., 2011) and fir wood (Eletskii, et al., 2011), etc. Certainly, AC can be also prepared from various carbon-contained materials (principally plant wastes (Hameed&Daud, 2008, Kula, et al., 2008) in past decades.

It should be noted that high initial cost and costly maintenance of AC including high regeneration temperature and tremendous carbon consumption have greatly restricted to its extensive use in industries. To greatly decease the cost of AC, the various ACs prepared from non-conventional sources such as coir pith (Agrawal, et al., 2011), sawdust and rice husk (Chen, Y., et al., 2011), pinewood (Tseng, et al., 2003) applied in removal of some compounds have also been investigated and better results have been attained.

The reason why adsorption capacity of adsorbents is high or low is mainly due to pore development and surface groups of AC, which is closely associated with its preparation methods. Here, preparation process of AC should be discussed because it has an important effect on pore structures and surface groups of AC. Preparation of AC includes a primary process and a secondary one which are denoted as carbonization and activation, respectively. Carbonization is followed by activation in sequence. In carbonization process, a large amount of gases, mainly small low molecular weight hydrocarbons-containing molecules or fragments possibly with oxygen (O), nitrogen (N) and sulfur (S), are released from materials surface and adhesion agents from 380 °C to 550 °C in an inert atmosphere due to decomposition of some groups which are readily destroyed. The carbonized materials have primary development of pore structure. In contrary to inert atmosphere used in carbonization process, some activating agents are frequently used in activation process (the secondary process). Gas activating agents consist of vapour, carbon dioxide and air used to further/promote development of pore properties at 600 °C to 980 °C. It is noting that chemical activating agents including phosphorus acid (H_3PO_4), phosphorus pentoxide (P_2O_5) (Izquierdo, et al., 2011), zinc chloride ($ZnCl_2$) (Cronje, et al., 2011), calcium chloride ($CaCl_2$) (Hu&Srinivasan, 2001), sodium hydroxide (NaOH) (Vargas, et al., 2011), potassium hydroxide (KOH) (Król, et al., 2011) and potassium sulfide (K_2S) (Huang, Z., 2006) seem to widely applied in preparation of wood-derived ACs which has only the carbonization process. The resulting AC after being cooled down to room temperature must be washed several times in order to eliminate residues of activating agents. After carbonization and activation, ACs have huge surface area, deep developed pore structures and plenty of different types of surface groups are produced. In recent years, combination of a chemical method and a physical one has been also introduced into the activation process and can yield high surface area and developed pore structures (Huang, Z., 2006).

It is well known that principal component of AC is carbon element and surface of AC itself is non-polar, and hence it is concluded that as a whole, AC has hydrophobic surface. Various activation conditions and more or less amount of oxygen AC contacts when it is stored/exposed in air will to some extent implant some/many oxygenated groups on AC surface. The surface groups will alter surface state of AC and subsequently has an influence on its surface chemical properties. This results in the difference in surface chemistry of ACs. When a significant effect of surface chemistry on adsorption and catalysis process is investigated, AC is frequently modified through different treatment methods in order to strengthen or attenuate the surface chemical properties, e.g. surface functional groups. Usually, the properties of the functional groups such as oxygen-containing groups and nitrogen-containing groups of AC surface can be modified by physical, chemical, and electrochemical treatment methods. These methods consist of gas phase oxidation via O_2 or N_2O at different temperatures, various contact times and partial/total pressures, liquid phase oxidation treatments through HNO_3 or H_2O_2-containing aqueous solutions at various temperatures, concentrations, and reaction times and heat treatment at various high temperatures and in different gaseous environments (N_2 and H_2 and some high pure inert gases such as Ar and He) in order to selectively eliminate some of the functional groups.

Functional groups of AC surface may be acidic groups and basic ones because the surface chemistry of modified AC is closely associated to the types of heteroatoms (oxygen, hydrogen, nitrogen, et al) other than carbon atom within the carbon matrix (Wibowo, et al., 2007). Some common heteroatoms mainly including oxygen (O), nitrogen (N), phosphorus (P), hydrogen (H), chlorine (Cl) and sulfur (S) introduced in the modification process are bound to the edges of graphite-like lays, forming organic functional groups such as carboxyl groups, lactonic groups, phenolic groups, carbonyl groups, aldehydes, ethers, animo groups and other N-containng groups and phophates (Salame&Bandosz, 2001). The surface functional groups of ACs determines their moisture content, catalytic properties, acid/base character, and adsorption capacity (Salame&Bandosz, 2001). Therefore, it is important that the above groups are characterized. However, the surface chemistry characterization of ACs is a complex/difficult task. At present, the characterization methods are better established. In characterization of ACs, it is possible not only to identify the surface chemical groups but also to quantify them.

The objective of this work was to investigate the pore structure and surface chemistry of activated carbons (ACs) due to its effect on adsorption and catalytic property. The present review compiles the work done over the last few decades on types, formation and characterization of surface functionalities of ACs. Special attention is paid to characterization of pore structure and surface groups on ACs by various technologies including N_2 adsorption, scanning electron micrograph, elemental analysis, Boehm Titration, potentiometric titration, Fourier transform infra red/diffuse Fourier infrared transform spectroscopy, X-ray photoelectron spectroscopy, temperature programmed desorption, thermal gravimetric analysis/differential scanning calorimetry, which are illustrated by some examples. Moreover, the most important aspects referring to comparison of different characterization methods are also overviewed in this paper. The present work carried out indicates that various methods have their advantages and disadvantages in investigating the surface groups of ACs and the methods mentioned above must be used to complement each other. As so far, relatively little work has been published and there is considerable scope for more detailed studies on the characterization of surface groups of ACs.

Characterization

Adsorption capacity of ACs is dependent on surface area, pore structure and surface groups, polarity, solubility and molecule size of adsorbates; solution pH and the presence of other ions in solution and so on (Radovic, et al., 2001). To investigate and obtain information pertaining to its physical structures and chemistry properties, AC needs to be characterized by various analysis apparatus.

Physical structure characterization

N_2 adsorption

Surface area and pore structure can be determined by N_2 adsorption (Li, et al., 2010b), mercury porosimetry (García, et al., 2011), capillary condensation (CC) (Liu, Y., et al., 2011c), X-ray Diffraction (XRD) (Tongpoothorn, et al., 2011) and scanning/transmission electron microscope (SEM) (Chingombe, et al., 2005), small angle X-ray scattering (SAXS) (Bradley, et al., 2011).

Here, physical structures of AC analyzed only by liquid Nitrogen adsorption will be introduced in this paper because Nitrogen adsorption are frequently used in investigating the texture of AC. Brunauer-Emmitt-Teller surface area (S_{BET}), Langmuir surface area ($S_{langmuir}$), pore volume and pore distribution of the adsorbents were determined through liquid Nitrogen adsorption at 77 K on a Micromertics ASAP 2020 surface area analyzer (USA). Micropore and mesopore volumes and surface areas of the adsorbents were calculated using Density Functional Theory (DFT). The pore size distribution of adsorbents was determined from adsorption isotherm data using DFT.

Adsorbent	V_{mic} (cm^3/g)	S_{mic} (m^2/g)	V_{mes} (cm^3/g)	S_{mes} (m^2/g)	$\bar{r}$ (nm)	S_{BET} (m^2/g)	$S_{langmuir}$ (m^2/g)
AC	0.305	689	0.069	51	1.103	977	1231±8
ACO1	0.304	777	0.071	50	0.996	903	1135±8
ACO2	0.302	779	0.063	45	0.968	890	1121±8
ACD	0.306	805	0.028	21	0.847	847	1066±5

Note: $\bar{r}$ is median pore radius

Table 2. Textural characterization of the adsorbents (Li, et al., 2009a)

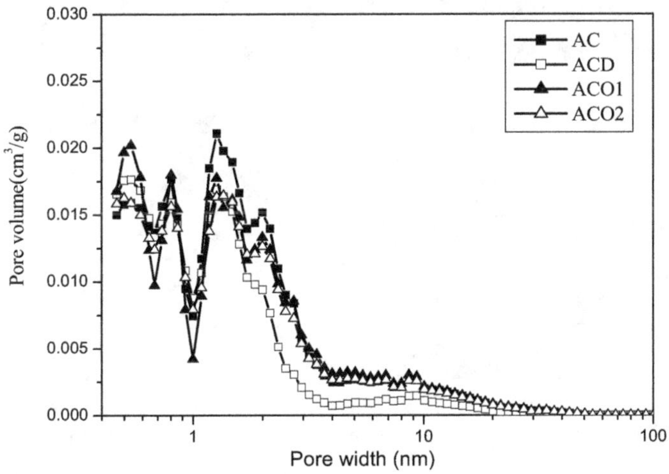

Fig. 1. Pore size distribution of adsorbents (Li, et al., 2009a)

Table 2 suggests that the difference in micro-pore properties, most relevant to adsorption of organic compounds, are small, with similar micro-pore volumes but different micro-pore surface areas, indicating the major difference between them is pore distribution within the micro-pore range as indicated by the mean pore diameter (Table 1). Among the four adsorbents, ACD contains more smaller micropores than the other adsorbents while AC contains less smaller pores in comparison to ACO1 and ACO2. This trend seems correlating well with the ash contents of the adsorbents (Table 4) and suggests that the acid leaching processes remove fine mineral matters from the adsorbents and result in formation of micro-pores, as depicted as Fig. 1. The meso-pore properties (V_{meso} and S_{meso}) of AC, ACO1 and ACO2 are very similar, but that of ACD are significantly smaller than others. These parameters again correlate well with ash content of the adsorbents, suggesting that there is removal of minerals from the mesopores, which converts them into macro-pores.

Scanning electron micrograph (SEM)

Preparation (Chingombe, et al., 2005)

The surface morphology of the carbons was analyzed using a Cambridge Instrument 360 scanning electron microscope at accelerating voltages of 10–20 kV. Prior to analysis, samples were dried at 373 K and stored in a desiccator overnight. The samples were then put on an aluminum platform for analysis.

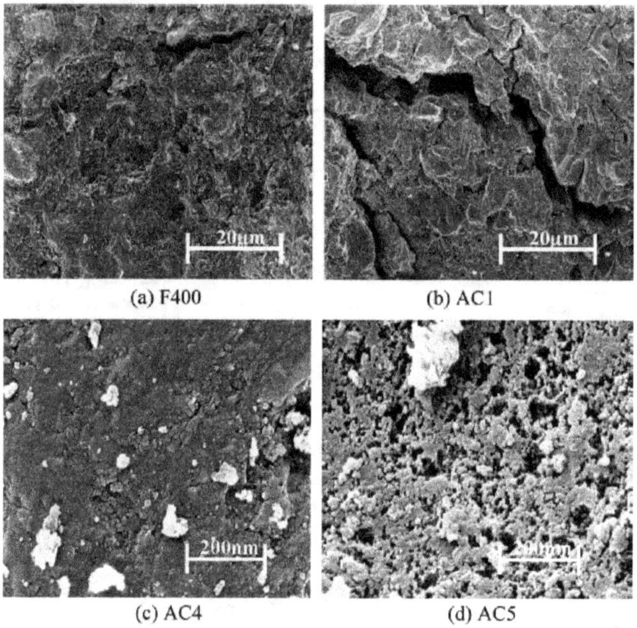

Fig. 2. SEM images of carbon samples before and after modification. (Chingombe, et al., 2005)

Chingombe, et al described the SEM of the samples in Fig. 2 (Chingombe, et al., 2005). There is little difference in the surface morphology of the samples except for some apparent pore widening on AC1 that could have occurred from the oxidation process. It was noticed that the oxidized samples disintegrated to small particles when compared to F400 sample particles. This observation could be linked to the cleavage of C–O bridging bonds on the carbon surface during the oxidation process. The SEM images of AC2 and AC3 were similar in appearance to AC1 image. This implies that post-oxidation treatments of the carbon samples did not make any apparent change in the surface morphology of the adsorbents. The surface morphology for AC4 was compared to that of AC5 and it is readily found that modification of AC4 to produce AC5 had a visual impact on surface morphology. The surface looks spongy and to some extent eroded.

Surface chemistry characterization

Prior to 1960s, researchers did not know how to characterize surface chemistry of AC. Fortunately, Boehm in 1966 began to make attempt to determinate surface chemistry of AC with acid/base titration theory/method, which makes researchers further analyze/ understand surface chemistry of AC. Thereafter, with the development of science and technologies, more and more characterization methods including elemental analysis, potentiometric titration, Fourier transform infrared/diffuse Fourier infrared transform spectroscopy, X-ray photoelectron spectroscopy, temperature programmed desorption, thermal gravimetric analysis/differential scanning calorimetry, X-ray diffraction (XRD), and so on have been proposed and utilized by researchers all over the world.

Element analyses

Sometimes, contents of elements contained in raw material of AC, e.g. characteristics of raw tobacco residue are provided in the literature (Kilic, et al., 2011). The characteristics of raw tobacco residue were shown in Table 3. From Table 3, it is obviously seen that content of carbon is high and content ash is low, indicating that tobacco residue is suitable for AC production.

Type	Value
Proximate analysis (%)	
Moisture	8.13
Ash	11.73
Volatiles	67.55
Fixed C	12.59
Ultimate analysis (%) (dry-ash basis)	
C	40.95
H	5.21
N	5.99
O*	47.85
H/C	1.52
O/C	0.88
HHV (MJ/kg)	15.07
Structural analysis (%)	
Lignin	8.75
Cellulose	42.30
Oil	3.30

Table 3. Characteristics of raw tobacco residue (Kilic, et al., 2011)

Li, et al determined ash amount and ultimate analyses of the adsorbents in the literature (Li, et al., 2009a). Carbon (C), hydrogen (H), and nitrogen (N) contents of the adsorbents were determined by an element analyzer (Analysensysteme Gmbh Elementar Vario EL), sulfur (S) content was measured by a sulfur analyzer (SC-132, LECO, USA), and oxygen (O) content was calculated by difference. Analytic results were shown in Table 4.

Adsorbent	Ash (wt.%)	C (wt.%)	H (wt.%)	S (wt.%)	N (wt.%)	O (wt.%)
AC	8.32	88.47	0.50	0.46	0.43	1.82
ACO1	7.11	86.89	2.11	0.38	0.83	2.67
ACO2	7.04	86.36	2.04	0.36	0.77	3.49
ACD	0.53	93.40	1.76	0.42	0.58	1.95

Table 4. Ash amount and ultimate analyses of the adsorbents

Table 4 shows the composition analysis of the four adsorbents. Among the four adsorbents, AC has the highest ash content at 8.32 wt%. The nitric acid treatment removed some of the ash and yielded adsorbents with medium ash contents, 7.11 wt.% for ACO1 and 7.04 wt.% for ACO2. The HCl/HF treatment removed most of the ash and yielded an adsorbent with the lowest ash content (ACD, 0.53 wt.%). Compared to that of the AC, the higher H, N and O contents for adsorbents ACO1 and ACO2 indicate that the main role of the nitric acid treatment is chemistry modification, with formation of many H, N and O containing groups on the surface; the slightly lower S contents for the acid treated adsorbents indicate that most of the sulfur in the adsorbents is organic.

Boehm titration (Ayranci&Duman, 2006, Boehm, 1966, Nevskaia, et al., 2004)

Boehm Titration proposed by Boehm in 1966 is one of many methods to characterize surface chemistry of AC. As so far, the titration based on acid/base titration theory is widely used because it can effectively determine oxygen-containing functional groups of AC surface and obtain information of groups. Boehm titration is based on the theory that acid and base in aqueous solution react with various basic sites and acidic sites of AC which are derived from different oxygenated groups, respectively.

Titration principle (Boehm, 1966)

Sodium hydroxide can centralize carboxylic groups, lactonic groups and phenolic groups of AC surface. The reaction processes of sodium hydroxide with carboxyl groups, lactonic groups, phenolic groups are shown in Equations 1-3, respectively.

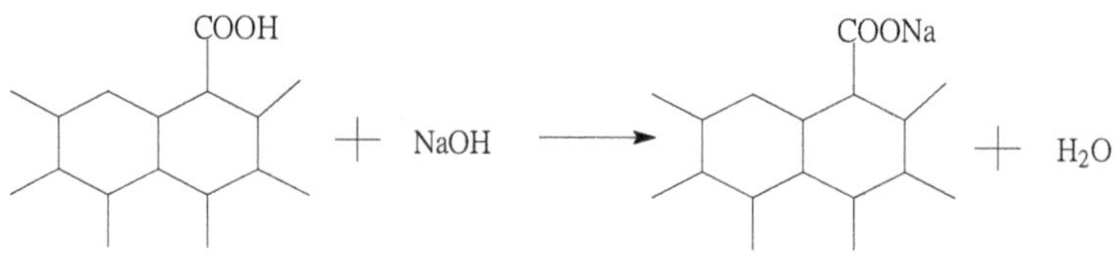

(1)

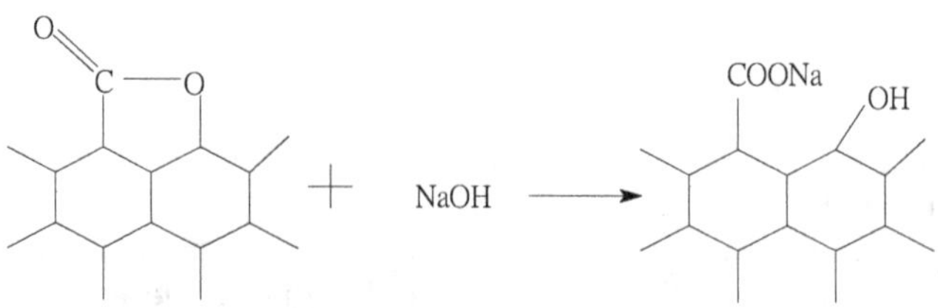

(2)

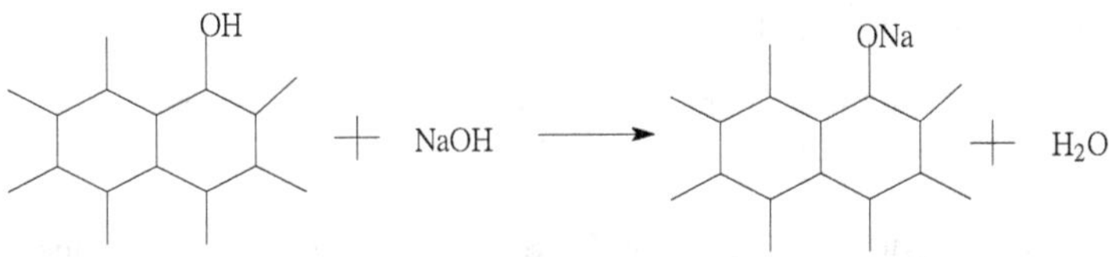

(3)

Hydrogen ion of carboxylic groups and phenolic groups is replaced by sodium cation of sodium hydroxide, forming sodium carboxylates and sodium phenolates when sodium hydroxide reacts with carboxylic groups, and phenolic groups of AC surface. In reaction of sodium hydroxide with lactonic groups, sodium carboxylates and phenolic groups are formed.

Owing to basic property of sodium carbonate poorer than that of sodium hydroxide, sodium carbonate centralizes only carboxyl groups and lactonic groups, of which reaction processes are shown in Equations 4-5.

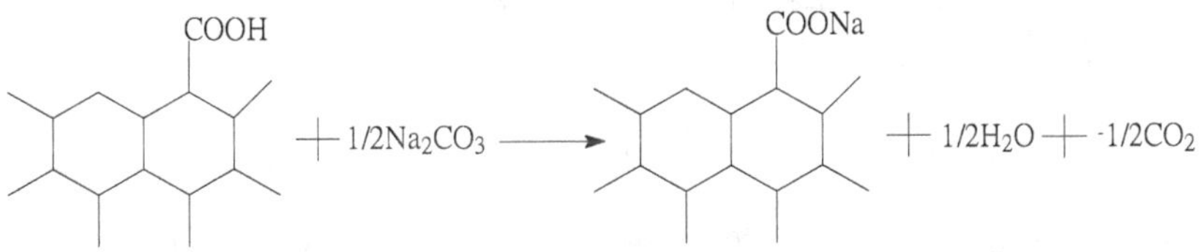

(4)

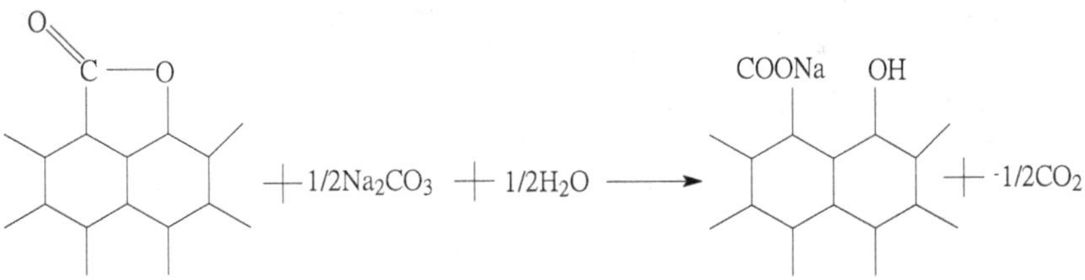

(5)

Sodium bicarbonate reacts only with carboxyl groups of AC surface, yielding sodium carboxylates due to its poorest basic property among the three bases mentioned above. The Equation 6 describes the above reaction.

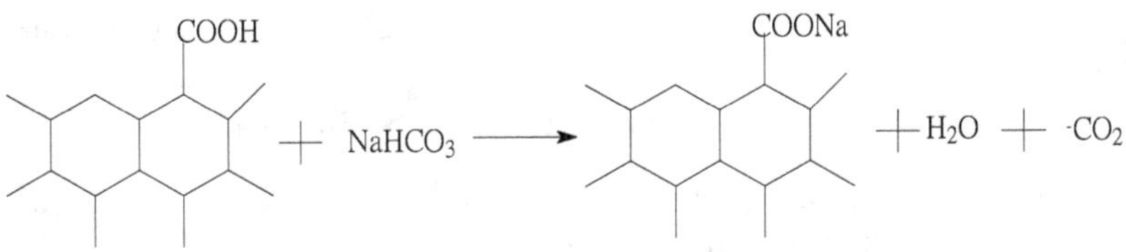

(6)

Hydrochloric acid can centralize surface base of AC, which is shown in Equation 7.

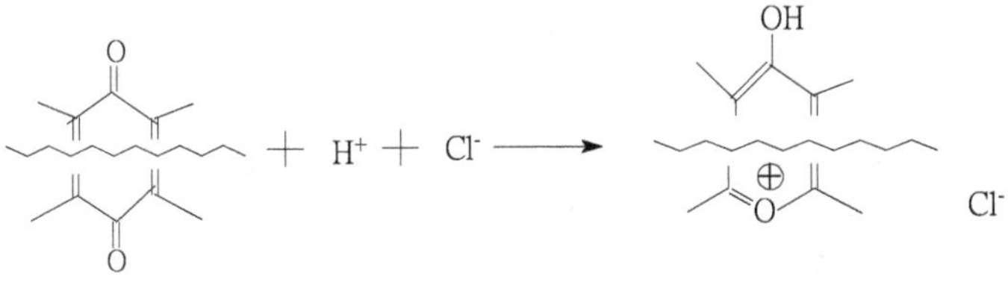

(7)

Preparation and test (Li, et al., 2010a)

Li, et al determined and analyzed the surface functional groups of activated carbon (AC) and activated carbon fiber (ACF). AC and ACF, used as the control adsorbents in the experiments, were commercial coal-based granular activated carbon from Xinhua Chemical Plant (Taiyuan, China) and novel cellulose activated carbon fiber from Shanxi Institute of Coal Chemistry (Taiyuan, China), respectively. AC was crushed into particles of 40–60 mesh size (0.3–0.45 mm) and the ACF was cut into pieces (40 mm × 40 mm); The carbon materials mentioned above were finally dried at 110 °C for 48 h.

To obtain adsorbents with similar textural properties and different surface chemistry, the chemical treatment of ACF by HNO_3 oxidation was carried out in a 500 mL Teflon bottle containing 16 g ACF and 200 mL 5 M nitric acid solution at 70 °C for 6 h. Once the oxidation was complete, the oxidized ACF was repeatedly filtered and washed with distilled water several times to remove the impurities. The oxidized and washed ACF was dried at 110 °C for 48 h. The resulting ACF was designated as ACFN.

An exact amount of adsorbent (0.200 g) was placed in a series of 100 mL well-sealed Teflon bottles containing 25 mL of 0.1 M NaOH, Na_2CO_3, $NaHCO_3$, and HCl solutions, respectively. After shaking at 150 rpm and 30 °C for 24 h in a thermostatic automatic shaker (HZQ-C, Haerbin, China), the adsorbent was separated from the solutions by filtration, and the filtrates were then titrated with a 0.1 M HCl or NaOH. The number of acidic groups was calculated based on the following assumptions: NaOH neutralizes carboxylic, lactonic, and phenolic groups; Na_2CO_3 neutralizes carboxylic and lactonic groups; and $NaHCO_3$ neutralizes only carboxylic groups. The number of basic sites was determined from the amount of HCl that reacted with the adsorbents.

Adsorbent	Surface Acidity (mmol/g)	Carboxylic (mmol/g)	Lactonic (mmol/g)	Phenolic (mmol/g)	Surface Basicity (mmol/g)
AC	0.489	0.033	0.327	0.129	0.646
ACF	1.279	0.068	0.911	0.300	1.605
ACFN	3.724	1.409	1.224	1.091	1.383

Note: Surface Acidity and Surface Basicity are the Adsorbent's surface acidic groups and surface basic groups, respectively.

Table 5. Surface chemistry of the adsorbents determined by Boehm titration (Li, et al., 2010a)

Table 5 showed the acid/base properties of the adsorbents characterized by Boehm titration. From table 5, it is easily seen that the surface property of AC and ACF is totally basic, but that ACFN is acidic. The total surface acidic groups, carboxyl groups, lactonic groups, phenolic groups and basic groups in ACF are much higher than the corresponding parameters in AC, indicating the difference in surface chemistry between AC and ACF resulting from the different preparation of the carbon materials. In addition, contents of total surface acidic groups, carboxyl groups, lactonic groups, phenolic groups but basic groups in ACFN are much higher than those in ACF. It indicates that the HNO_3 oxidation clearly increased the surface acidic groups (carboxylic, lactonic, and phenolic groups) and decreased the basic groups of ACF. It should be noted that other surface functional groups including anhydrides, pyrones, benzoquinones and benzofurans and nitrogen-containing groups of adsorbents can not be determined by Boehm titration.

It comes to conclusion that the incomprehensive information, instead of total surface information, of oxygen-containing groups of AC can be discovered by results analyses of Boehm titration due to ability of the titration to qualitatively and quantitatively detecting only some surface acidic groups (carboxylic, lactonic, and phenolic groups) and the basic groups.

Potentiometric titration (Zpc Titration) (Dabrowski, et al., 2005, Martin, et al., 2003, Ramrakhiani, et al., 2011)

Potentiometric titration was proposed by Y.Matsura when the effect of surface charge of AC on its adsorption/catalysis abilities began to be concerned about by researchers. At the beginning of potentiometric titration proposed, acidic functional groups of carbon black were only approximately classified into strong acid type (pK<7) and weak acid one (7<pK<11) due to the researcher's poor understanding of acidic functional groups. Potentiometric titration to some extent has been developed since Y.Matsura,et al. proposed a novel analysis technology, potentiometric titration, to estimate the surface acidity of carbon black with in 1970s, but the potentiometric titration method has been greatly developed since constant distribution of AC surface was investigated by Teresa J., et al, in 1990s. At present, potentiometric titration has been a convenient analysis method of surface chemistry of AC through determining surface charge (Chen, J. P., et al., 2003, Noh&Schwarz, 1990).

Owing to the amphoteric character of the carbon surface derived from the acidic and/or basic functional groups, the surface properties may be influenced by the pH value of the coexisting liquid bulk phase. The effects of surface functionalities on adsorption of organic electrolytes including weak electrolytes such as phenols are usually significantly more complex than porosity effects, and so is their assessment. The attention to the surface charge of the carbon as well as the extent of ionization of the solute should be paid.

It is commonly assumed, that for pH < pKa adsorption of non-ionized organics does not depend on the surface charge of AC. However, for pH > pKa the phenolic compound is dissociated, and adsorption of its ionic form depends on the surface charge.

Principle

Potentiometric titration designed to determinate surface chemistry is based on the principle that surface charge of AC is a function of pH value of aqueous solutions. Potentiometric titration measurements are carried out by an automatic titrator.

Determination of potentiometric titration (Dabrowski, et al., 2005)

The pH mode of the titrator was set, then 0.1000 g of AC and 50ml of solution containing 0.1 M $NaNO_3$ were placed in the reaction flask, and finally the mixture including AC and electrolyte was balanced in 298K for 24h. To eliminate the effect of CO_2 in the air, the volume above liquid was protected with N_2. In measurements with 0.1M NaOH or HCl standard solution, solution was titrated to pH = 3~10. The pH of the solution adjusted to the corresponding potential of the solution is Zpc potential

It should be noted that C1, C2 and C3 represents ACs with different surface chemistry.

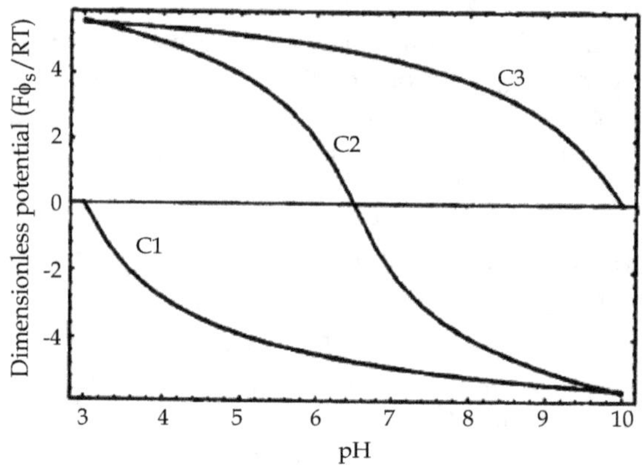

Fig. 3. the relation of surface potential with pH (hypothetical) ACs: C1, acidic carbon (pH_{zpc}=8.0); C2, amphoteric carbon (pH_{zpc}=6.5); C3, basic carbon (pH_{zpc}=10.0) (Radovic, et al., 2001)

Fig. 3 showed the dependence of surface potential on pH for hypothetical ACs supposed by Radovic. The pH value, at which the surface charge is zero, is called the point of zero charge (PZC). For typical amphoteric carbons, the surface is positively charged at pH < pH_{PZC} and negatively charged at pH > pH_{PZC}. The so-called isoelectric point, IEP, is denoted as the pH value of zero potential. In practice, pH_{IEP} is usually close to pH_{PZC}, but it is lower than pH_{PZC} of ACs. Potential of C1 is less than zero, indicating the surface of C1 is acidic; Potential of C3 is more than zero, indicating the surface of C3 is basic. It is commonly assumed that for pH < pKa adsorption of non-ionized organics does not depend on the surface charge of AC. However, for pH > pKa the phenolic compounds is dissociated, and adsorption of its ionic form depends on the surface charge. In accordance with pKa, it is obtained that C1 surface has carboxylic groups and C3 has carbonyls and benzoquonones (Dabrowski, et al., 2005).

From Fig. 3, various acidic or basic groups can not be differentiated by potentiometric titration and specific amount of groups such as carboxylic group, anhydrides, and so on, can not be given by potentiometric titration.

Diffuse reflectance infrared Fourier transform spectroscopy (DRIFTS) (Azargohar&Dalai, 2011, Chingombe, et al., 2005, Li, et al., 2010a)

The method pertaining to Fourier infrared transform spectroscope sprang up at the beginning of 20th century and was extremely developed due to the rapid development of technologies on computer and Fourier transform. It, combining Michelson interferometer with computer technology, has many advantages such as short time of measurement, high sensibility and resolution, broad range of measurement spectrum. In 1970s, Kortuum and GriffihsIn narrated the fundamental of principle of diffuse reflectance infrared Fourier transform spectroscopy (DRIFTS). DRIFTS is applied in determining the samples with scattering property and strong adsorption capacity, especially AC. Owing to its ability of decreasing or even eliminating the effect of scattering factor derived from preforming process on detecting chemistry, loose powder of some samples is also effectively analyzed. In addition, organic functional groups of AC surface not being detected by common FTIR due to special property of AC, which belongs to a type of blank body, can be determined by DRIFTS (Brown, 1990, Liang, Xiaotian, 1988, Wang, K., et al., 2006, Zhong, 1984).

Principle

Usually, infrared absorption spectra of molecules are vibrational/rotational spectrums of molecules. Infrared absorption spectrums are based on the absorption resulted from transition of vibrational/rotational energy level of molecules. Compared with standard spectrums, information of functional groups of samples can be determined by frequency of absorption band.

Sample preparation (Chingombe, et al., 2005)

F400 is a commercial granular coal-based AC from Chemviron (USA). In the surface modification, F400 was washed with distilled water, and dried at 383 K for 24 h, followed by various treatments including HNO_3, amination and annealing in H_2. All other reagents but HNO_3 were analytically pure.

The pre-treated F400 was stirringly oxidized by about 35 wt.% of HNO_3 at 363 K for 6 h. After the oxidation, a fresh HNO_3 solution at the same concentration oxidized the above treated sample for 3 h. The oxidized carbon was then washed with distilled water until no further change in pH could be detected. The resultant water-washed sample (after oxidation) was divided into three portions. The first portion was denoted as AC1, and this is a sample that was washed with water after the oxidation reaction. AC2 was the sample that was washed with 0.1 M NaOH to remove humus that is a byproduct of the oxidation process. The washing was continued until no further coloration from the humic substances could be detected. Reconditioning of the sample to the hydrogen form was performed using 0.1 M hydrochloric acid and further washing with distilled water was done until the pH of the supernatant solution stabilized to a pH of about 4.0. The third sample, denoted as AC3, was as a result of heating the oxidized sample at a temperature of 580 K under vacuum of 2 mm Hg for 12 h to remove humic substances. Prior to the annealing process the sample was conditioned in a flow of hydrogen (3.3×10^{-6} m^3/s) for 45 min at ambient temperature. The furnace temperature was then raised at a rate of 10 °C /min until it reached 1173 K where it was maintained for 3 h. After the annealing cycle, the furnace was allowed to cool down and the flow of hydrogen was maintained even during the cooling process.

Operation conditions

Chingombe P., et al. investigate surface groups of ACs through DRIFTS. Fourier Transmission infrared experiments, which were conducted on a Nicolet DXC20 FTIR spectrometer with liquid nitrogen-cooled mercury–cadmium–telluride detector and a Spectra Tech diffuse reflectance accessory (Chingombe, et al., 2005). The preparation process was drying of samples of particle size <45 μm for 24 h at a 383 K and mixture of the dried samples with fine KBr at a ratio of 1:100. In the preparation process, FTIR spectra of samples were recorded at a resolution of 4 cm^{-1} and with 200 scans and an aperture setting of 15. A previously recorded background spectra of water vapour was subtracted from the spectrum of each sample (Dabrowski, et al., 2005).

Results (Chingombe, et al., 2005)

DRIFTS spectra for a series of F400 and oxidized samples were shown in Fig. 4. It is seen that some characterization peaks were at wave numbers 1240, 1610 and 1750 cm^{-1}. The spectra for AC1, AC2 and AC3 have some similar marked curves, indicating the three samples could possess similar surface groups. F400 does not have a characterization peak at 1750 cm^{-1}, and its peak at 1240 cm^{-1} is also less pronounced than that for the other oxidized samples, suggesting that oxidation increase surface groups.

It is usually difficult to ascribe the peak at 1240 cm^{-1} (a superimposed peak) to a specific groups because the peak of a certain bands is overlapped by that of other bands. The superimposed peaks can be resulted from ether, epi-oxide and phenolic structures in various chemical environments. Beniak et al. reported that tertiary C–N stretching vibration may also result in the contribution of the peak in the region of 1240 cm^{-1}. The C–N contribution can probably explain the lack of a pronounced peak on F400 sample which did not undergo HNO_3 oxidation. The peak at 1610 cm^{-1} can be ascribed to quinone-like structures, and this peak appears in all the samples. Sutherland et al. and Shim et al. have also observed such a peak. As mentioned earlier, the peak at 1750 cm^{-1} appears on the oxidized samples only, but its position on the spectra of the other samples makes it difficult for positive identification. However, Lopez et al. obtained peaks in this region when they modified carbon samples by air oxidation and they assigned it to free carboxyl groups, lactonic groups, esters and carbonyl groups near the hydroxyl groups. The most logical explanation for the peak at 1750 cm^{-1} would be the existence of carboxyl groups that are formed as a result of nitric acid oxidation. This is also supported by the high sodium capacity results on the oxidized samples as opposed to F400 sample.

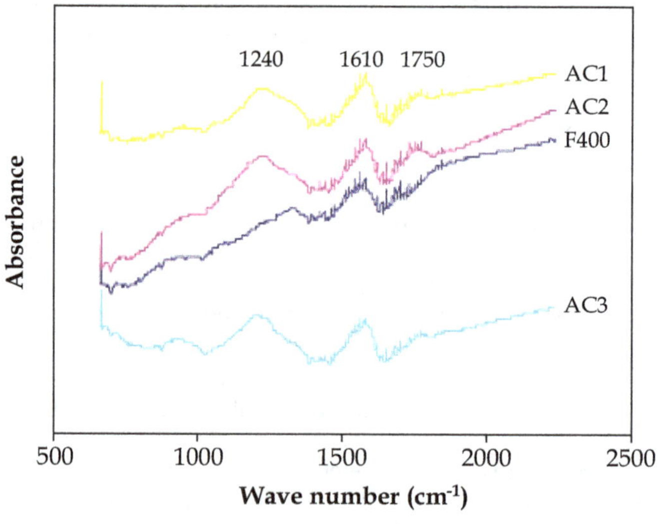

Fig. 4. DRIFTS curves for F400 with different treatments. (Chingombe, et al., 2005)

Owing to the similarity of DRIFTS curves of samples in Fig. 4, another example pertaining to characterization of surface chemistry by DRIFTS below is given to better describe the role of DRIFTS.

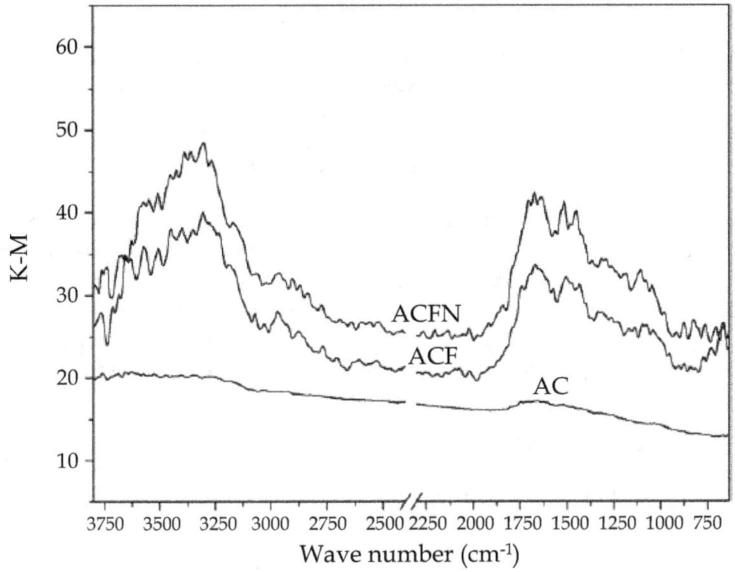

Fig. 5. DRIFTS curves of the adsorbents (Li, et al., 2010a)

The DRIFTS spectra of the adsorbents were shown in Fig. 5. As can be seen, among the three adsorbents, the lowest intensity of bands is in AC, indicating that AC has fewer groups than both ACF and ACFN. Compared with ACF, ACFN has stretching vibrations similar to those of ACF, with the exception of the stronger intensity of bands of the surface functionalities of ACFN with OH of around 3384cm^{-1}, C=O (carboxylic groups, anhydride, lactone and ketene) at 1674cm^{-1}, C–O (lactonic groups, ether, phenol, etc.) including υ_{as} (C–O) at 1300cm^{-1}, and υ_s (C–O) at 1110cm^{-1}. This indicates that more oxygen-containing groups have been introduced on ACF surface after HNO_3 oxidation. Although the DRIFTS spectra do not provide quantitative information of the carbon surface chemistry, it can provide other important results, such as nitrogen-containing groups, that could not be detected using the titration method due to their inability to dissociate. Clearly, ACFN has the strong band with the stretching vibrations around 1519cm^{-1}, which can be ascribed to the presence of more nitro groups on the ACF surface.

From the analysis, DRIFTS does not accurately quantify oxygen-containing functional groups, but can acquire carboxyl groups, hydroxyl groups, carbonyls and nitro group produced from HNO_3 oxidation. Furthermore, some absorption peaks can not be ascribed to a specific group due to the overlapping peaks.

X-ray photoelectron spectroscopy (XPS)

The chemical shift effect of inner electron energy level originated from copper oxidation was discovered by Siegbahn K. in 1958, which represented the birth of XPS. XPS technology was fast developed during 1970s and 1980s. From then on, XPS has been not only an important measurement method in research, but also an effective means of analysis and test in controlling industrial quality.

Principle (Wang, K., et al., 2006)

When samples to determine is being exposed by X-ray with enough energy, an inner-shell electron will be bombarded from the sample atoms, producing an ion in an excited state (Equation 8) and XPS can detect the kinetic energy of the bombarded electron.

$$S + hv_1(Xray) \rightarrow S^{+^*} + e^- \tag{8}$$

Kinetic energy of each electron is related with the orbital energy of emission electron, and orbital energy is characteristic of atoms or molecules, so the XPS can be used to qualitatively analyze samples. In a given experimental condition, number of emission electron is generally proportional to the concentration of emitters, so the XPS can also quantify the samples. The depth of samples studied in electron spectrum is less than 5 nm, so XPS is a surface analysis method. Specific binding energy of each electron corresponds to a Gaussian peak, representing a type of functional group.

Experiments (Wang, S.&Zhu, 2007)

To obtain the surface chemistry of AC, samples can be also characterized by XPS. Here, an example of XPS of the samples done by Wang is given below.

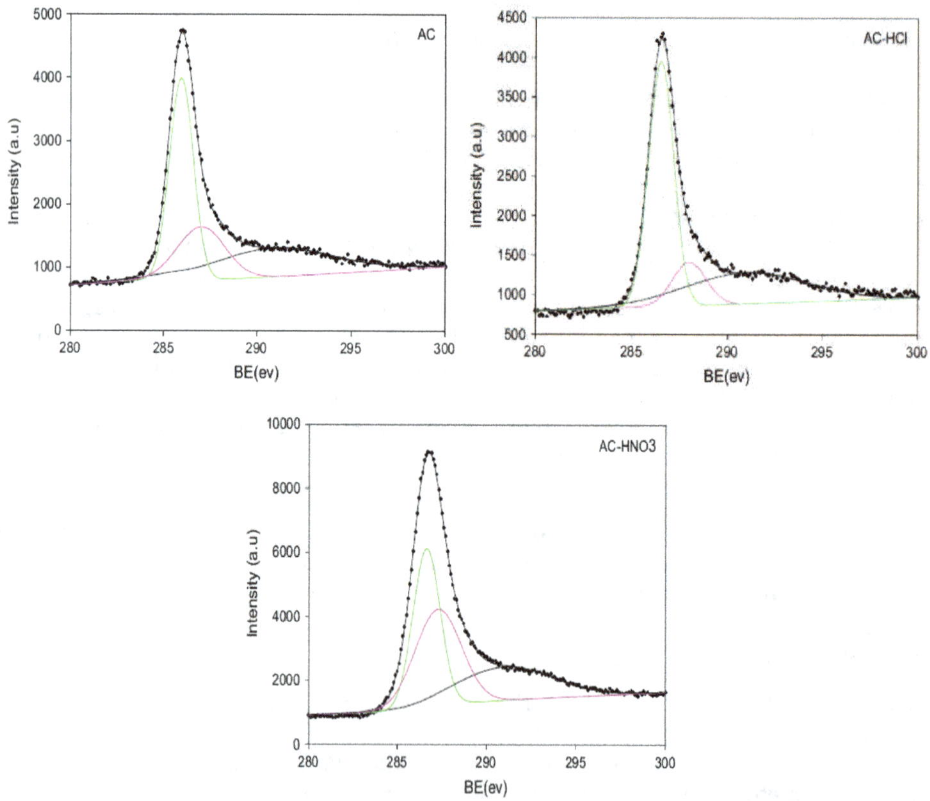

Fig. 6. XPS spectra of C1s on AC surface (Wang, S.&Zhu, 2007)

Wang used the three samples including AC (A commercial activated carbon), AC-HCl (AC treated with 2 M HCl at room temperature) and AC-HNO_3 (2 M HNO_3 for about 24 h at room temperature), and determined the surface chemistry of the above samples by resolution analysis of XPS peaks (Fig. 6).

The XPS measurements were carried out in PHI-560 ESCA system (Perkin Elmer), conditions of which were set at a basic pressure of 2×10^{-7} Torr with Mg Ka excitation at 15 kV and recorded in ΔE constant mode, pass energy 50 and 100 eV.

It should be noted that in measuring functional groups, the content percent (%) of other elements of functional groups to determine can be identified by XPS if the elements (such as C or O) have been given. The distribution of C and O structures can be derived from C1s and XP O1s spectra.

Fig. 6 showed the C1s signals of ACs. It is readily seen that C1s signals of samples are different. The maximum C1s signal on acid-treated carbons are shifted to higher binding energy due to an increase in oxidic species (alcohols, carbonates, or carboxylic groups) and ethers on the carbon surface after acid treatment. The resolution of peak indicates that three functional groups (C-C, C-O and C=O) can be obtained. For AC-HNO_3, C-O and C=O functional groups are much higher than AC and AC-HCl and C=O on AC-HCl is also higher than AC (Wang, S.&Zhu, 2007). Therefore, XPS results confirm that more acidic groups such as carboxylic groups and ethers were produced by acid treatment (Wang, S.&Zhu, 2007).

From analysis of Fig. 6, XPS can monitor oxygen-containing groups, but inaccurately quantify the groups.

Temperature programmed desorption (TPD) (Li, et al., 2009a, Li, et al., 2010a)

TPD is one of temperature-programmed analysis technology to measure structures and properties of catalysts and interaction between reaction molecules. It has been frequently used to identify and quantify oxygen-containing groups on AC through chemical adsorption apparatus or mass spectrum.

Principle

Surface oxygen-containing groups on carbon materials decompose upon heating, releasing CO_2 and/or CO at different characteristic temperatures (Mahajan&Moreno-Castilla, 1980), so it is readily concluded that a certain functional group decomposes at its special/corresponding temperature. Usually, decomposition law in TPD is as follows. For example, the CO_2 produced at temperatures lower than 400 °C may be ascribed to the decomposition of carboxylic acids, and at around 650 °C to the decomposition of lactones. CO_2 and CO released at temperatures about 550 °C account for the decomposition of anhydrides; the CO released at around 700 °C can be attributed to the decomposition of phenolic groups, and beyond 850 °C to the decomposition of carbonyl groups and quinones.

Experiments

To obtain the surface chemistry of AC, samples can be also characterized by TPD experiments. Here, an example of TPD of the samples treated by different methods is given to describe the analysis process (Li, et al., 2009a).

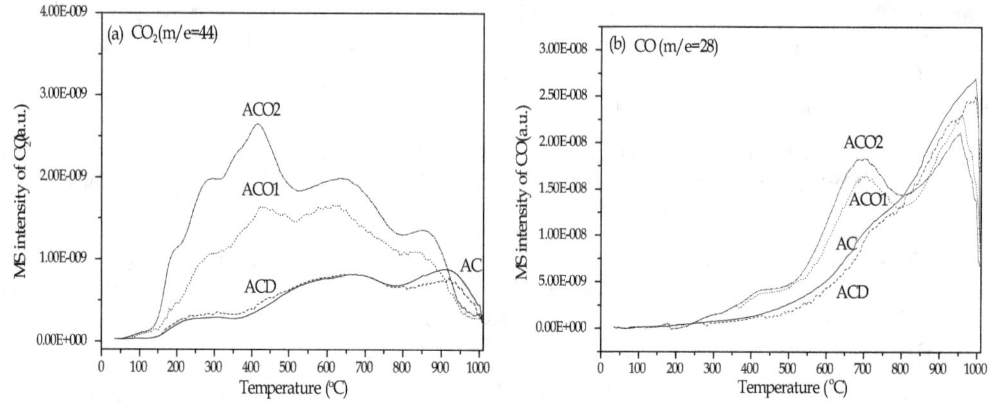

Fig. 7. CO_2 (a) and CO (b) evolution during TPD of the four adsorbents (Li, et al., 2009a)

Fig. 7 showed curves of CO_2 and CO released from the adsorbents during TPD. In Fig. 7, The similar CO_2 and CO release profiles for AC and ACD indicate that the HCl/HF treatment resulted in little change in acid and basic groups on the adsorbents' surface and the main groups on AC and ACD are basic, i.e. carbonyl and quinones. More CO_2 released from ACO1 and ACO2 at temperatures lower than 400 °C and around 650 °C is higher than that AC, indicating carboxylic acids and lactones of AC surface are increased by HNO_3 oxidationt; more CO_2 and CO released from ACO1 and ACO2 around 550 °C account for decomposition of more anhydrides; more CO released from ACO1 and ACO2 than that AC at around 700 °C and beyond 850 °C shows that more phenolic groups, carbonyl groups and quinones of AC were formed by HNO_3 treatment. Carboxylic acids, lactones, anhydrides, phenolic groups, carbonyl groups and quinones introduced by HNO_3 treatment can be analyzed by TPD.

It should be noted that some nitrogen-containing functional groups were also produced during HNO_3 oxidationt. However, these nitrogen-containing groups can not be better monitored and analyzed due to disturbance of other gases produced in heating. For example, mass charge ratio of 46 can represent NO_2 and CO_2, so NO_2 curve of AC is frequently disturbed by CO_2 released in heating. Furthermore, the information of surface groups of samples obtained by TPD is an indirect analysis result because groups to determine are ascribed only to the CO_2 and CO released from decomposition of groups at different temperature.

Therefore, it draws a conclusion that part of surface groups (the acidic groups such as carboxylic, lactonic, phenolic groups and the basic groups) of AC are determined by TPD, but it does not directly reflect surface chemical properties of AC and only analyze the decomposition behavior of oxygen-containing functional groups. Characterization of surface chemistry through TPD method should be further investigated.

Water vapor adsorption by thermal gravimetric analysis/differential scanning calorimetry (Li, et al., 2009a)

To identify hydrophilic properties of the adsorbents, water vapor adsorption of the adsorbents was measured by a thermal gravimetric analyzer (TGA 409 PC, Netzsch). The

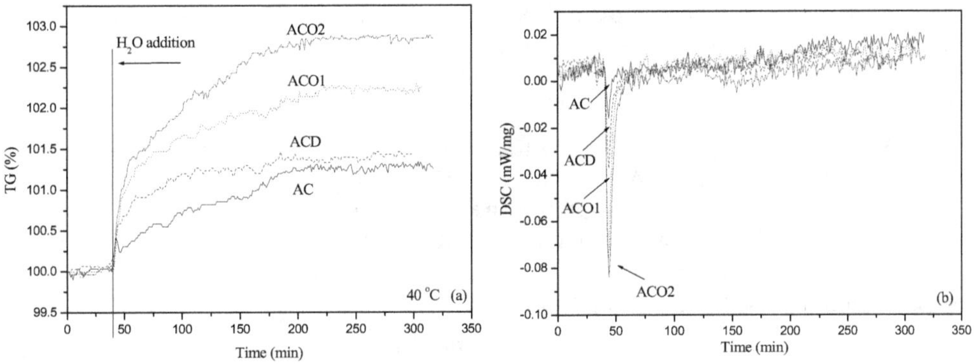

Fig. 8. Water adsorption on adsorbents (AC, ACD, ACO1 and ACO2) (Li, et al., 2009a)

Type	Advantages	Disadvantages
N_2 adsorption	Determination of micro/meso pore structure, surface area, average pore diameter and pore distribution	Long process and high expense
SEM	Determination of Pore structure and morphology	Inability of detecting pore distribution
Element analysis	Determinating element C, H, O, N, S, ash content	Inability of specific functional groups
Boehm Titration	Determination of acidic groups (carboxylic groups, lactonic groups, anhydrides, phenolic groups and carbonyl groups) and the basic groups, accurate quantification, large amount of sample, low expense and direct results	Inability of indentifying anhydrides, benzoquinones pyrones and other groups, complex operation, long process. Sample was destroyed
Pzc Titration	less samples, low expense, process slow, indirect results	Inability of indentifying types of some groups, Inability of quantification. Sample was destroyed
DRIFTS	Part of functional groups, nitrogen-containing groups to determine, less samples, medium expense, easy operation, intact samples and direct results	Inability of some absorption to be attributed and approximate quantification
XPS	Medium mass, more samples, high expense and direct results	Inner surface layer in depth less than 5 nm, approximate quantification, high expense and easy operation. Some peaks is difficult to disintegrate and fit.
TPD	Analysis of functional groups decomposing at temperatures lower than 1100 °C, accurate quantitative, a small amount of sample	Medium expense, medium easy and indirect results. Sample was destroyed

Table 6. Comparisons of characterization methods of AC surface chemistry

measurement started with flowing Ar (99.99%, 50 mL/min) over 8.0 mg adsorbent at 40 °C, followed by replacing the Ar stream by water vapour-containing Ar stream (P_{H_2O} of 7.381 kPa) at steady state. The weight increase in the adsorbent after the introduction of water vapour is attributed to water adsorption.

The TG and DSC data in Fig. 8 all show that water adsorption of the adsorbents increases from AC to ACD and then to ACO1 and to ACO2. Clearly, this trend corresponds to the changes in surface acid/basic properties; a higher surface acidity of an adsorbent is consistent with a higher water adsorption capability. Adsorption heat also verifies the trend.

Comparison of characterization methods

Form the discussion in many examples, characterization methods mentioned above can be used to analyze physical texture and surface chemistry of AC. To further understand the methods, it is important to investigate the advantages and disadvantages between the methods, which were listed below.

Comparisons of characterization methods of AC surface chemistry were shown in Table 6. N_2 adsorption and SEM can be used to better analyze physical structures of AC. It should be noted that in analyzing surface chemistry of AC by Boehm, Pzc Titration and TPD, carboxylic, anhydride, lactonic groups and phenolic groups can be determined, but ketones, aldehydes, ethers, esters, pyrones and other functional nitrogen-containing groups can not be detected, which can be characterized by DRIFTS and XPS. DRIFTS and XPS have the greater advantage of determining nitrogen-containing functional groups than other methods mentioned above.

Conclusion

There is the difficulty in analyzing its chemical properties of surface derived from the special nature of activated carbon itself, which is a black body and has complex components on surface, so the different characterization methods are proposed and used. Given the above characterization methods are not perfect in characterizing activated carbon surface chemistry, a variety of characterization methods complement each other to acquire more accurate and more comprehensive surface information. It is predicted that with the development of science and technology, these methods will be continually developed and improved and the nature of surface chemistry of AC will be further understood.

Acknowledgment

The author expresses his grateful appreciation for the financial support from the Doctorate Science Fund of Taiyuan University of Science & Technology (20102001)

References

Agrawal, S. G., King, K. W., Fischer, E. N. & Woner, D. N. (2011). PO_4^{3-} Removal by and Permeability of Industrial Byproducts and Minerals: Granulated Blast Furnace Slag, Cement Kiln Dust, Coconut Shell Activated Carbon, Silica Sand, and Zeolite. *Water Air Soil Poll*, Vol. 219, No. 1-4, pp. (91-101), 0049-6979

Al-Asheh, S., Banat, F. & Abu-Aitah, L. (2003). Adsorption of phenol using different types of activated bentonites. *Sep. Purif. Technol.*, Vol. 33, No. 1, pp. (1-10), 1383-5866

Andersson, K. I., Eriksson, M. & Norgren, M. (2011). Removal of lignin in wastewater generated by mechanical pulping using activated charcoal and fly ash: adsorption isotherms and thermodynamics. *Ind. Eng. Chem. Res.*, Vol. 50, No. 13, pp. (7722 - 7732), 0888-5885

Ayranci, E. & Duman, O. (2006). Adsorption of aromatic organic acids onto high area activated carbon cloth in relation to wastewater purification. *J. Hazard. Mater.*, Vol. 136, No. 3, pp. (542-552), 0304-3894

Azargohar, R. & Dalai, A. K. (2011). The direct oxidation of hydrogen sulphide over activated carbons prepared from lignite coal and biochar. *Can. J. Chem. Eng.*, Vol. 89, No. 4, pp. (844-853), 1939-019X

Bassin, J. P., Pronk, M., Kraan, R., Kleerebezem, R. & van Loosdrecht, M. C. M. (2011). Ammonium adsorption in aerobic granular sludge, activated sludge and anammox granules. *Water Res.*, Vol. 45, No. 16, pp. (5257-5265), 0043-1354

Boehm, H. P. (1966). Chemical identification of surface groups. *Advances in catalysis,* Vol. 16, No. pp. (179-274), 0360-0564

Bradley, R. H., Smith, M. W., Andreu, A. & Falco, M. (2011). Surface studies of novel hydrophobic active carbons. *Appl Surf Sci*, Vol. 257, No. 7, pp. (2912-2919), 0169-4332

Brown, R. D. (1990). *New Instrument Analysis Technique Book* Chemical Indutry Press,

Bundschuh, M., Pierstorf, R., Schreiber, W. H. & Schulz, R. (2011). Positive Effects of Wastewater Ozonation Displayed by in Situ Bioassays in the Receiving Stream. *Environ. Sci. Technol*, Vol. 45, No. 8, pp. (3774 - 3780), 0013-936X

Chakravarty, S., Mohanty, A., Sudha, T. N., Upadhyay, A. K., Konar, J., Sircar, J. K., Madhukar, A. & Gupta, K. K. (2010). Removal of Pb (II) ions from aqueous solution by adsorption using bael leaves (Aegle marmelos). *J Hazard Mater*, Vol. 173, No. 1-3, pp. (502-509), 0304-3894

Chen, J. P., Wu, S. & Chong, K. H. (2003). Surface modification of a granular activated carbon by citric acid for enhancement of copper adsorption. *Carbon*, Vol. 41, No. 10, pp. (1979-1986), 0008-6223

Chen, Y., Zhu, Y., Wang, Z., Li, Y., Wang, L., Ding, L., Gao, X., Ma, Y. & Guo, Y. (2011). Application studies of activated carbon derived from rice husks produced by chemical-thermal process--A review. *Adv Colloid Interfac*, Vol. 103, No. 1, pp. (39-52), 0001-8686

Chingombe, P., Saha, B. & Wakeman, R. J. (2005). Surface modification and characterisation of a coal-based activated carbon. *Carbon*, Vol. 43, No. 15, pp. (3132-3143), 0008-6223

Çoruh, S., Geyikçi, F. & Ergun, O. N. (2011). Adsorption of basic dye from wastewater using raw and activated red mud. *Environ Technol*, Vol. 32, No. 11, pp. (1183-1193), 0959-3330

Cronje, K. J., Chetty, K., Carsky, M., Sahu, J. N. & Meikap, B. C. (2011). Optimization of chromium (VI) sorption potential using developed activated carbon from sugarcane bagasse with chemical activation by zinc chloride. *Desalination*, Vol. 275, No. 1-3, pp. (276-284), 0011-9164

Dabrowski, A., Podkoscielny, P., Hubicki, Z. & Barczak, M. (2005). Adsorption of phenolic compounds by activated carbon--a critical review. *Chemosphere*, Vol. 58, No. 8, pp. (1049-1070), 0045-6535

Dos Santos, V. C. G., De Souza, J. V. T. M., Tarley, C. R. T., Caetano, J. & Dragunski, D. C. (2011). Copper ions adsorption from aqueous medium using the biosorbent sugarcane bagasse in natura and chemically modified. *Water Air Soil Poll*, Vol. 216, No. 1-4, pp. (351-359), 0049-6979

Eletskii, P. M., Yakovlev, V. A. & Parmon, V. N. (2011). Modern approaches to the production of carbon materials from vegetable biomass. *Theor. Exp. Chem.*, Vol. 47, No. 3, pp. (139-154), 0040-5760

Feng, N., Guo, X., Liang, S., Zhu, Y. & Liu, J. (2011). Biosorption of heavy metals from aqueous solutions by chemically modified orange peel. *J Hazard Mater*, Vol. 185, No. 1, pp. (49-54), 0304-3894

Foley, J., De Haas, D., Hartley, K. & Lant, P. (2010). Comprehensive life cycle inventories of alternative wastewater treatment systems. *Water Res.*, Vol. 44, No. 5, pp. (1654-1666), 0043-1354

García, A., Izquierdo-Barba, I., Colilla, M., de Laorden, C. L. & Vallet-Regí, M. (2011). Preparation of 3-D scaffolds in the SiO_2–P_2O_5 system with tailored hierarchical meso-macroporosity. *ACTA Mater*, Vol. 7, No. 3, pp. (1265-1273), 1742-7061

Ghafari, S., Aziz, H. A., Isa, M. H. & Zinatizadeh, A. A. (2009). Application of response surface methodology (RSM) to optimize coagulation-flocculation treatment of leachate using poly-aluminum chloride (PAC) and alum. *J Hazard. Mater.*, Vol. 163, No. 2-3, pp. (650-656), 0304-3894

Gokel, G. W. & Dean, J. A. (2004). *Dean's handbook of organic chemistry* McGraw-Hill, 0071375937,

Hadi, M., Samarghandi, M. R. & McKay, G. (2011). Simplified Fixed Bed Design Models for the Adsorption of Acid Dyes on Novel Pine Cone Derived Activated Carbon. *Water Air Soil Poll*, Vol. 218, No. 1-4, pp. (197-212), 0049-6979

Hameed, B. H. & Daud, F. B. M. (2008). Adsorption studies of basic dye on activated carbon derived from agricultural waste: Hevea brasiliensis seed coat. *Biochem. Eng. J.*, Vol. 139, No. 1, pp. (48-55), 1385-8947

Hu, Z. & Srinivasan, M. P. (2001). Mesoporous high-surface-area activated carbon. *Microporous Mesoporous Mater.*, Vol. 43, No. 3, pp. (267-275), 1387-1811

Huang, L., Yang, Z., Li, B., Hu, J., Zhang, W. & Ying, W. C. (2011). Granular activated carbon adsorption process for removing trichloroethylene from groundwater. *AIChE J*, Vol. 57, No. 2, pp. (542-550), 1547-5905

Huang, Z. (2006). *Fundmentals of technology of activated carbon* Ordnance industrial press, 7-80172-648-0,

Izquierdo, M. T., Martínez de Yuso, A., Rubio, B. & Pino, M. R. (2011). Conversion of almond shell to activated carbons: Methodical study of the chemical activation based on an experimental design and relationship with their characteristics. *Biomass Bioenerg*, Vol. 35 No. 3, pp. (1235-1244), 0961-9534

Kagalkar, A. N., Jagtap, U. B., Jadhav, J. P., Govindwar, S. P. & Bapat, V. A. (2010). Studies on phytoremediation potentiality of Typhonium flagelliforme for the degradation of Brilliant Blue R. *Planta*, Vol. 232, No. 1, pp. (271-285), 0032-0935

Khan, A. R., Al-Bahri, T. A. & Al-Haddad, A. (1997a). Adsorption of phenol based organic pollutants on activated carbon from multi-component dilute aqueous solutions. Vol. 31, No. 8, pp. (2102-2112), 0043-1354

Kilic, M., Apaydin-Varol, E. & Pütün, A. E. (2011). Adsorptive removal of phenol from aqueous solutions on activated carbon prepared from tobacco residues: Equilibrium, kinetics and thermodynamics. *J. Hazard. Mater.*, Vol. 189, No. 1-2, pp. (397-403), 0304-3894

Król, M., Gryglewicz, G. & Machnikowski, J. (2011). KOH activation of pitch-derived carbonaceous materials effect of carbonization degree. *Fuel Process Technol*, Vol. 92, No. 1, pp. (158-165), 0378-3820

Kula, I., Ugurlu, M., Karaoglu, H. & elik, A. (2008). Adsorption of Cd (II) ions from aqueous solutions using activated carbon prepared from olive stone by ZnCl2 activation. *Bioresour. Technol.*, Vol. 99, No. 3, pp. (492-501), 0960-8524

Lee, Y. & von Gunten, U. (2010). Oxidative transformation of micropollutants during municipal wastewater treatment: Comparison of kinetic aspects of selective (chlorine, chlorine dioxide, ferrateVI, and ozone) and non-selective oxidants (hydroxyl radical). *Water Res.*, Vol. 44, No. 2, pp. (555-566), 0043-1354

Li, B., Lei, Z. & Huang, Z. (2009a). Surface-Treated Activated Carbon for Removal of Aromatic Compounds from Water. *Chem. Eng. Technol.*, Vol. 32, No. 5, pp. (763-770), 1521-4125

Li, B., Lei, Z., Zhang, X. & Huang, Z. (2010a). Adsorption of simple aromatics from aqueous solutions on modified activated carbon fibers. *Catal. Today*, Vol. 158, No. 3-4, pp. (515-520), 0920-5861

Li, B., Liu, Z., Lei, Z. & Huang, Z. (2009b). Catalytic dry oxidation of aniline, benzene, and pyridine adsorbed on a CuO doped activated carbon. *Korean J. Chem. Eng.*, Vol. 26, No. 3, pp. (913-918), 0256-1115

Li, B., Liu, Z., Lei, Z. & Huang, Z. (2010b). Effect of structure and properties of simple aromatic compounds on adsorption behavior of activated carbon. *J. Fuel Chem. Technol.*, Vol. 38, No. 2, pp. (252-256), 0253-2409

Liang, X. (1988). *The application of modern laboratory instruments* Beijing Science and Technology Press,

Liang, Z., Wang, Y., Zhou, Y. & Liu, H. (2009). Coagulation removal of melanoidins from biologically treated molasses wastewater using ferric chloride. *Chem Eng J*, Vol. 152, No. 1, pp. (88-94), 1385-8947

Liu, B., Wang, D., Li, H., Xu, Y. & Zhang, L. (2011a). As (III) removal from aqueous solution using [alpha]-Fe2O3 impregnated chitosan beads with As (III) as imprinted ions. *Desalination*, Vol. 272, No. 1-3, pp. (286-292), 0011-9164

Liu, X. L., Wang, L. J., Chen, Y. L., Chen, N. & Wang, S. F. (2011b). Adsorption of Heavy Metal Ions by Chelate-Fiber Prepared by Chemical Surface Modification. *Adv. Mater. Res.*, Vol. 308-310, No. pp. (178-181), 1022-6680

Liu, Y., Liu, Z., Wang, Y., Dai, J., Gao, J., Xie, J. & Yan, Y. (2011c). A surface ion-imprinted mesoporous sorbent for separation and determination of Pb (II) ion by flame atomic absorption spectrometry. *Microchim ACTA*, Vol. 172, No. 3-4, pp. (309-317), 0026-3672

Lorenc-Grabowska, E. & Gryglewicz, G. (2007). Adsorption characteristics of Congo Red on coal-based mesoporous activated carbon. *Dyes Pigments*, Vol. 74, No. 1, pp. (34-40), 0143-7208

Lu, S., Xu, J., Zhang, C. & Niu, Z. (2011). Adsorption and desorption of radionuclide europium (III) on multiwalled carbon nanotubes studied by batch techniques. *J Radioanal Nucl Ch*, Vol. 287, No. 3, pp. (893-898), 0236-5731

Mahajan, O. P. & Moreno-Castilla, C. (1980). Surface-treated activated carbon for removal of phenol from water. *Sep. Sci. Technol*, Vol. 15, No. 10, pp. (1733-1752), 0149-6395

Mane, V. S. & Babu, P. V. (2011). Studies on the adsorption of brilliant green dye from aqueous solution onto low-cost NaOH treated saw dust. *Desalination*, Vol. 273, No. 2-3, pp. (321-329), 0011-9164

Martin, M. J., Artola, A., Balaguer, M. D. & Rigola, M. (2003). Activated carbons developed from surplus sewage sludge for the removal of dyes from dilute aqueous solutions. *Chem. Eng. J.*, Vol. 94, No. 3, pp. (231-239), 1385-8947

Matilainen, A., Vepsäläinen, M. & Sillanpää, M. (2010). Natural organic matter removal by coagulation during drinking water treatment: A review. *Adv Colloid Interfac*, Vol. 159, No. 2, pp. (189-197), 0001-8686

Montazer-Rahmati, M. M., Rabbani, P., Abdolali, A. & Keshtkar, A. R. (2011). Kinetics and equilibrium studies on biosorption of cadmium, lead, and nickel ions from aqueous solutions by intact and chemically modified brown algae. *J Hazard Mater*, Vol. 185, No. 1, pp. (401-407), 0304-3894

Namasivayam, C., Prabha, D. & Kumutha, M. (1998). Removal of direct red and acid brilliant blue by adsorption on to banana pith. *Bioresour. Technol.*, Vol. 64, No. 1, pp. (77-79), 0960-8524

Naumova, L. B., Minakova, T. S., Chernov, E. B., Gorlenko, N. P. & Ekimova, I. A. (2011). Adsorption-desorption of water vapor on initial and modified peat samples. *Russian Journal of Applied Chemistry*, Vol. 84, No. 5, pp. (792-797), 1070-4272

Nevskaia, D. M., Castillejos-Lopez, E., Guerrero-Ruiz, A. & Mu oz, V. (2004). Effects of the surface chemistry of carbon materials on the adsorption of phenol-aniline mixtures from water. *Carbon*, Vol. 42, No. 3, pp. (653-665), 0008-6223

Noh, J. S. & Schwarz, J. A. (1990). Effect of HNO_3 treatment on the surface acidity of activated carbons. *Carbon*, Vol. 28, No. 5, pp. (675-682), 0008-6223

Okolo, B., Park, C. & Keane, M. A. (2000). Interaction of phenol and chlorophenols with activated carbon and synthetic zeolites in aqueous media. *J. Colloid Interf. Sci.*, Vol. 226, No. 2, pp. (308-317), 0021-9797

Padoley, K. V., Mudliar, S. N., Banerjee, S. K., Deshmukh, S. C. & Pandey, R. A. (2011). Fenton oxidation: A pretreatment option for improved biological treatment of pyridine and 3-cyanopyridine plant wastewater. *Chem Eng J*, Vol. 166, No. 1, pp. (1-9), 1385-8947

Parab, H., Joshi, S., Sudersanan, M., Shenoy, N., Lali, A. & Sarma, U. (2010). Removal and recovery of cobalt from aqueous solutions by adsorption using low cost lignocellulosic biomass coir pith. *J Environ Sci Heal A*, Vol. 45, No. 5, pp. (603-611), 1093-4529

Radovic, L. R., Moreno-Castilla, C. & Rivera-Utrilla, J. (2001). Carbon materials as adsorbents in aqueous solutions. *A series of advances*, Vol. 27, No. pp. (227-405), 0069-3138

Ramrakhiani, L., Majumder, R. & Khowala, S. (2011). Removal of hexavalent chromium by heat inactivated fungal biomass of Termitomyces clypeatus: Surface characterization and mechanism of biosorption. *Chem Eng J*, Vol. 171, No. 3, pp. (1060-1068), 1385-8947

Salame, I. I. & Bandosz, T. J. (2001). Surface chemistry of activated carbons: combining the results of temperature-programmed desorption, Boehm, and potentiometric titrations. *Journal of Colloid and Interface Science*, Vol. 240, No. 1, pp. (252-258), 0021-9797

Saoudi, F. & Hamdaoui, O. (2011). Innovative technique for 4-chlorophenol desorption from granular activated carbon by low frequency ultrasound: Influence of operational parameters. *Micropor. Mesopor. Mat*, Vol. 141, No. 1-3, pp. (69-76), 1387-1811

Saratale, R. G., Saratale, G. D., Chang, J. S. & Govindwar, S. P. (2011). Bacterial decolorization and degradation of azo dyes: A review. *J. Taiwan. Inst. Chem. E.*, Vol. 42, No. 138-157 pp. 1876-1070

Seifi, L., Torabian, A., Kazemian, H., Bidhendi, G. N., Azimi, A. A. & Charkhi, A. (2011). Adsorption of Petroleum Monoaromatics from Aqueous Solutions Using Granulated Surface Modified Natural Nanozeolites: Systematic Study of Equilibrium Isotherms. *Water Air Soil Poll*, Vol. 217, No. 1-4, pp. (611-625), 0049-6979

Sevim, A. M., Hojiyev, R., Gül, A. & Çelik, M. S. (2011). An investigation of the kinetics and thermodynamics of the adsorption of a cationic cobalt porphyrazine onto sepiolite. *Dyes Pigments*, Vol. 88, No. 1, pp. (25-38), 0143-7208

Silva, E. L., Roldan, P. S. & Giné, M. F. (2009). Simultaneous preconcentration of copper, zinc, cadmium, and nickel in water samples by cloud point extraction using 4-(2-pyridylazo)-resorcinol and their determination by inductively coupled plasma optic emission spectrometry. *J Hazard. Mater.*, Vol. 171, No. 1-3, pp. (1133-1138), 0304-3894

Tabrez, S. & Ahmad, M. (2009). Effect of wastewater intake on antioxidant and marker enzymes of tissue damage in rat tissues: Implications for the use of biochemical markers. *Food Chem. Toxicol.*, Vol. 47, No. 10, pp. (2465-2478), 0278-6915

Tanyildizi, M. S. (2011). Modeling of adsorption isotherms and kinetics of reactive dye from aqueous solution by peanut hull. *Chem Eng J*, Vol. 168, No. 3, pp. (1234-1240), 1385-8947

Tongpoothorn, W., Sriuttha, M., Homchan, P., Chanthai, S. & Ruangviriyachai, C. (2011). Preparation of activated carbon derived from Jatropha curcas fruit shell by simple thermo-chemical activation and characterization of their physico-chemical properties. *Chem Eng Res Des*, Vol. 89, No. 3, pp. (335-340), 0263-8762

Tseng, R. L., Wu, F. C. & Juang, R. S. (2003). Liquid-phase adsorption of dyes and phenols using pinewood-based activated carbons. Vol. 41, No. 3, pp. (487-495), 0008-6223

Vargas, A. M. M., Cazetta, A. L., Garcia, C. A., Moraes, J. C. G., Nogami, E. M., Lenzi, E., Costa, W. F. & Almeida, V. C. (2011). Preparation and characterization of activated carbon from a new raw lignocellulosic material: Flamboyant (Delonix regia) pods. *J Environ Manage*, Vol. 92, No. 1, pp. (178-184), 0301-4797

Vinod, V. T. P. & Sashidhar, R. B. (2011). Bioremediation of industrial toxic metals with gum kondagogu(Cochlospermum gossypium): A natural carbohydrate biopolymer. *Ind J Biotechnol*, Vol. 10, No. 1, pp. (113-120), 0972-5849

Vinodh, R., Padmavathi, R. & Sangeetha, D. (2011). Separation of heavy metals from water samples using anion exchange polymers by adsorption process. *Desalination*, Vol. 267, No. 2, pp. (267-276), 0011-9164

Viraraghavan, T. & de Maria Alfaro, F. (1998). Adsorption of phenol from wastewater by peat, fly ash and bentonite. *J. Hazard. Mater.*, Vol. 57, No. 1-3, pp. (59-70), 0304-3894

Wang, J. C., Liu, Q. Y., Liu, Z. Y. & Huang, Z. G. (2008). Heterogeneity of V2O5 supported cylindrical activated coke used for SO2 removal from flue gas. *Chem. Eng. Technol.*, Vol. 31, No. 7, pp. (1056-1061), 1521-4125

Wang, K., Zhang, Y., Qi, R., Yang, M. & Deng, R. (2006). Effects of activated carbon surface chemistry and pore structure on adsorption of HAAs from water. *Journal of Chemical Industry and Engineering*, Vol. 57, No. 007, pp. (1659-1663),

Wang, S. & Zhu, Z. H. (2007). Effects of acidic treatment of activated carbons on dye adsorption. *Dyes Pigments*, Vol. 75, No. 2, pp. (306-314), 0143-7208

Wang, X., Hu, X., Hu, C. & Wei, D. (2011). Sequential use of ultraviolet light and chlorine for reclaimed water disinfection. *J Environ Sci*, Vol. 23, No. 10, pp. (1605-1610), 1001-0742

Wibowo, N., Setyadhi, L., Wibowo, D., Setiawan, J. & Ismadji, S. (2007). Adsorption of benzene and toluene from aqueous solutions onto activated carbon and its acid and heat treated forms: Influence of surface chemistry on adsorption. *Journal of Hazardous Materials*, Vol. 146, No. 1-2, pp. (237-242), 0304-3894

Zahrim, A. Y., Tizaoui, C. & Hilal, N. (2011). Coagulation with polymers for nanofiltration pre-treatment of highly concentrated dyes: A review. *Desalination*, Vol. 266, No. 1, pp. (1-16), 0011-9164

Zhang, C., Tezel, U., Li, K., Liu, D., Ren, R., Du, J. & Pavlostathis, S. G. (2011a). Evaluation and modeling of benzalkonium chloride inhibition and biodegradation in activated sludge. *Water Res.*, Vol. 45, No. 3, pp. (1238-1246), 0043-1354

Zhang, Y., Xu, S., Luo, Y., Pan, S., Ding, H. & Li, G. (2011b). Synthesis of mesoporous carbon capsules encapsulated with magnetite nanoparticles and their application in wastewater treatment. *J. Mater. Chem*, Vol. 21, No. pp. (3664-3671), 0959-9428

Zhong, H. (1984). *Introduction to Infrared Spectroscopy* Chemical Industry Press

7

Bioleaching of Galena (PbS)

E. R. Mejía[1], J. D. Ospina[1], M. A. Márquez[1] and A. L. Morales[2]
[1]*Materials Engineering School, Applied Mineralogy and Bio-process Group (GMAB)*
National University of Colombia, Medellín AA 1027,
[2]*Solid State Group, University Research Centre*
Antioquia University, Medellín AA 1226,
Colombia

Introduction

Microbial leaching of metals from sulfide minerals has been widely studied (Marsden & House, 1992; Brierley e Luinstra, 1993; Partha & Nataraja, 2006; Watling, 2006; Al-Harahssheh *et al.*, 2006). However, few studies have reported on the bacterial oxidation of galena (PbS) to insoluble lead sulfate (Santhiya *et al.*, 2000; Da Silva, 2004a and b). Formation of lead sulfate prevents the recovery of lead from traditional solvent extraction via electrowinning routes (Da Silva, 2004b).

Galena is a mineral of vast industrial importance, not only for being the world's main source of lead, but also for being a semiconducting material with a band gap around of 0.4 eV (Muscat *et al.*, 2003). Sulfide materials are also of interest from an environmental perspective as they are a major cause of water system acidification in mining operations.

In contrast to studies on galena, great attention has been paid to the bioleaching of sphalerite ((Zn,Fe)S) (Muscat & Gale, 2003). This interest stems from the increasing need to process grade ores of mixed mineralogy (Da Silva, 2004b; Muscat & Gale, 2003). One particular problem is the common association of sphalerite with galena, especially at fine particle sizes, which could complicate the differential flotation of the two minerals (Da Silva, 2004; Liao & Deng, 2004; Bolorunduro *et al.*, 2003). Although the kinetics and mechanism of sphalerite bioleaching are well known, they are not completely understood (Da Silva, 2004a,b; Boon *et al.*, 1998; Paar *et al.*, 1984; Rodrigez *et al.*, 2003; Zapata *et al.*, 2007).

Two different minerals can be selectively bioleached with galvanic interactions where the mineral with lower rest potential is sacrificed and that with higher potential is passivated (Das *et al.*, 1999; Suzuki, 2001; Da Silva, 2004a,b; Abraitis *et al.*, 2004; Cruz *et al.*, 2005; Urbano *et al.*, 2007).

The mechanism of galena oxidation is important in flotation processes, where mineral oxidation, through the grinding/flotation circuit, can affect its hydrophobicity and, therefore, its interaction with surfactants (Da Silva, 2004; Jañezuk *et al.*, 1993; Nowak *et al.*, 2000; Peng *et al.*, 2002). However, the pretreatment of refractory ores to recover metals from lower-grade sulfide ores or refractory minerals is unusual in Colombia (Muñoz *et al.*, 2003).

This leads to economic losses in mining processes, especially in subsistence mining. Their implementations in mining and metallurgical industries are also very attractive (Flower *et al.*, 1999; Rohwerder *et al.*, 2003; Olson *et al.*, 2003).

In this work, biological oxidation of galena was performed using *Acidithiobacillus ferrooxidans*-like bacteria and a mixed culture. Characterization techniques such as scanning electron microscopy (SEM), X-ray diffraction (XRD), and Fourier transform infrared spectroscopy (FTIR) were used to follow the morphologic and chemical changes occurring during the process.

Materials and methods

Minerals

All experiments were carried out using a galena sample from an El Silencio miner, property of Frontino Gold Mines Company (Segovia, Antioquia, Colombia). The mineral was subjected to crushing and milling processes followed by gravimetric separation in a Wilfley table. Manual concentration using stereographic microscopy was then performed. The mineralogical composition of the concentrate measured by countdown points was 93.3% galena (PbS), 6.2% sphalerite (ZnS), and 0.5% chalcopyrite ($CuFeS_2$) for -200 Tyler and 90% galena (PbS), 7.5% sphalerite (ZnS), 0.7% chalcopyrite ($CuFeS_2$), and 1.8% gangue (SiO_2), for -325 Tyler. An agate mortar was used to obtain two particle sizes: a pass through 200 Tyler mesh (~75 μm) and a pass through 325 Tyler mesh (~45 μm). XRD results confirmed that galena was the principal mineral phase in both sizes (Fig. 1). The mineral was sterilized in a furnace at 80°C for 90 min.

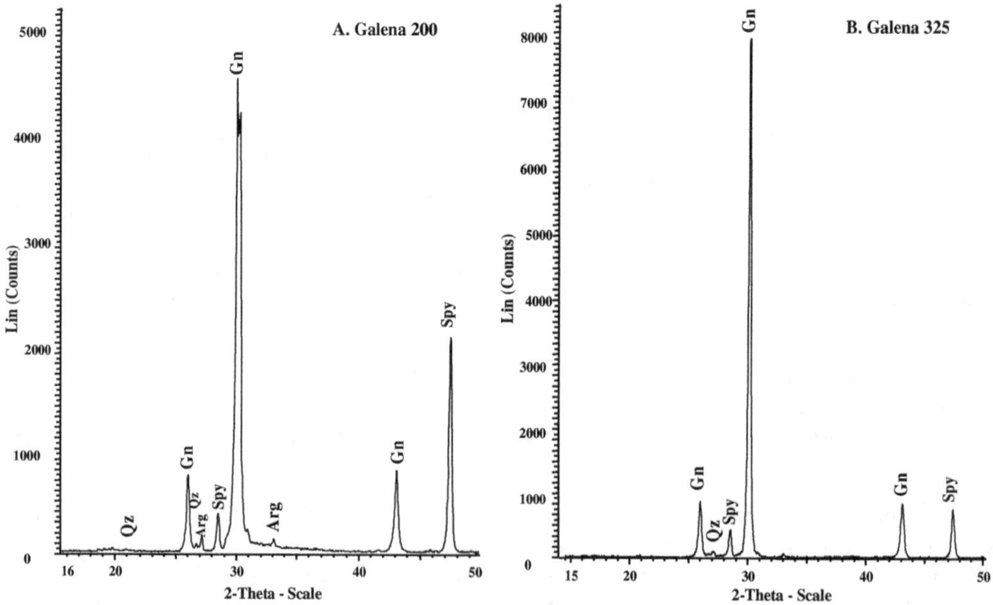

Fig. 1. X-ray spectra of the concentrates. Mineral pass through A) 200 and B) 325 Tyler mesh

Bioleaching experiments

A. ferrooxidans- and *Acidithiobacillus thiooxidans*-like strains were used in the bioleaching experiments. The strains were isolated by Cardona (2008). The microorganisms were previously grown in T&K medium by successive replacement of the ferrous sulfate with galena. The medium was acidified to pH 1.8 using H_2SO_4. The flasks were sterilized by autoclaving for 20 min at 120°C and 18 psi. The experiments were inoculated with *A. ferrooxidans* 10% (v/v) for the single culture and *A. ferrooxidans* 5% (v/v) and *A. thiooxidans* 5%(v/v) for the mixed culture. The experiments were carried out for 30 days in 500 mL shake flasks containing 300 mL of medium with 10% (w/v) galena at 180 rpm and 30°C. All conditions were duplicated and the respective abiotic control was included.

Chemical analysis

Measurements of pH (HACH HQ40d multi PHC30103) and redox potential (Shot Handylab 1 Pt 6880) *in situ* (reference electrode Ag^0/AgCl) were performed daily. Samples were aseptically withdrawn from the flasks after 24 h and then every five days. The samples were separated in a DIAMOND IEC DIVISION centrifuge for 15 min at 3000 rpm. Iron and sulfate concentrations were measured using an ultraviolet-visible spectrophotometer GENESYS™ 10. The methods employed included 3500-FeD (O-phenanthroline) for ferrous and total iron according to the standard methods for water analysis.

Mineralogical analysis

Combinations of analytical techniques were used in the mineralogical characterization of the samples. The FTIR spectra of the solid samples were recorded by an FTIR spectrophotometer (Shimadzu Advantage 8400) using KBr pellets in transmission mode. A sample KBr mixture at a ratio of 1:200 was used. The total number of scans was 20, and a spectral resolution of 4 cm^{-1}, a range of 400–4000 cm^{-1}, and Happ-Henzel correction were used.

The biooxidation samples were mounted in epoxy resin and polished with sequentially finer SiC grit paper followed by a final polish with 0.05-μm sized alumina powder. Analysis of the polished sections was performed with a JEOL JSM 5910 LV SEM in backscattering electron mode with an energy dispersive X-ray (EDX) detector (Oxford Instruments), using a beam voltage of 18kV. XRD analyses of the samples were conducted on a Bruker D8ADVANCE diffractometer with Cu λ= 1.5406 Å radiation generated at 35 kV and 30 mA. XRD data were obtained using computer-controlled X-ray Diffractometer Panalytical X'Pert Pro MPD. The initial characterizations of the mineral-polished sections were performed by optical microscopy of reflected light.

Results

Galena leaching experiments using *A. ferrooxidans*

Variation on pH and redox potential (Eh) values for the inoculated systems and the abiotic controls are presented in Fig. 2. In order to prevent inhibition of the bacteria, H_2SO_4 was added to maintain the pH values around 2.0 until day 15. The pH values first increased and then decreased over time to levels around 1.1. The pH values in the abiotic controls stabilized at around 2.0 after day 15.

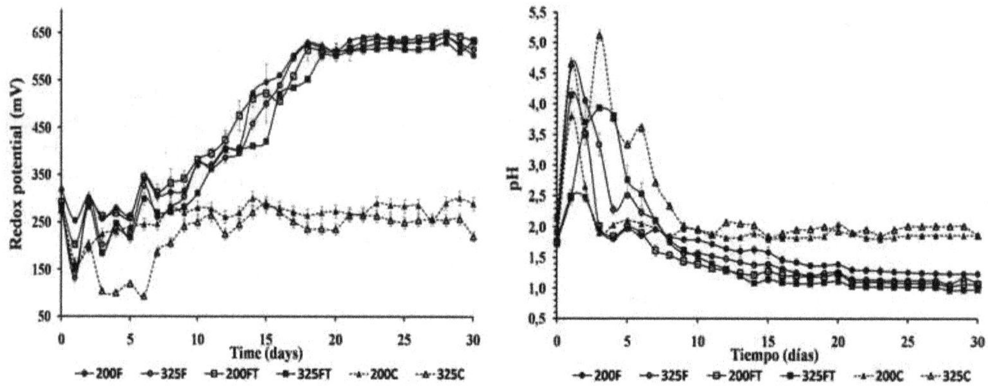

Fig. 2. Changes in redox potential and pH during the bacterial oxidation process. Graph legend: 200F, test with *Acidithiobacillus ferrooxidans* and 200 Tyler mesh; 325F, test with *A. ferrooxidans* and 325 Tyler mesh; 200FT, test with consortium and 200 Tyler mesh; 325FT, test with consortium and 325 Tyler mesh; 200C, inoculate test with 200 Tyler mesh; and 325C, inoculate test with 325 Tyler mesh

The variation in redox potential in both cultures with different grain sizes presented a relatively low redox potential at the beginning of the test until day 11 followed by an increase and finally a stationary phase with a small decrease with time. Eh values in the abiotic controls were around 284 mV throughout the processes. The SO_4^{2-} concentration in the solution as well as in the solid phase gradually increased with time. However, the increase of sulfate concentration was greater in the solid phase than in the solution (Fig. 3).

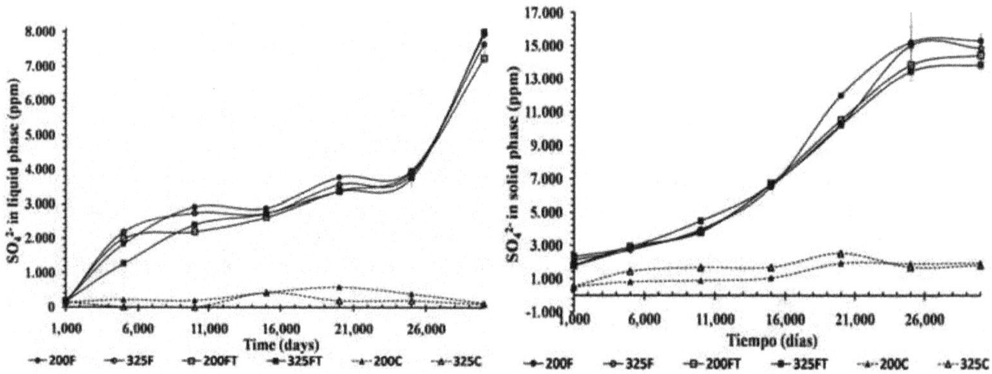

Fig. 3. Changes in SO_4^{2-} during the bacterial oxidation process. Graph legend: 200F, test with *Acidithiobacillus ferrooxidans* and 200 Tyler mesh; 325F, test with *A. ferrooxidans* and 325 Tyler mesh; 200FT, test with consortium and 200 Tyler mesh; 325FT, test with consortium and 325 Tyler mesh; 200C, inoculate test with 200 Tyler mesh; and 325C, inoculate test with 325 Tyler mesh

The dissolution of Fe^{2+} increased and reached a maximum during days 6–15. It then decreased sharply, becoming somewhat stationary with a small decrease on day 25 (Fig. 4).

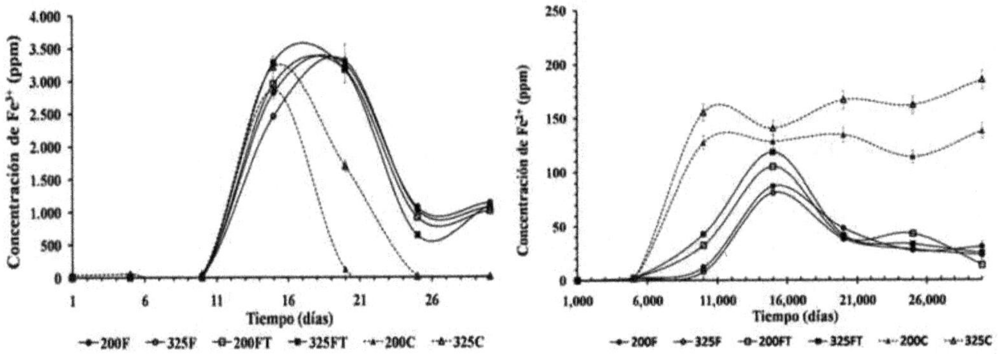

Fig. 4. Changes in Fe^{2-} and Fe^{3+} during the bacterial oxidation process. Graph legend: 200F, test with *Acidithiobacillus ferrooxidans* and 200 Tyler mesh; 325F, test with *A. ferrooxidans* and 325 Tyler mesh; 200FT, test with consortium and 200 Tyler mesh; 325FT, test with consortium and 325 Tyler mesh; 200C, inoculate test with 200 Tyler mesh; and 325C, inoculate test with 325 Tyler mesh

In the abiotic experiments, ferrous iron increased between day 5 and 10 and then remained stationary until the end of the process. The Fe^{3+} increased from day 10 to day 15, became stationary until day 25, and finally decreased sharply to about 1 ppm, where it remained until the end of the experiment. Lead extraction was around 57% for all texts. On the other hand, <5% of Pb was solubilized in the chemical controls (Fig. 5).

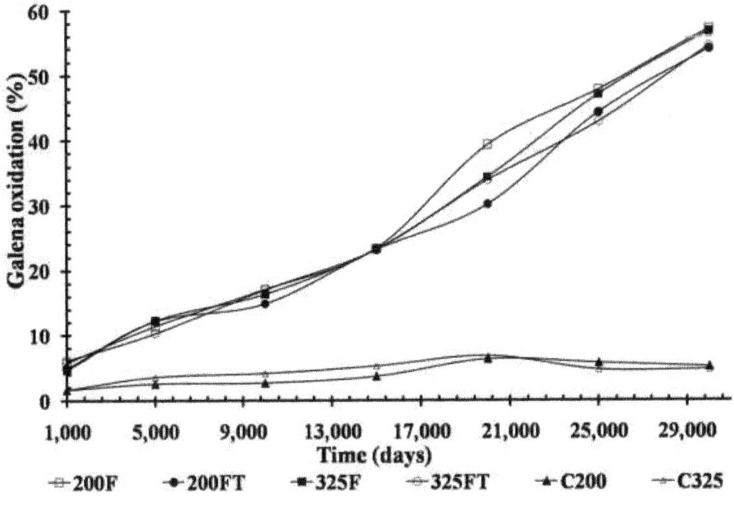

Fig. 5. Galena oxidation during the bacterial oxidation process. Graph legend: 200F, test with *Acidithiobacillus ferrooxidans* and 200 Tyler mesh; 325F, test with *A. ferrooxidans* and 325 Tyler mesh; 200FT, test with consortium and 200 Tyler mesh; 325FT, test with consortium and 325 Tyler mesh; 200C, inoculate test with 200 Tyler mesh; and 325C, inoculate test with 325 Tyler mesh

FTIR

Results obtained by FTIR for the bioleached galena samples showed typical bands of anglesite, the main mineral product of the process, with absorption bands at 950–1000, 1165–1765, 1115–1125, 1050–1060, and 592–620 cm^{-1} (Chernyshova, 2003). Scotlandite ($PbSO_3$) was also identified by bands at 920, 870, 970, and 600–620 cm^{-1} (Paar *et al.*, 1984; Chernyshova, 2003). The bands around 2935 cm^{-1} related to the total carbon present on the cell surface increased permanently (Naumann & Helm, 1995; Sharma & Hynumantha, 2005; Xia *et al.*, 2008).The FTIR spectra showed at the beginning of the process, for all tests, an increased numbers of anglesite and scontlandite bands.

The tests conducted using *A. ferrooxidans* with different grain sizes presented a continuous increase in the bands of the anglesite and scontlandite until day 10, followed the strong band increases. This was possibly due to the strong anglesite and scontlandite precipitate. It then increased continuously until day 20. Finally, the typical bands of anglesite and scontlandite showed a strong increase (Figs. 6a and 6b). However, the FTIR spectra for the test conducted using the mixed culture with different grain sizes presented continuous increases in the anglesite and scontlandite bands throughout the process (Figs. 6c and 6d).

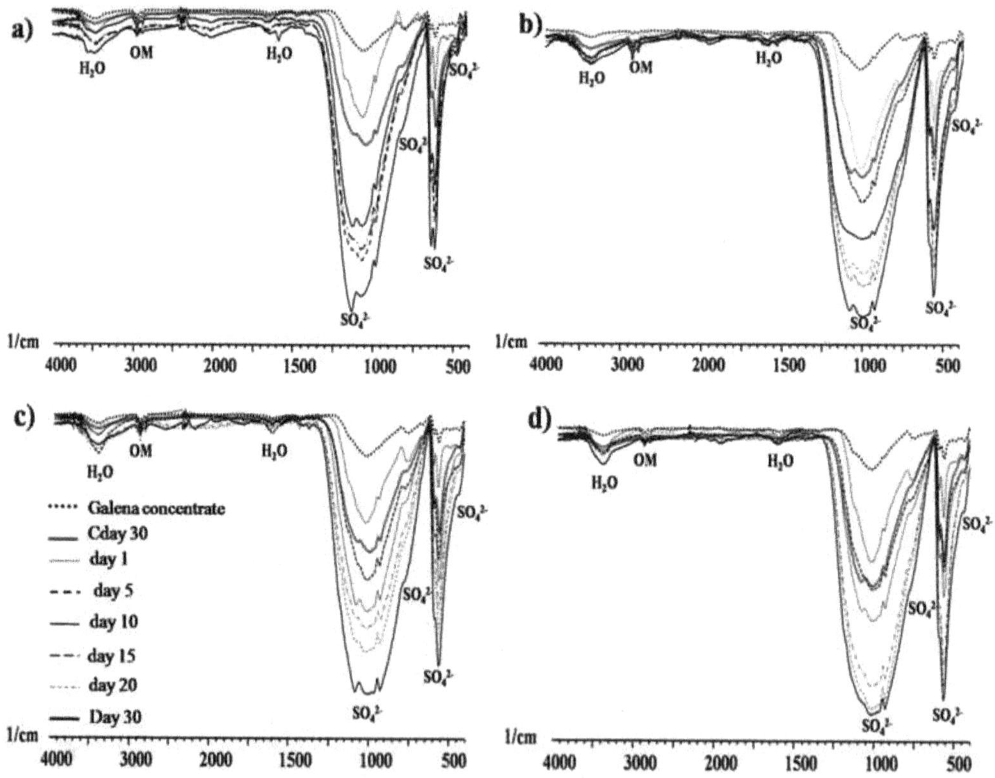

Fig. 6. Fourier transform infrared spectra of solid residues after galena bioleaching. OM, organic matter; SO_4^{2-}: anglesite

SEM/EDX

SEM images of leached galena are shown in Figs. 7, 8, and 9. All of the samples had corrosion features such as pits and gulfs on the grain surfaces (Figs. 7H and 7I). Moreover, coarse particle size porous films coating grains (Figs. 7C, 7F, and 7L), and aciculate precipitates of anglesite (Figs. 7D and 7L) were observed. These characteristics became observable on day 7 of the process (Figs. 7A, 7B, 7G, and 7H) and were more evident with time. After day 15, the galena grains were coated with anglesite film with porous texture (Figs. 7C and 7K). The anglesite aciculate formations were more evident at the end of the process (Fig. 7L). It is important to note the passivating effect that galena has on sphalerite and pyrite in which the pyrite and sphalerite grains did not show evidence of oxidation (Figs. 7C, 7D, 7J, 7F, and 7L).

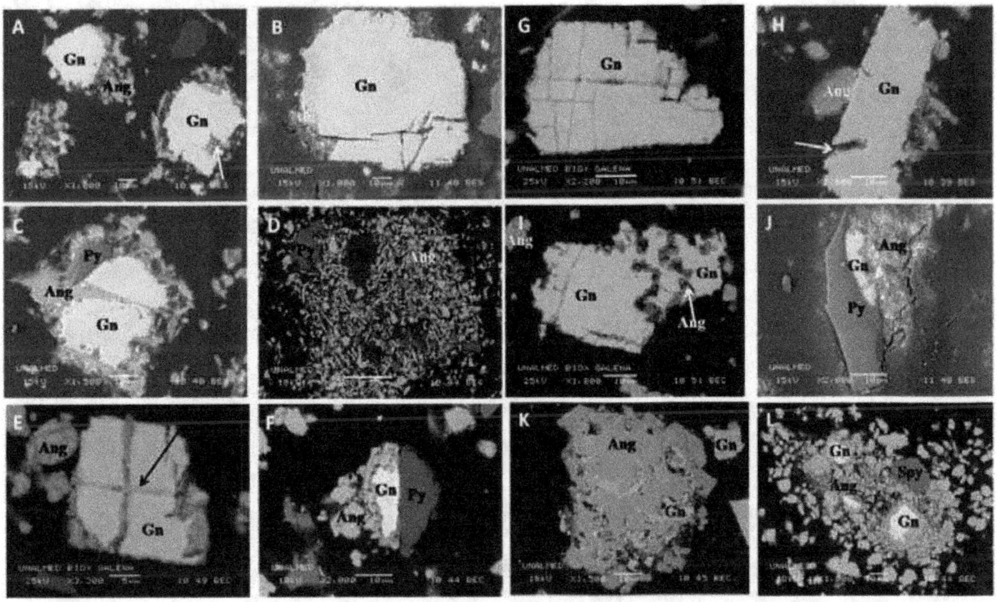

Fig. 7. Scanning electron microscopy micrograph of the residues after galena bioleaching by passing through 200 (A, B, C, D, E, and F) and 325 (G, H, I, J, K, and L) Tyler mesh. Gn, galena; Ang, anglesite; Py, pyrite. A) Galena grain and aciculate anglesite precipitates (arrows indicate corrosion gulfs on day 5). B) Galena grain in an incipient oxidation state (day 5). C) Galena grain covered and coated with anglesite porous film and pyrite grain without apparent oxidation (day 15). D) Anglesite grains aciculate and anhedrals, quartz grains without apparent oxidation and pyrite grains in the insipient oxidation state (day 15). E) Galena grain oxidant along the cleavage plane (arrows indicate cleavage plane) and anglesite grain (day 30). F) Galena grain covered and coated with anglesite porous film and pyrite grain without apparent oxidation (day 30). G) Galena grain corroded in gulfs and cleavage plane (day 5). H) Galena grain in an incipient oxidation state showed gulfs of corrosion (day 5). I) Galena grain corroded with anglesite cavity formation (day 15). J) Galena grain covered and coated with anglesite porous film and pyrite grain without apparent oxidation (day 15). K) The anglesite was porous and had a remaining galena nucleus. L) Anglesite grains aciculate and anhedrals and galena grains coated with anglesite porous film and sphalerite grain without apparent oxidation

However, in some cases, we observed sphalerite and pyrite grains with oxidation grooves at the end of the process (Figs. 8A and 8B). Moreover, we observed remaining galena and anglesite nuclei with porous texture (Figs. 7K and 8B). On other hand, some galena grains showed oxidation along the cleavage planes (Figs. 7E and 7G). SEM images of the uninoculated samples for all tests (Figs. 8a and 8b) showed surfaces with a few alterations such as small oxidation in the cleavage planes.

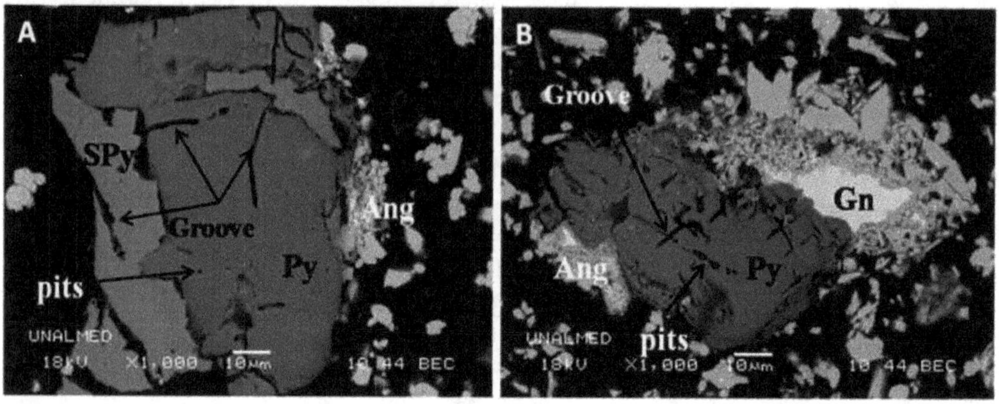

Fig. 8. Scanning electron microscopy micrograph of the residues after galena bioleaching. Gn, galena; Ang, anglesite; Py, pyrite; and Spy, sphalerite. A) Sphalerite and pyrite grains showed a typical corrosion groove and pits and porous anglesite. B) Pyrite grain with corrosion grooves, remaining galena nucleus, and porous anglesite

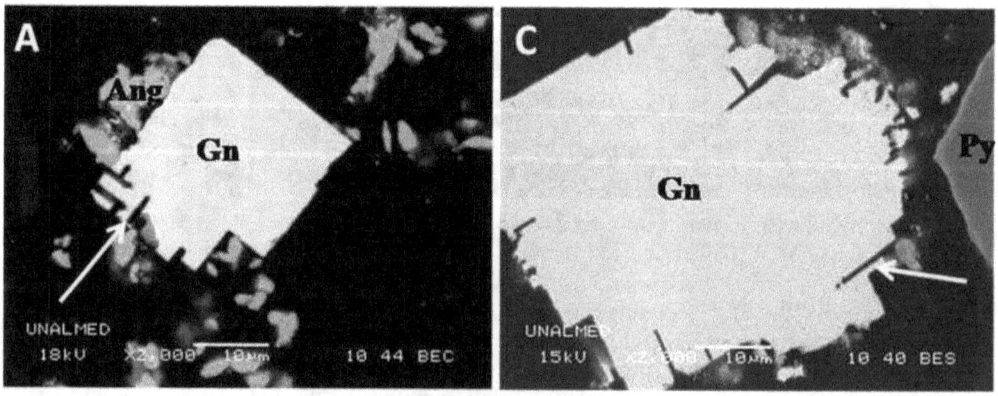

Fig. 9. Scanning electron microscopy micrographs of uninoculated residues after galena bioleaching. A) Galena grain with insipient oxidation state along the cleavage plane (pass through 200 Tyler mesh) and anglesite aciculate on the galena surface. B) Galena grain with an incipient oxidation state along the cleavage plane (pass through 325 Tyler mesh), and pyrite without an oxidation state. Cpy, chalcopyrite; Qz: quartz

EDX analysis of the grains showed the galena, anglesite, pyrite, and sphalerite stoichiometric composition (Table 1).

EDS Sphalerite

Element	Weight%	Atomic%
S	28,20	44,88
Fe	8,60	7,72
Zn	60,79	46,59
Cd	1,90	0,85

EDS Anglesite

Element	Weight%	Atomic%
O	17,83	61,66
S	11,24	19,40
Pb	70,93	18,94

EDS Pyrite

Element	Weight%	Atomic%
S	49,42	62,99
Fe	50,48	37,01

EDS Galena

Element	Weight%	Atomic%
S	12,97	49,06
Pb	87,03	50,94

Table 1. Energy dispersive spectrography (EDS) analysis of the residues after bioleaching of galena

XRD analysis

Initial XRD analyses revealed that galena was the main mineral phase present in the original samples, with small quantities of quartz (SiO_2), sphalerite (ZnS), chalcopyrite ($CuFeS_2$), and aragonite (Fig. 1). XRD spectra of bioleached samples are shown in Fig. 11. Mineralogical evolution of the mineral phases consists of a gradual reduction of galena peaks and the appearance of anglesite ($PbSO_4$). Anglesite peaks were observed from day 5 onward for all samples. However, peak intensity was higher in concentrate passed through -325 Tyler mesh. The sphalerite peaks remained unchanged throughout the process. XRD for uninoculated controls showed minimal anglesite formation around day 30 (Fig. 10).

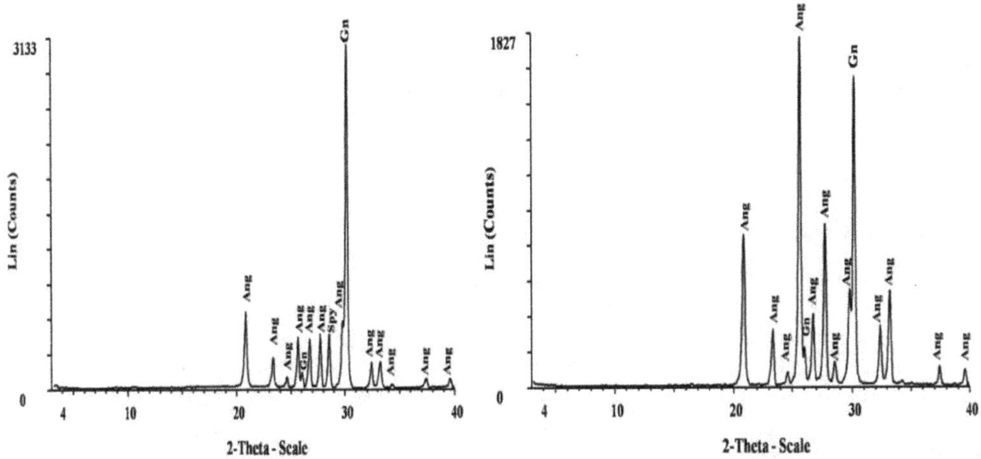

(A) (B)

Fig. 10. X-ray diffractograms of inoculated samples after 30 days of the biooxidation process. A) Sample passed through -200 Tyler mesh; B) Sample passed through -325 Tyler mesh. Gn, galena; sph, sphalerite; and Ang, anglesite

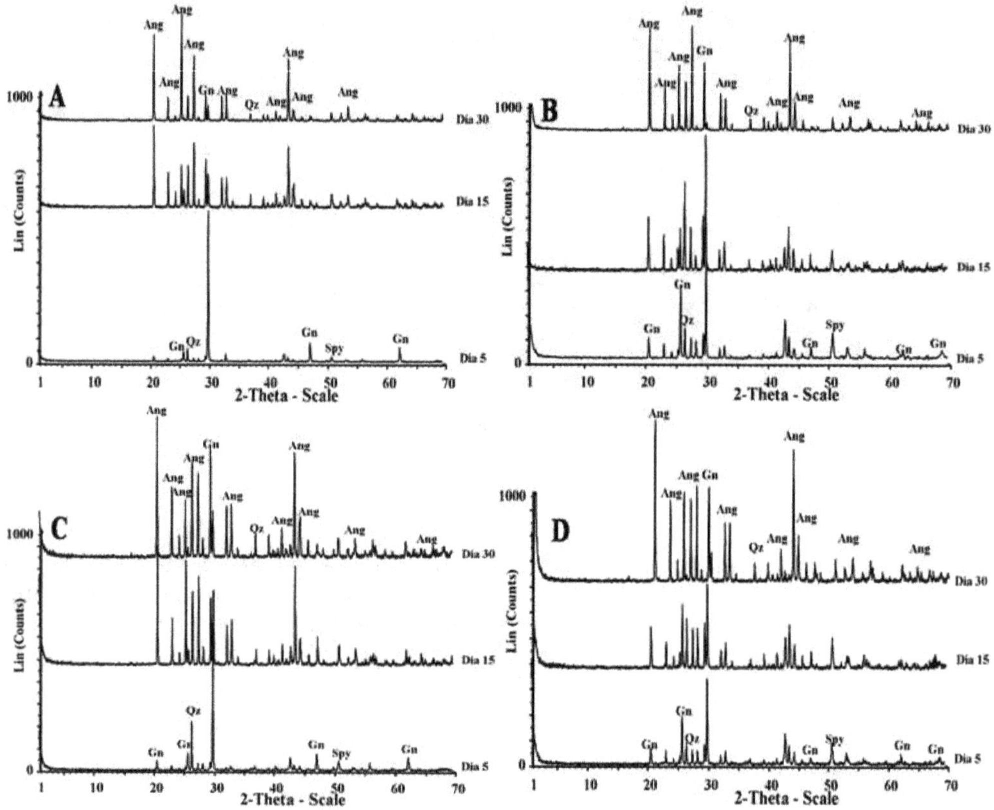

Fig. 11. X-ray diffraction spectra for galena before the bioleaching process. A. Particle size using 200 Tyler mesh for pure culture. B. Particle size using 325 Tyler mesh for the pure culture. C. Particle size using 200 Tyler mesh for the mixed culture. D. Particle size using 325 Tyler mesh for the mixed culture. Gn, galena; sph, sphalerite; Qz, quartz; and Ang, anglesite

Discussion

Galena leaching experiment

A. ferrooxidans-like bacteria showed good adaptation on galena with a high oxidative capacity (SEM, FTIR, XRD, and chemical data) since the microorganism was grown in a mineral concentrate that was the only source of energy. Jiang *et al.* (2008) stated that the bacteria may directly oxidize galena by taking energy from it. However, several authors suggest that *A. ferrooxidans* does not have a direct effect on galena oxidation; rather, it only indirectly acts on it via oxidizing hydrogen sulfide (H_2S) and sulfur (Da Silva 2004; Muscat & Gale, 2003; Garcia *et al.*, 1995). Thus, *A. ferrooxidans* utilizes H_2S dissolved in the solution as an energy source (Dutrizac & Chen, 1995; Mizoguchi & Habashi, 1981) according to the following equations:

$$PbS + H_2SO_4 \rightarrow PbSO_4 + H_2S \tag{1}$$

$$2H_2S + O_2 \xrightarrow{bacteria} 2S + 2H_2O \quad (2)$$

$$2S + 2H_2O + 3O_2 \xrightarrow{bacteria} 2SO_4^{2-} + 4H^+ \quad (3)$$

Furthermore, this study suggests that galena biooxidation also produces anglesite ($PbSO_4$) by reacting with sulfuric acid as shown in equation 1.

This finding is in agreement with the results obtained in this work in which Pb^{2+} was released from galena and precipitated as lead sulfate. Nevertheless, in accordance with the results obtained by FTIR, where there was evidence of the presence of scontlandite ($PbSO_3$), it is possible that the anglesite was not the only sulfate mineral phase and was able to generate other PbS mineral phases as accessories.

On other hand, the solubility of Pb^{2+} is very low, around 45 ppm (Mousavi *et al.*, 2006), and the SO_4^{2-} increased gradually in the solid and liquid phases (Fig. 3), being higher in solids (Fig. 4). This finding indicates that Pb^{2+} and SO_4^{2-} or SO_3^{2-} ions react to form anglesite ($PbSO_4$) and scontlandite ($PbSO_3$) as shown in equations 6 and 7. Moreover, anglesite ($PbSO_4$) was detected in the residual solid and increased over time according to FTIR, XRD, and SEM analysis. This phenomenon can be represented by the following equations:

$$PbS + H_2SO_4 + 0.5O_2 \rightarrow PbSO_4 + H_2O + S^0 \quad (4)$$

$$2PbS + H_2SO_4 + 3/2O_2 \rightarrow 2PbSO_3 + H_2O + S^0 \quad (5)$$

$$Pb^{2+} + SO_4^{2-} \rightarrow PbSO_4 \quad (6)$$

$$Pb^{2+} + SO_3^{2-} \rightarrow PbSO_3 \quad (7)$$

However, elemental sulfur was not detected by SEM, XRD, or FTIR. Moreover, the pH (Fig. 2) decreased after six days around to 1.3 for all tests after it was 1.8, indicating that the increased H^+ concentration was produced by bacterial activity. This behavior was due to the galena dissolution in an acid environment occurring as a result of the protonation of the mineral surface. The only protonation mechanism that has been proven to be energetically favorable in aqueous solution consists of the attachment of three H^+ onto three surface S atoms surrounding a central Pb atom, which is then replaced by a fourth H^+ (equation 1).

$$PbS + H_2SO_4 \rightarrow PbSO_4 + H_2S \quad (1)$$

This work detected few changes in chemical controls in galena oxidation and SO_4^{2-} in solid and liquid concentration at the beginning of the process, indicating that galena dissolution was favored in acid media via a purely chemical mechanism. This finding is in agreement with those of Gerson & O'Deo (2003) and Acero *et al.* (2007).

The bacteria oxidize H_2S, generating elemental sulfur and water (equation 1). Elemental sulfur is also oxidized by bacteria producing sulfates and H^+ (equation 2).

$$2H_2S + O_2 \xrightarrow{bacteria} 2S + 2H_2O \quad (2)$$

The H^+ reattach to the mineral and, thus, generate a cycle. This resulted in great dissolution in the inoculate tests (around 57%) compared with the uninoculated test.

On the other hand, on day 10, when the concentration of Fe^{3+} was high probably due to biooxidation of the minor quantities of sphalerite and pyrite (Fig. 4), galena oxidation was favored, possibly indicating that this ion contributes to the overall process efficiency. Jiang *et al.* (2007) found that ferric can oxidize galena and generate elemental sulfur according to the following reaction:

$$PbS + 2Fe^{3+} \rightarrow Pb^{2+} + 2\ Fe^{2+} + S^0 \quad (8)$$

It is important to emphasize that the behaviors of the kinetic parameters analysis were similar for both culture types. This finding was probably due to the fact that the pH value was highly unstable at the beginning of the process, increasing to around 4.0, inhibiting *A. thiooxidans.* This suggested that this type of microorganism was unable to obtain energy from galena and required elemental sulfur addition as an additional energy source. On the other hand, the particle size used did not generate test differences.

Mineralogical analysis

Mineralogical studies indicated that anglesite was the predominant mineral phase of galena biooxidation. The presence of anglesite was not clear, and evidence of passivation of the galena biooxidation process was seen in the galena oxidation curve, which was linear and increased throughout the process (Fig. 5), and the SEM images, where anglesite grains with remaining galena nuclei were observed on day 30 (Figs. 7K and 8B). The formation of this mineral phase was confirmed by the FTIR spectra, which showed increased numbers of anglesite and scontlandite bands at the beginning of the process for all tests, probably because anglesite and scontlandite present precipitate pulse (Fig. 6). This sharp increase was observed on days 10 and 20 for the tests using *A. ferrooxidans,* and its further slow increases on the other days of the process were consistent with sharp increases in the chemical sulfate data (Fig. 3). The FTIR spectra for the test conducted using the mixed culture presented continuously increased numbers of bands of anglesite and scontlandite throughout the process (Figs. 6C and 6D).

On the other hand, the band increases at 2935 and 2847 cm^{-1} organic matter was possibly due to an increase in bacterial population, indicating bacterial activity (Naumann *et al.*, 1995; Sharma *et al.*, 2005; Xia *et al.*, 2008). Furthermore, SEM analysis indicated that galena dissolution in the presence of the microorganisms occurred on the surface due to the presence of grain roughening, dissolution gulf formation (Figs. 7A and 7H), and preferential cleavage planes dissolution (Fig. 7E), all of which increase with time. The anglesite precipitation phenomenon occurs because the solubility limit of lead and sulfates in the medium is exceeded and could be mitigated by a sulfate concentration reduction in the medium. Moreover, SEM analysis revealed anglesite film precipitation on the galena grain from day 15 onward (Fig. 7C).

The preferential oxidation of the galena cleavage planes was probably due to the fact that this region was more potentially favorable or more chemically reactive because this zone has higher surface energy and, therefore, was easily oxidized as demonstrated by Bennett and Tributsch (1978).

XRD analysis (Fig. 10) indicated that anglesite ($PbSO_4$) was formed at the expense of galena dissolution. In contrast, the galena in the control reaction system showed a few alterations. Nevertheless, dissolution of the minority phases as sphalerite can be observed.

Finally, mineralogical data showed the passivating effect of galena on pyrite and sphalerite (Figs. 7F and 7L, respectively), in which the latter, with a higher rest potential, is provided at the expense of galena oxidation, which acts as the sacrificial anode in agreement with earlier studies (Das *et al.*, 1999; Suzuki, 2001; Da Silva 2004b; Abraitis *et al.*, 2004; Cruz *et al.*, 2005; Urbano *et al.*, 2007). However, in some cases, dissolution was observed in pyrite and sphalerite grains (Figs. 8A and 8B). These data confirmed that Fe^{2+} lixiviation likely originates from the minor quantities of sphalerite and pyrite that are present in the sample concentrate. However, the rest potential of sphalerite is less than that of pyrite, which possibly indicates that iron leached into sphalerite, but high iron concentration in solution (Fig. 4) apparently indicates that the pyrite made an important contribution. Moreover, the iron content of sphalerite (around 8.6% weight) was smaller than that of pyrite (around 50.48% weight) (Table 1).

Conclusions

Examinations of the bioleaching of natural galena concentrate in T&K medium by *A. ferrooxidans*-like bacteria and mixed culture resulted in the following conclusions:

- The bacteria have an impact on higher yield in the course of reaction of oxidizing PbS into $PbSO_4$ in acid medium (1.8), and experiments were inoculated with *Acidithiobacillus ferrooxidans* 10(v/v)% and 5%(v/v) for the single culture and *Acidithiobacillus thiooxidans* 5%(v/v) with 10%(w/v) galena. Where there was released the lead from galena and precipitated as lead sulfate. The level of lixiviation after 30 days of bioleaching was 57%, whereas that of the control examination was only 6%.
- The Fe^{3+} favored the biolixiviation of galena because when its concentration increased, galena dissolution was favored.
- The predominant new mineral phase was anglesite, a porous film that was formed on galena, but this film did not limit the access of the leaching agent and microorganisms inside the grain.
- In the presence of bacteria, the XRD peaks corresponding to galena decreased and, at the same time, new peaks appeared, anglesite, during the bioleaching process. This signal became more intense over time.
- In both cultures, the microorganisms gradually modified the original galena surface, increasing the rest potential and the SO_4^{2-} of the solid and liquid states.
- The particle size was apparently not a determining factor in the process.
- The galena was initially dissolved by acidic medium.

Acknowledgement

The authors would like to thank the biotechnology program of Colciencias, Colombia, the laboratory of biomineralogical of the National University of Colombia, Medellínd, Professor Diego Hernan Giraldo of the University of Antioquia, and the laboratory and group of molecular studies of the University of Antioquia.

References

Abraitis, P. K., Pattrick, R. A. D., Kelsall, G. H., Vaughan, D. J. 2004. Acid leaching and dissolution of major sulphide ore minerals: processes and galvanic effects in complex systems. Mineralogical Magazine. 68(2) 343–351

Acero, P., Cama, J., Ayora, C. 2007. Rate law for galena dissolution in acidic environment. Chemical Geology. 245 219–229.

Al-Harahsheh, M., Kingman, S., Rutten, F., Briggs, D. 2006. ToF-SIMS and SEM study on the preferential oxidation of chalcopyrite. International Journal of Mineral Processing. 80 2-4.

Bennett, J.C. and Tributsch, H. J. 1978. Bacterial leaching patterns on pyrite crystal surfaces. Bacteriol. 134:310-317.

Bolorunduro, S.A., Dreisinger, D.B., Van Weert, G. 2003. Zinc and silver recoveries from zinc-lead-iron complex sulphides by pressure oxidation. Minerals Engineering. 16 375 - 389.

Boon, M., Snijder, M., Hansford, G.S., Heijnen, J.J. 1998.The oxidation kinetics of zinc sulphide with Thiobacillus ferrooxidans. Hydrometallurgy. 48. 171 - 186.

Brierley J.A. & Luinstra, L. 1993. Biooxidation-heap concept for pretreatment of refractory gold ore. In: Biohydrometallurgical Biohydrometallurgical Technologies, A.E.

Cardona I.C. 2008. Mineralogía del proceso de biodesulfurización de carbones provenientes de la zona río Aguachinte – río Asnazú (valle del cauca y cauca). Tesis de Maestría. Universidad Nacional de Colombia, sede Medellín.

Chernyshova., I.V. 2003. An in situ FTIR of galena and pyrite oxidation in aqueous solution. Vol 558. Pp. 83-98.

Cruz, R., Luna-Sánchez, R.M., Lapidus, G.T., González, I., Monroy, M. 2005. An experimental strategy to determine galvanic interactions affecting the reactivity of sulfide mineral concentrates. Hydrometallurgy 78 198– 208.

Da Silva, G. 2004a. Relative impoortance of diffusion and reaction control during the bacterial and ferric sulphate leaching of zinc sulphide. Hydrometallurgy. Vol. 73. pp. 313 - 324.

Da Silva, G. 2004b. Kinetics and mechanism of the bacterial and ferric sulphate oxidation of galena. Hydromatallurgy. 75 99 - 110.

Das T, Ayyappan S, Chaudhury G.R. 1999. Factors affecting bioleaching kinetics of sulfide ores using acidophilic micro-organisms. BioMetals, 12. 1-10.

Dutrizac, J.E., Chen, T.T. 1995. The leaching of galena in ferric sulfate media. Metallurgical and Materials Transactions. 26 219–227.

Fowler, T.A., Holmes, P.R., Crundwell, F.K. 1999. Mechanism of pyrite dissolution in the presence of Thiobacillus ferrooxidans. Applied Environmetanl Microbiology. 65 2987-2993.

Garcia, Jr, O., Bigham, J.M., Tuovinen, O.H. 1995. Oxidation of galena by Thiobacillus ferrooxidans and Thiobacillus thiooxidans. Canadian Journal Microbiology. 41. 508 - 514.

Gerson, A., O'Dea, A. 2003. A quantum chemical investigation of the oxidation and dissolution mechanisms of Galena. Geochim. Cosmochim. Acta 67 (5) 813–822.

Jańczuk, B., Perea, R., González-Caballero., F. 1993.The influence of oxidation degree of galena surface and of ethyl xanthate on the stability of galena-air aggregates. Powder Technology. 75. 43 - 48.

Lei Jiang, Huaiyang Zhou, Xiaotong Peng, Zhonghao Ding. 2008. Bio-oxidation of galena particles by *Acidithiobacillus ferrooxidans.* Particuology 6 99–105.

Liao, M.X., Deng, T.L. 2004. Zinc and lead extraction from complex raw sulfides by sequential bioleaching and acidic brine leach. Minerals Engineering. 17 17 - 22.

Marsden J. and House I. 1992. The chemistry of gold extraction. Ed. Ellis Horwood Limited, England.

Mizoguchi, T., Habashi, F. 1981. The aqueous oxidation of complex sulfide concentrates in hydrochloric acid. International Journal of Mineral Processing. 8 177–193.

Mousavi, S.M., Jafari, A., Yaghmaei, S., Vossoughi, M., Roostaazad, R. 2006. Bioleaching of low-grade sphalerite using a column reactor. Hydrometallurgy. 82 75–82

Muñoz, A., Márquez, M.A., Montoya, O.I., Ruiz, O., Lemehsko, V. 2003. Evaluación de oxidación bacteriana de sulfuros con Acidithiobacillus ferrooxidans mediante pruebas de FTIR y difracción de rayos X. REVISTA COLOMBIANA DE BIOTECNOLOGÍA. V. 73 - 81.

Muscat, J., Gale, J.D. 2003. First principles of the surface of galena PbS. Geochimica et Cosmochimica Acta. 67. (5) 799 - 805.

Naumann, D., Helm, D. 1995. FEMS Microbiology Letters. 126 75

Nowak, P., Laajalehto, K., Kartio, I. 2000. A flotation related X-ray photoelectron spectroscopy study of the oxidation of galena surface. Colloids Surface. A 161. 447- 460.

Olson, G. J., Brierley, J.A., Brierley, C.L. 2003. Bioleaching review part B: Progress in bioleaching: applications of microbial processes by the minerals industries. Appl Microbiol Biotechnol. 63 249-257.

Paar, W.H., Braithwaite, R.S.W., Chen, T.T., Keller, P. 1984. A new mineral, scotlandite (PbSO3) from Leadhills Scotland; the first naturally occurring sulphite. Mineralogical Magazine. 48 283 - 288.

Patra, P., Natarajan, K.A. 2006. Surface chemical studies on selective separation of pyrite and galena in the presence of bacterial cells and metabolic products of Paenibacillus polymyxa. Journal of Colloid and interface Science. 298 (2) 720 - 729.

Peng, Y., Grano, S., Ralston, J., Fornasiero, D. 2002.Towards prediction of oxidation during grinding: I. Galena flotation. Minerals Engineering. 15 493 - 498.

Rodriguez, Y., Ballester, A., Blázquez, M.L., Gonzalez, F., muñoz. J.A. 2003. New information on the sphalerite bioleaching mechanism at low and high temperature. Hydrometallurgy. 71 57 - 66.

Rohwerder, T.T., Kinzler, G.K. Sand, W. 2003. Bioleaching review part A: Progress in bioleaching: fundamentals and mechanisms of bacterial metal sulfide oxidation. Appl Microbiol Biotechnol. 63 239-248.

Santhiya, D., Subramanian, S., Natarajan, K.A. 2000. Surface chemical studies on galena and sphalerite in the presence of *Thiobacillus thiooxidans* with reference to mineral beneficiation. Minerals Engineering. 13(7) 747 - 763.

Sharma, P.K., Hanumantha, R. K. 2005. Miner. Metal. Process. 22 31

Suzuki., I. 2001. Microbial laching of metals from sulfide minerals. Biotechnology Advances. 19. 119-132.

Urbano, G., Meléndez, A.M., Reyes, V.E., Veloz, M.A., Gonzáles, I. 2007. Galvanic interactions between galena - sphalerite and their reactivity. International Journal of Mineral Processing. 82148 - 155.

Waltling. H.R. 2006. The bioleaching of sulphide minerals with emphasis on copper sulphides—a review. Hydrometallurgy. 84 1-2 81-108.

Xia L., Liu J., Xiao L., Zeng J., Li B., Geng M. and Qiu G. 2008. Single and cooperative bioleaching of sphalerite by two kinds of bacteria—*Acidithiobacillus ferriooxidans* and *Acidithiobacillus thiooxidans*. Trans. Nonferrous Met. Soc. China. 12, pp 190-195.

8

Fourier Transform Spectroscopy of Cotton and Cotton Trash

Chanel Fortier
United States Department of Agriculture, Agricultural Research Service, Southern Regional Research Center, New Orleans, LA, USA

Introduction

The application of Fourier Transform (FT) spectroscopy instruments has broadened the knowledge and capabilities of textile researchers to study a host of projects including surface modifications of cotton fibers to impart flame retardant capabilities, monitoring aging effects on ancient cellulosic artifacts (Kavkler et al., 2011; Lojewski et al., 2010), determining the amount of dye fixation to fabrics (Choi, H-M, et al., 1994), and characterization of cotton byproducts during the processing of cotton (Cheng and Biswas, 2011), to name a few. As can be observed from the above examples, the power and versatility of analyzing textiles with FT spectroscopy is proven.

In the textile community, it is very important to characterize cotton quality parameters since this directly affects its potential profitability. Specifically, cotton strength, length, micronaire, fineness, color, and trash amount are conventionally monitored using an Uster® High Volume Instrument (HVI). In regards to the characterization of trash comingled with the cotton fiber, the identity of trash components evaluated with the current HVI system is not possible.

Recently, FT spectroscopy has been instrumental in identifying cotton trash components and foreign matter that are present with cotton fiber (Fortier et al., 2010). The ability to classify, quantify, and ultimately remove cotton trash and foreign matter present with cotton lint has the potential to increase the market value and durability of cotton from its conversion of fiber to yarn. When compared to dispersive instruments, some advantages of FT spectroscopy are higher signal-to-noise capabilities, higher throughput, negligible stray light, continuous spectra, and higher resolution. In addition, FT spectroscopy affords for frequencies in spectra to be measured all at once with more precise wavelength calibrations. FT spectroscopy clearly enhances the data garnered from near-infrared (NIR) and mid-infrared (MIR) spectroscopy studies of cotton fiber and cotton trash components. Since Fourier Transform spectroscopy is a broad topic, a brief introduction of its application to mid-infrared and near infrared spectroscopy will be included.

Theory of FT-MIR

FT-MIR spectroscopy, commonly known as Fourier Transform infrared (FT-IR) spectroscopy, is a branch of vibrational spectroscopy and is a primary method that yields structural information about the chemical makeup of a sample. Group absorption frequencies defined by wavelength regions over the range of 400-4000 cm^{-1} (25000-2500 nm) can be used to get qualitative information about the functional groups in a sample. Typically, organic functional groups are further defined by molecular vibrations, such as symmetric or antisymmetric stretching, which refer to an oscillation in bond length, or deformation, and bending, which describes an oscillation in bond angle (Pelletier et al., 2010).

The atoms in molecules display three types of energy: translational, rotational, and vibrational. In terms of FT-IR spectroscopy, the degrees of freedom for linear and non-linear molecules are defined by 3n-5 and 3n-6, respectively. The 3n term is defined by the three dimensions of the Cartesian coordinates for both types of molecules. However, the linear molecule has 2 degrees associated with rotational energy and 3 degrees are associated with translational energy. For the non-linear molecules, 3 of the degrees are rotational and 3 are translational while the remaining corresponds to fundamental vibrations. Therefore, the net degrees of freedom for non-linear molecules are 3n-6, and for linear molecules it is 3n-5 (Jiskoot, & Crommelin, 2005). Recently, an attenuated total reflection FT-IR accessory has been widely used. The ATR accessory greatly simplifies or removes sample preparation compared to traditional FT-IR samples employing KBr pellets. Briefly, the ATR accessory operates on the principle that a highly reflective crystal such as diamond or germanium is used as the internal reflective element (IRE). This IRE has a higher index of refraction compared to the sample. The sample and IRE must be in close contact, usually accomplished with a high pressure clamp, to achieve best data acquisition results. This accessory also yields the advantage of the ability to study both solids and liquids (Mirabella, Jr., 1993).

Instrumentation for FT-IR

Since a detailed description of an FT-IR instrument is beyond the focus of this chapter, the reader is referred to reference (Griffiths & de Haseth, 2007). Briefly, the FT-IR is based on the Michelson interferometer furnished with a stationary and moving mirror and beamsplitter angled by bisecting the planes of these two mirrors. There is also a source, sensitive detector, and computer (Colthup et al, 1975). The moving mirror is controlled by a transducer. In FT-IR spectroscopy, the interferogram in the time domain is converted to the frequency domain through a Fourier transformation, from which the IR spectrum is derived (Perkampus, 1995).

Theory of FT-NIR

The NIR spectral region encompasses 700-2500 nm (4000-12000 cm^{-1}) where the regions are defined as first, second, and third overtones or combination bands. In terms of the electromagnetic spectrum, it is situated between the visible and mid-infrared spectral regions. The primary absorbencies observed in the NIR spectral region are for the chemical species CH_i, NH_i, and OH. However, absorbencies for the carbonyl, sulfydryl, and amide groups can also be detected. In addition, the first, second, and third overtones in the NIR region can be related to fundamental frequencies in the FT-MIR region (Perkampus, 1995).

The wide and overlapping combination and overtone spectral bands typically found using FT-NIR, can complicate the interpretation of its data. In addition, the non-fundamental spectra of FT-NIR, compared to FT-IR, require that calibrations are made using chemometric software preferably over a wide range of samples. Chemometrics involves the application of statistics and mathematics to define chemical processes (Moros et al., 2010). With the evolution of chemometrics and high performance computing over the last 40 years, the popularity and use of FT-NIR spectroscopy has increased. Moros and co-workers defined three of the most traditional types of chemometric approaches to vibrational spectroscopy: Pre-processing techniques, classification methods, and regression methods. Pre-processing techniques encompasses pre-processing of the spectral data including derivative math, normalization, and baseline correction. Classification methods involve the application of pattern recognition, be it supervised or unsupervised. Multivariate regression methods include applications of principal component analysis (PCA) and partial least squares (PLS).

Instrumentation of FT-NIR

A detailed description of a FT-NIR system has been previously reported (Griffiths & Haseth, 2007; Bell, 1972). Briefly, a typical FT-NIR spectrometer is composed of a source, interferometer, beamsplitter, laser, detector, and optical components. A halogen source is conventionally used. The interferometer includes a moving and stationary mirror perpendicular to each other which modulates the NIR source. A beamsplitter, made up of quartz, calcium fluoride or potassium bromide, splits the NIR signal between the moving and stationary mirror, then recombines the signal and directs it to the detector. The laser, commonly made of helium and neon, controls the moving mirror and aligns the interferometer. NIR spectrometers can be selected based on their response, speed, and detection capabilities. Other optical components, such as mirrors or lenses, can focus or collimate the NIR signal (McCarthy & Kemeny, 2008).

Comparison of the FT-IR and FT-NIR

FT-IR affords the use of fundamental frequencies to identify molecular functional groups in both solid and liquid media. Also, when used in conjunction with the ATR accessory, little or no sample preparation is required. However, FT-NIR spectroscopy offers distinct advantages, such as flexibility of multiple sampling systems (e.g. fiber optic probe, rotating sphere) and the option of analyzing powder-size, pepper-size, raw samples (e.g. "sticks"), and large cotton trash samples, which is particularly useful for analyzing a heterogeneous sample such as cotton. Like the ATR/FT-IR sampling system, the FT-NIR sampling system is also non-destructive, easy to use, and has a short analysis time. In addition, FT-NIR instruments are capable of carrying out measurements on samples with longer path lengths than that of FT-IR spectrometers (Perkampus, 1995). These advantages make the study of textiles using FT-NIR, specifically for cotton, very attractive (Rodgers, 2002; Thomasson and Shearer, 1995; Camjani and Muller, 1996; Thibodeaux, 1992; Rodgers and Beck, 2005; Rodgers and Beck, 2009; Rodgers and Ghosh, 2008; Montalvo, et al., 1991).

Cotton fibers

Cotton fiber is composed mostly of cellulose with a small percentage being made up of waxes, proteins, pectic substances, organic acids, ash, and sugars (Lewin, 1998). The percentage of each non-cellulosic content can vary based on the maturity, cotton variety, and environmental conditions. This desirable chemical makeup along with its physical properties makes cotton a very important agricultural crop in both domestic and international markets. The popularity of this natural fiber for many years stems in part from its versatility (used in apparel, bedding, home furnishings, and industrial products), appearance, and performance (Wakelyn et al., 2007).

Cotton trash

Impact of cotton trash on the industry

Cotton trash present with cotton fiber affects the quality, appearance, price paid by textile mills to producers, yarn and fabric properties, as well as the performance of ginning and spinning equipment (Himmelsbach et al., 2006). In addition, cotton trash identification is important to the textile industry because certain types of trash have different effects on the textile processing of cotton and the marketability of the final fabric. Previous reports have shown the deleterious effects of cotton trash on yarn quality. Yarn breakage during spinning increased nearly 60% with a 1% increase in bark content when cotton fiber was processed with standard cleaners and carding equipment (Brashears et al, 1992). Veit and co-workers found a correlation between yarn imperfections and seed coat content present with cotton fiber (Veit et al., 1996). It was also determined that seed coat trash could be implicated in the onset of yarn breaks, deposits in rotor spinning equipment, and nep formation (Frey & Schneider, 1989). Thus, identifying cotton trash components present with the fiber may lead to more effective instrumental techniques that could be designed to remove the trash content co-mingled with cotton fiber.

Cotton trash from field to fabric

Initially, raw cotton has both fiber and trash material present. In the U.S., cotton is conventionally machine-stripped or machine-picked during harvesting. It is then sent to a gin where the cotton fiber is separated from the cottonseed. The cotton fiber separated from the cottonseed is referred to as the cotton lint. The cotton lint is then cleaned to remove the cotton trash. During the ginning and lint cleaning methods, cotton trash present with the lint tends to become smaller and smaller in size. Next, the cotton lint is classified by the Agricultural Marketing Service (AMS), and sold to textile mills where the yarn and fabric is developed.

It is necessary to clean the raw cotton because the value of cotton lint is strongly influenced by the amount of cotton trash present. Samples from each bale of American cotton are visually examined. Human "classers" give each bale a leaf grade and extraneous matter description. The leaf grade goes from 1-8 with leaf grade 1 having the least amount of trash and leaf grade 8 having the most amount of trash. Obviously, this method's accuracy and precision along with being time-consuming and labor intensive is limited by the subjective evaluation of a human "classer". Thus, instrumental techniques which can reproducibly classify cotton and the trash found with it are desired.

An instrumental method employing the Uster® High Volume Instrument (HVI) yields information on key quality parameters such as length, strength, color, micronaire, length uniformity, as well as trash content. The Shirley Analyzer is another conventional method to analyze cotton trash through and aero-mechanical technique which separate cotton trash from lint. However, these methods do not yield individual trash component identification present with the cotton lint. Thus, being able to accurately identify individual cotton trash components would lead to more efficient trash removal and favorable yarn and fabric production.

Non-fourier transform techniques to analyze cotton trash

Geometric/imaging approaches

Cotton trash has been previously analyzed by a variety of methods. A clustering analysis method was used to identify cotton trash components using sum of squares, fuzzy and neural networks (Xu et.al, 1999). The clustering approach was based on color, shape and size attributes. When comparing the different clustering approaches, the color features were more reliable than the shape and size parameters since trash particles can become smaller during the processing of cotton lint. The neural network clustering method yielded an accurate trash type classification greater than 95% of the time. However, a limitation of this method was the time required for computation.

Siddaiah and co-workers also classified cotton trash components based on geometry. Their approach involved defining simple geometric shapes (disks, points, arcs, and straight lines) to identify bark leaf, and pepper trash. This technique, which incorporated imaging and intelligent pattern recognition, was highly accurate with 98% correct classification of the trash types (Siddaiah, 1999).

A machine vision system comparing a camera- and scanner-based imaging technique was developed (Siddaiah et.al, 2006). The Cotton Trash Identification System (CTIS) was compared to conventional instrumentation such as the Uster® HVI, Advanced Fiber Information System (AFIS), and the Shirley Analyzer for measuring cotton trash. It was determined that surface based measurements (CTIS, HVI) had a higher correlation compared to volumetric measurement systems (AFIS and Shirley Analyzer). The CTIS system, calibrated using "classer" calls (human visual identification) was later found to accurately identify trash categories (bark/grass, stick, leaf, and pepper) 97% of the time.

UV-Vis spectroscopy

Fortier and co-workers investigated the capability of identifying botanical cotton trash types using UV-Vis spectroscopy (Fortier et al., 2011a). Figure 1 shows the result of applying first derivative math to classify cotton and some botanical trash types (hull, leaf, seed coat, and stem). The chemometric software package employed in their study did not allow the use of sub-libraries to distinctly identify the highly similar trash spectra. Thus, 67% of the individual trash samples were correctly identified as shown in Table 1.

Applications of fourier transform spectroscopy to cotton and cotton trash

Trash effect on spinning efficiency

Cotton trash has been reported to reduce the efficacy of textile processing (Bargeron et al. 1988, Verschraege, 1989). After cotton fiber is separated from the cottonseed, the fiber is dried, cleaned and converted into yarn through different spinning techniques. The three most commonly used spinning methods are ring, open-end rotor, and vortex spinning. Ring spinning is characterized by a twist inserted into a yarn by a circulating traveler with the

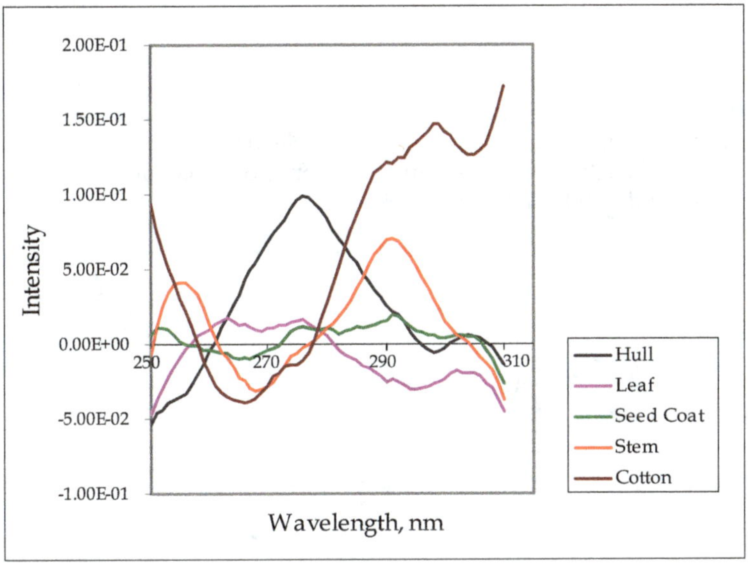

Fig. 1. Average first derivative UV-Vis spectra of cotton, hull, leaf, seed coat and stem botanical trash. Source: Fortier, C., et al., 2011a, *Journal of Cotton Science*, Vol. 15, pp.174.

Trash Type	No. of Samples	No. Correct	% Correct
Hull	27	14	52%
Leaf	30	16	53%
Seed Coat	12	9	75%
Stem	27	25	93%
Total	96	64	67%

Table 1. UV-Vis identification results of botanical trash types Source: Fortier, C., et al., 2011a, *Journal of Cotton Science*, Vol. 15, p.173

winding action made possible by a rotating spindle. In open-end rotor spinning, cotton fiber bundles from the sliver (loose piece of cotton fiber) are separated into individual fibers by a roller which opens and an air stream. Vortex spinning has become available commercially in recent years. This spinning type is characterized by cotton fibers being introduced into a spindle orifice by way of air jet technology. One of the three spinning types, open-end rotor spinning efficiency has been found to be greatly influenced by the presence of cotton trash (Baker, et al., 1994).

The presence of cotton contamination in the rotor groove has been associated with the onset of yarn breakage, thick places in the yarn, and yarn entanglement (Thibodeaux & Baril, 1981; McCreight et al., 1997). Compounding this issue is the tendency for trash types to become smaller and smaller in size making it virtually impossible to visibly identify individual cotton trash components. Foulk and co-workers carried out an FT-IR classification study where cotton trash components (e.g. hull vein, hull outside, hull stem, hull inside, shale vein, shale) were evaluated based on color differences (Foulk et al., 2004). Pima and upland cotton varieties having widely different trash amounts and moisture levels were analyzed using a diverse set of ginning methods. The objective of this study was to identify the type of cotton and trash type or types built up in the rotor groove. The rotor dust was collected and analyzed by FT-IR and compared to the spectral database (calibration set). In the spectral database, the top five matches were provided for each sample. Classification results were determined by counting the number spectral matches for a particular type of trash as shown in Figure 2. The trash collected in the rotor grove was determined to be largely composed of shale and hull trash types (Foulk, 2004).

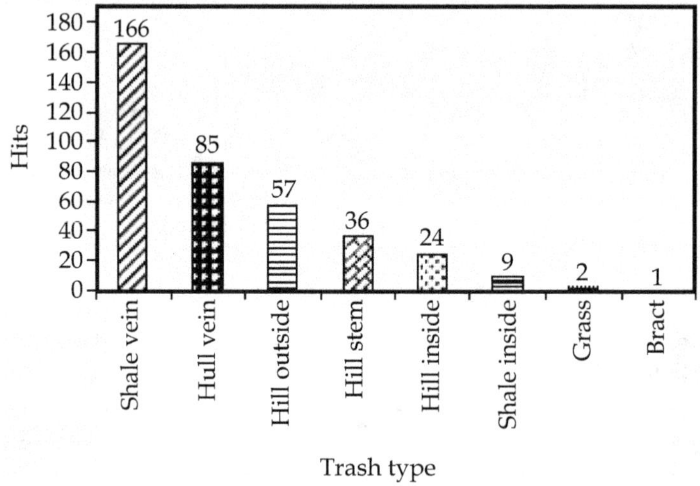

Fig. 2. FT-IR spectra matched hits for open end spinning rotor dust. Source: Foulk, J. et al, 2004, Journal of Cotton Science, Vol. 8, p.249

Size effects and heat treatment

Allen and co-workers conducted an ATR/FT-IR spectroscopy study to analyze botanical cotton trash (leaf, seed coat, stem, and hull) subjected to different temperatures and particle sizes typically observed in a ginning or textile mill environment (Allen et al., 2007). During ginning the cotton fiber temperature is raised to aid in the separation of cotton trash from the lint. Also, raw trash components become smaller and smaller due to mechanical treatment of the cotton fiber. This effect makes it very challenging to determine the origin of the trash particles.

The cotton trash components were evaluated at elevated (149 °C) and standard room temperature conditions (22 °C). The FT-IR spectral data was obtained at 21 °C using a benchtop FT-IR instrument fitted with an ATR accessory. To determine the effects of the two temperature settings, the ratio of the area under the total spectrum to the area under each of the IR bands of interest were calculated. The assigned ATR/FT-IR bands included in this study were based on the O-H stretch at 3300 cm^{-1}, the C-H asymmetric stretch at 2900 cm^{-1}, the C-H symmetric stretch at 2850 cm^{-1}, the C=O stretch of a carboxylic acid and ester at 1700 cm^{-1}, and the C=O stretch of an acid salt at 1600 cm^{-1} (Chung et al., 2004; Silverstein and Webster, 1998). All heated samples were allowed to return to room temperature to observe any irreversible changes in the FT-IR spectra. No change in trash color was observed in pepper-sized trash (0.841mm standard diameter). However, over the O-H and C-H stretching frequencies (2900 and 3300 cm-1), the stem and hull trash had an appreciable change (Figures 3 and 4) compared to the leaf and seed coat spectra (data not shown). For the IR bands representing the C=O stretching frequency (1600 and 1700 cm-1), the seed coat trash component has the most noticeable spectral difference (Figure 5) compared to the leaf, hull, and stem components. The relevance of this observation is that the seed coat may have volatile aliphatic esters which are generated with heat.

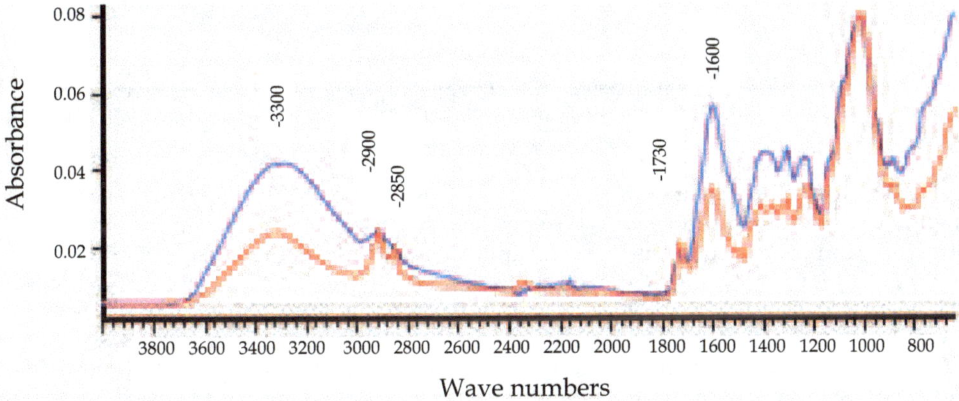

Fig. 3. The effects of heat treatment on the ATR/FT-IR spectra of the pepper-size stem samples. The blue curve represented the spectra at standard conditions and the red curve represebts samples heated to 149 °C for 20 minutes then allowed to return to standard conditions prior to analyzing. Source: Allen et al., 2007, Journal of Cotton Science, Vol. 11, p.72

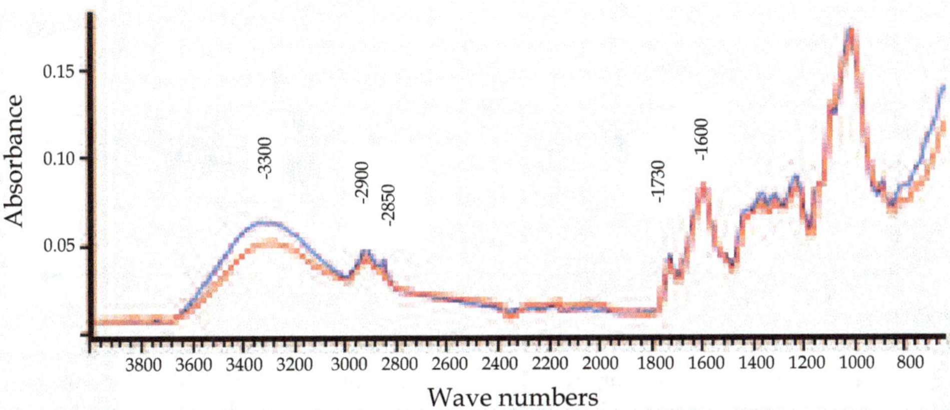

Fig. 4. The effects of heat treatment on the ATR/FT-IR spectra of the pepper-size hull samples. The blue curve represented the spectra at standard conditions and the red curve represents samples heated to 149 °C for 20 minutes then allowed to return to standard conditions prior to analyzing. Source: Allen et al., 2007, Journal of Cotton Science, Vol. 11, p.72

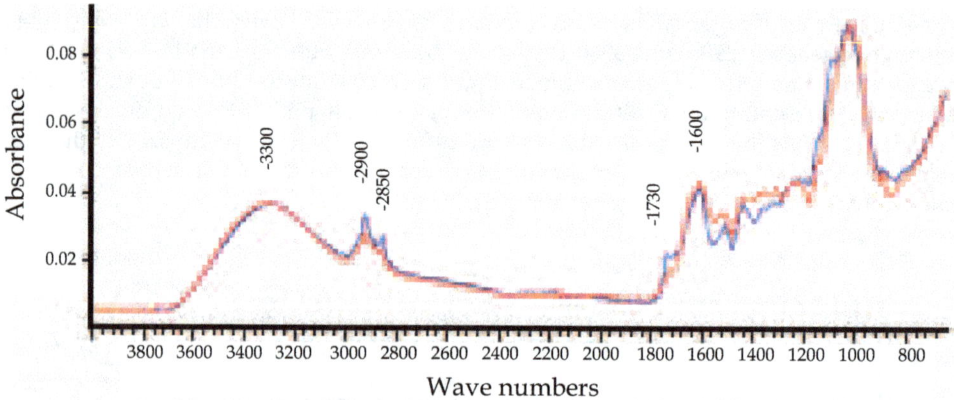

Fig. 5. The effect of heat treatment on the ATR/FT-IR spectra of the pepper-size seed coat samples. The blue curve represented the spectra at standard conditions and the red curve represents samples heated to 149 °C for 20 minutes then allowed to return to standard conditions prior to analyzing. Source: Allen et al., 2007, Journal of Cotton Science, Vol. 11, p.72

In terms of the effect of particle size (powder, pepper, and raw), distinct spectral FT-IR differences over the spectral range of 3500 to 2800 cm^{-1} were observed for the stem, (Figure 6) leaf and hull samples (data not shown), but for the seed coat trash type (Figure 7) no observable differences were found. Pepper and powder-sized trash samples were generated using 20 (0.841 mm std. diameter) and 80 (0.177 mm std. diameter) meshes, respectively.

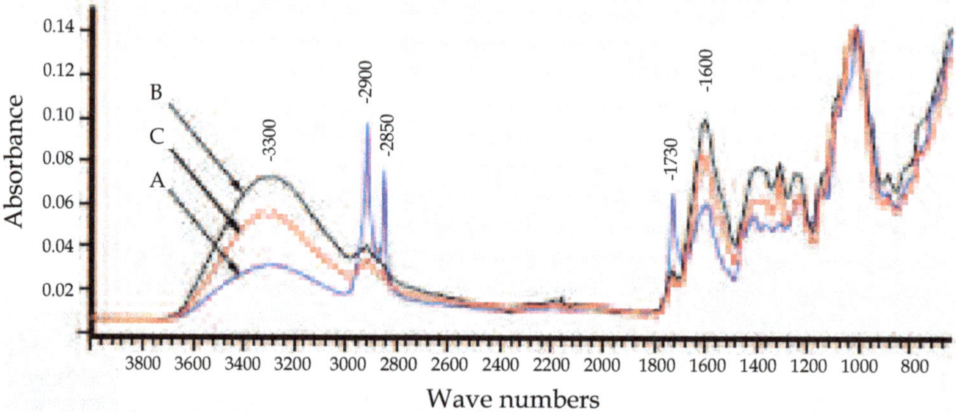

Fig. 6. The effect of size reduction on the ATR/FT-IR spectra of the stem sample, including ground raw size (blue, A), pepper-size (black, B), and powder size (red, C) Source: Allen et al., 2007, Journal of Cotton Science, Vol. 11, p.71

Overall, it was concluded that size and heat treatment can affect the FT-IR spectra of cotton trash types. Thus, spectral databases used for classifying these components should include spectra accounting for these physical applications.

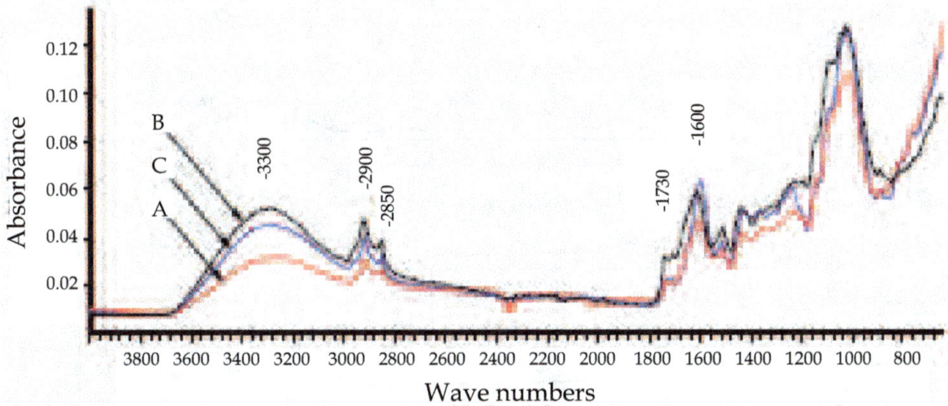

Fig. 7. The effect of size reduction on the ATR/FT-IR spectra of the seed coat sample, including ground raw size (blue, A), pepper-size (black, B), and powder size (red, C) Source: Allen et al., 2007, Journal of Cotton Science, Vol. 11, p.71

Botanical and synthetic cotton trash

To expand on the proven utility of applying ATR/FTIR as a technique to specifically identify cotton trash, Himmelsbach and co-workers did an extensive study on botanical cotton trash and other foreign matter both organic (other fibers, yarns, etc.) and synthetic (plastic bags, film, etc.) that can become co-mingled with cotton lint during its harvesting, ginning, and processing (Himmelsbach et al., 2006). The development of an FT-IR spectral database yielded highly accuracte identification of cotton trash and foreign matter serving as a "proof of concept" of the utility of this instrumental method. Fortier and co-workers later compared their FT-IR results to that of this preliminary FT-IR study by Himmelsbach and co-workers (Fortier et al., 2010). The two studies yielded comparable results. Representative FT-IR absorbance spectra results of cotton, hull, leaf, seed coat, and stem are shown in Figure 8. The spectra are based on three replicates averaged with a resolution of 8 cm^{-1} and 128 scans.

Fourier-transform near-infrared spectroscopy of botanical cotton trash

More recently, the application of FT-NIR spectroscopy has also been employed to identify individual botanical trash components (hull, leaf, seed coat, and stem) in a spectral library. Pepper-sized and powder-sized pure trash components were included in the calibration set. While pepper-sized, powder-sized and raw samples of different varieties were included in the prediction set. Specifically, the calibration set was different from the prediction set in that the corresponding powder and pepper samples in the calibration set were represented as the opposite powder and pepper samples in the prediction set. There were 9 different varieties of trash samples taken from Mississippi, New Mexico, and South Carolina. The cotton trash varieties are denoted by the first two letters from the state the samples were acquired in and the last letter as the sample variety as shown in Table 2. Sample measurements were taken by directly placing a solid probe in contact with the trash samples. When creating the calibration set (spectral library), multiple wavelength regions

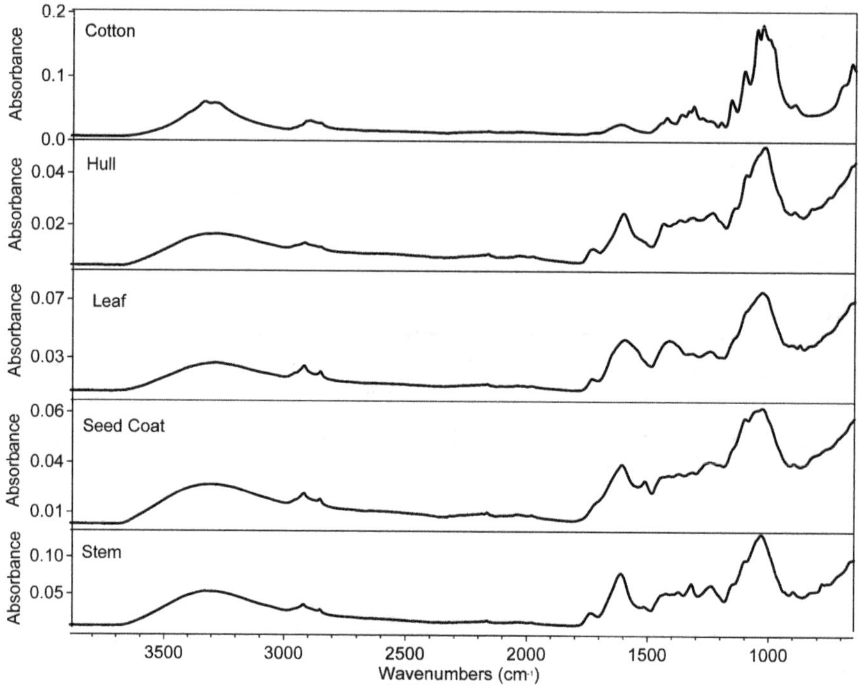

Fig. 8. Average FT-MIR absorbance spectra of cotton and cotton trash components at a resolution of 8 cm^{-1} with 128 scans Source: Fortier et al., 2010, Textile Research Journal, Vol. 81, p. 233

(full spectra range, and small spectral ranges) and pre-processing methods (vector normalization and derivative math) were investigated to optimize the correct identification of the pure botanical trash types in the prediction set.

Figure 9 show the results of comparing the average FT-NIR absorbance spectra for a clean cotton reference to pure botanical trash components. The "clean" cotton has specific bands represented at 1490 nm, 1930 nm, and 2100 nm. Over the spectral region of 1490 nm to 1600 nm which is referred to as the first overtone of the hydroxyl, the cotton band is more uniquely defined compared to the cotton trash. The band at 1450-1490 nm represents the OH stretch first overtone due to moisture and indicates the presence of OH groups in cotton. There was considerable overlap of cotton and cotton trash spectra at the very intense moisture band present at 1930 nm which is also the OH bend stretch and water deformation combination band. At 2100 nm, which is the OH bend/CO stretch combination, the band representing OH groups present in cotton, and to a lesser extent moisture, had considerable overlap with the cotton trash spectra even though the intensity of the cotton band is more pronounced. In terms of the cotton trash, there was considerable overlap throughout the entire spectral region with a defining peak over the spectral region of 1400 nm to 1500 nm which did not include a cotton peak. Even though cotton had distinctive spectral bands, the investigation was continued to gain non-overlapping spectral identity for all the trash components with a validation report showing no confused components (Fortier et al., 2010).

Thus, the investigation was expanded to include different wavelength ranges and other pre-processing methods. First derivative math was then employed to uniquely identify the botanical trash types, and cotton as shown in Figure 10. This pre-processing was able to distinctly identify cotton, leaf, and stem. However, a sub-library consisting of hull and seed coat was necessary to distinguish these trash types due to their very similar spectra as depicted in Figure 11. Overall, result in Table 3 reveals the trash types were correctly identified 98% of the time. The 2 misidentifications for the seed coat was a direct effect due to the spectral similarity of the hull and seed coat trash types. Raw samples of each botanical trash type was also investigated to determine the sample-size effect on correct identification of hull, leaf, seed coat, and stem pure trash types. Table 4 reveals the results of this experiment. The result for the raw stem trash misidentifications could be explained by the irregular shape of the stems during sampling with the solid fiber optic probe.

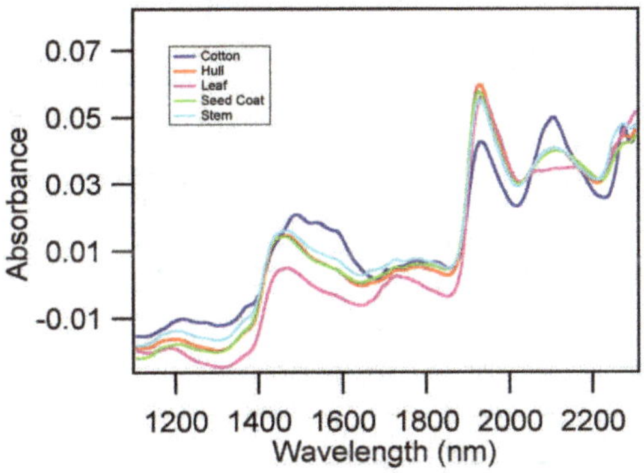

Fig. 9. Average FT-NIR absorbance spectra for "clean" cotton and cotton trash samples over entire spectral range 1100-2400 nm where cotton can be identified but cotton trash components are overlapping. Vector normalization and standard method pre-processing was applied Source: Fortier et al., 2010, Textile Research Journal, Vol. 81, p. 234

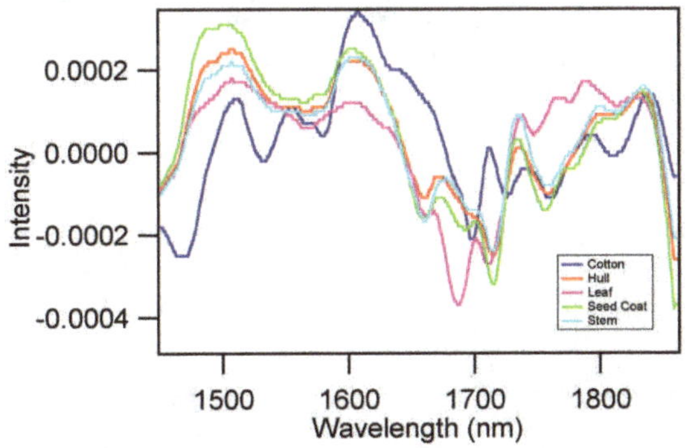

Fig. 10. Average FT-NIR spectra for cotton and cotton trash over the narrow spectral range of (1427-1867 nm) where cotton and cotton trash components are uniquely identified except for hull and seed coat. A factorization and first derivative pre-processing method was applied Source: Fortier et al., 2010, Textile Research Journal, Vol. 81, p. 236

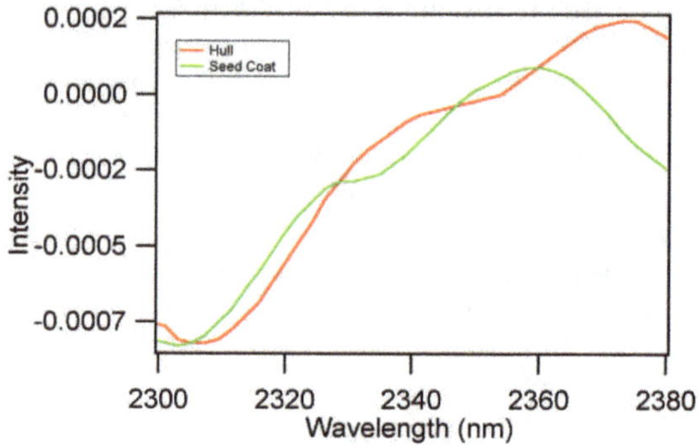

Fig. 11. Average first derivative spectra from sub-library for hull and seed coat. Source: Fortier et al., 2010, Textile Research Journal, Vol. 81, p. 236

Cotton Trash Variety and Size	Trash Types			
MSA (powder)	Hull	Leaf	Seed Coat	Stem
NMA (pepper)	Hull	Leaf	NS	Stem
NMB (powder)	Hull	Leaf	Seed Coat	Stem
NMC (pepper)	Hull	Leaf	Seed Coat	Stem
SCA (pepper)	Hull	Leaf	Seed Coat	Stem
SCB (powder)	Hull	Leaf	Seed Coat	Stem
SCC (pepper)	Hull	NS	Seed Coat	Stem
SCD (pepper)	Hull	Leaf	Seed Coat	Stem
SCE (powder)	Hull	Leaf	Seed Coat	Stem
MSA, NMB,NMA,NMC,SCA	Raw seed meat			

NS = no sample. (The cotton trash varieties are denoted by the first two letters from the state the samples were acquired in and the last letter as the sample variety (MSA = Mississippi DP555, NMA = New Mexico DP555, NMB = New Mexico AC151799, NMC = New Mexico Unknown, SCA = South Carolina DP458, SCB = South Carolina DP555, SCC = South Carolina DP555a, SCD = South Carolina FM989, SCE = South Carolina PM1218).

Table 2. Calibration set of cotton trash samples. Source: Fortier et al., 2010, Textile Research Journal, Vol. 81, p. 231

Prediction Set	Individual Powder and Pepper Samples[a]		
	%Correct	Number of samples	Number Correct
Hull	100%	27	27
Leaf	100%	27	27
Seed Coat	91.7%	24	22
Stem	100%	27	27
Total	98.1%	105	103

Table 3. NIR Identification by Cotton Trash Type for Powder and Pepper samples. Source: Fortier et al., 2010, Textile Research Journal, Vol. 81, p. 236

Prediction Set	Average Raw Samples[a]		
	%Correct	Number of samples	Number Correct
Hull	100%	9	9
Leaf	100%	9	9
Seed Coat	100%	8	8
Stem	77.78%	9	7
Total	94.3%	35	33

Table 4. NIR Identification by Cotton Trash Type for Raw samples. Source: Fortier et al., 2010, Textile Research Journal, Vol. 81, p. 236

Seed meat addition to botanical trash library

Fortier and co-workers added seed meat to the FT-NIR spectral library to make it more robust (Fortier et al., 2011b). The seed meat spectra (data not shown) was quite different from the original trash components in the FT-NIR spectral library (hull, leaf, seed coat, and stem) Thus, simply adding the seed meat spectra to the existing library caused some misidentifications by making the other botanical trash components' thresholds overlap. As a result, a different approach was necessary to improve the number of correct identifications in the prediction set. A "top down" approach was employed where the seed meat was isolated first in the library. Next, the cotton and botanical trash types were included in the spectral library, and a second sub-library was made to separate the hull and seed coat, due to their highly similar spectra. The end result was that 100% of the trash types were identified correctly in the prediction set (data not shown).

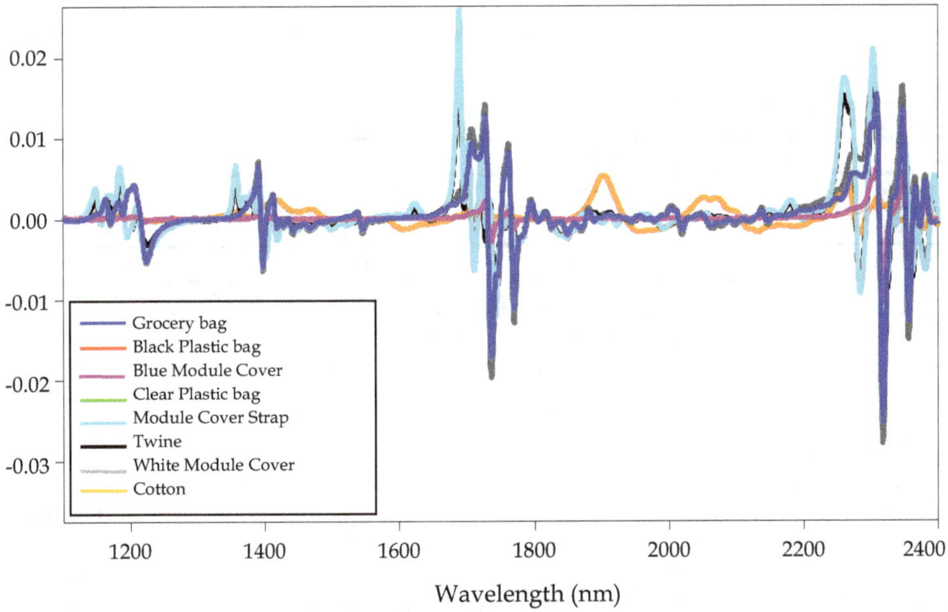

Fig. 12. Average first derivative spectra of field trash and cotton. Source: Fortier et al., 2011b, Journal of Cotton Science, in press

Field trash addition to FT-NIR spectral library

Field trash refers to any trash found in a cotton field which is not originating from the cotton plant. Recently, FT-NIR spectroscopy was used to identify field trash samples including a grocery bag, black plastic bag, blue module cover, clear plastic bag, module cover strap, twine, and white module cover (Fortier et al., 2011b). Three replicates were acquired for each trash type. Figure 12 shows the average first derivative spectra of cotton and field trash types. Applying a "top down" approach, Fortier and co-workers added field trash samples, to the FT-NIR spectral library (Fortier et al., 2011b). First, the seed meat and field trash were isolated from the rest of the botanical trash library. Next, a sub-library of the cotton and botanical trash types was designed. Finally a second sub-library to separate hull and seed coat was developed. The identification results can be found in Table 5. Overall, greater than 98% correct identifications were achieved.

Prediction Set	**Individual Powder, Pepper, and Field Trash Samples**		
Trash Type	%Correct	Number of Samples	Number Correct
Hull	100%	27	27
Leaf	100%	27	27
Seed Coat	95.24%	21	20
Seed Meat	91.67%	12	11
Stem	100%	27	27
Total Botanical Trash	98.25%	114	112
Grocery Bag	100%	2	2
Black Plastic Bag	100%	2	2
Blue Module Cover	100%	2	2
Clear Plastic Bag	100%	2	2
Module Cover Strap	100%	2	2
Twine	100%	2	2
White Module Cover	100%	2	2
Total Field Trash	100%	14	14
Overall Total Trash	98.44%	128	126

Table 5. NIR Identification by Cotton Trash Type for Individual Powder and Pepper samples adding seed meat and field trash. Source: Fortier et al., 2011b, Journal of Cotton Science, in press

Summary

This chapter has summarized the application of FT spectroscopy towards the identification and classification of cotton trash components which are found present with cotton lint. The successes reported herein were not possible with conventional techniques such as using the HVI and Shirley Analyzer. In the future, the feasibility of using FT techniques can benefit those in the textile industry such as farmers, ginners, spinners and merchants by providing a well-defined cotton product which has a minimal amount of trash and a high market value.

References

Allen, A., Foulk, J. & Gamble, G. 2007. Preliminary Fourier-Transform Infrared Spectroscopy Analysis of Cotton Trash, *Journal of Cotton Science*, Vol.11, pp.68-74, ISSN 15236919.

Baker, R., Price, J., & Robert, K. 1994, Gin and mill cleaning for rotor spinning, *Transactions of the American Society of Agricultural Engineers*, Vol. 37, No.4, pp. 1077-1082, ISSN 0001-2351.

Bargeron, J., Rayburn, & S. Griffith, S. 1988. Effects of grass contaminated with cotton on yarn manufacturing, *Transactions of the American Society of Agricultural Engineers*, Vol.31, No.1, pp.2-4, ISSN 0001-2351.

Bell, R.J. 1972. *Introductory Fourier Transform Spectroscopy*, Academic, ISBN 0120851504, New York, New York.

Brashears, A., Baker, R., Bragg, C., & Simpson, C., 1992. Effect of bark on spinning efficiency of cotton, *Proceedings of Beltwide Cotton Conference*, ISSN 10592644, Nashville, TN, January 1992, pp.1218-1219.

Camajani, N., & Muller, D., 1996. Textile Analysis by FT-NIR: Fourier Transform Near Infrared Spectroscopy, In: *Textile Applications of Near Infrared Technology: American Association of Textile Chemists and Colorists Symposium*, Rodgers, J., Beck, K., Howell, H., & Hunter, W., pp. 1-12, Asheville, North Carolina.

Cheng, H. & Biswas, A. 2011. Chemical modification of Cotton-Based Natural Materials: Products from Carboxymethylation. *Carbohydrate Polymers*, Vol 84, No. 3, pp.1004-1010, ISSN 01448617.

Choi, H-M., Srinivasan, M., & Morris, N. 1994. Single-Step Dyeing and Finishing Treatment of Cotton with 1,2,3,4-Butanetetracarboxylic acid. *Journal of Applied Polymer Science*, Vol. 54, pp.2107-2118, ISSN 00218995.

Chung, C., Myunghee, L. & Choe, E. 2004. Characterization of cotton fabrics scouring by FT-IR ATR spectroscopy. *Carbohydrate Polymers*, Vol. 58, No.4, pp.417-420, ISSN 0144-8617.

Colthup, N., Daly, L., & Wiberley, S. 1975. *Introduction to Infrared and Raman spectroscopy*, Academic Press, ISBN 0-12-182552-3, New York, New York.

Fortier, C., Rodgers, J., Cintron, M. S., Cui, X. & Foulk, J. 2010. Identification of Cotton and Cotton Trash Components by Fourier Transform Near-Infrared Spectroscopy. *Textile Research Journal*, Vol.81, pp. 230-238, ISSN 0040-5175.

Fortier, C., Rodgers, J., & Foulk, J. 2011a. Investigation of the Impact of Instrumental and Software Applications on Cotton and Botanical Trash Identification by Ultraviolet-Visible and Near-Infrared Spectroscopy, *Journal of Cotton Science*, Vol. 15, pp.170-178, ISSN 15236919.

Fortier, C., Rodgers, J., Foulk, J., & Whitelock, D. 2011b. Near-Infrared classification of cotton lint, botanical and field, *Journal of Cotton Science*, in press, ISSN 15236919.

Foulk, J., McAlister, D., Himmelsbach, D. & Hughs, E. 2004. Mid-Infrared Spectroscopy of Trash in Cotton Rotor Dust. *Journal of Cotton Science*, Vol.8, pp.243-253, ISSN 15236919.

Frey, M., & Schneider, U. 1989. Possibilities of Removing Seed Coat Fragments and Attached Fibers in the Spinning Process, *Melliand Textilberichte International Textile Reports*, Vol. 70, No. 5, pp. 315-317, ISSN 0341-0781.

Griffiths, P., De Haseth, J. 2007. *Fourier Transform Spectrometry*, 2nd edition, John Wiley and Sons, ISBN 978-0-471-19404-0, Hoboken, New Jersey.

Himmelsbach, D., J. Hellgeth, D. McAlister. 2006. Development and Use of an Attenuated Total Reflectance/Fourier Infrared (ATR/FT-IR) Spectral Database to Identify Foreign Matter in Cotton. *Journal of Agricultural and Food Chemistry*, Vol.54, pp.7405-7412, ISSN 0021-8561.

Jiskoot, W., Daan, J., & Crommelin, A. 2005. *Methods for structural analysis of protein pharmaceuticals*, Library of Congress, ISBN 0-9711767-2-8, Arlington, VA.

Kavkler, K., Smit, Z., Jezersek, D., Eichert, D., & Demsar, A. 2011. Investigation of Biodeteriorated Historical Textiles by Conventional and Synchrotron Radiation FTIR spectroscopy. *Polymer Degradation & Stability*, Vol. 96, pp.1081-1086, ISSN 0141-3910.

Lewin, M. & Pearce, E. 1998. *Handbook of Fiber Chemistry*, 2nd edition, Marcel Dekker, Inc., ISBN 0-8247-9471-0, New York, New York.

Lojewski, T., Miskowiec, P., Missori, M., Lubanska, A., Proniewicz, L., & Lojewska, J. 2010. FTIR and UV/Vis as Methods for Evaluation of Oxidative Degradation of Model Paper: DFT Approach for Carbonyl Vibrations. Carbohydrate Polymers, Vol. 82, pp.370-375, ISSN 0144-8617.

McCarthy, W. & Kemeny, G. 2008. Fourier Transform Spectrophotometers in the Near-Infrared, In: *Handbook of Near-Infrared Analysis, 3rd Edition*, Burns, D., Ciurczak, E., pp. 79-91CRC Press, ISBN 978-0-8493-7393-0, Boca Raton, Florida.

McCreight, D., Feil, R., Booterbaugh, J., & Backe, E. 1997, *Short Staple Yarn Manufacturing*, Carolina Academic Press, ISBN 9780890898536, Durham, NC.

Mirabella, Jr., F. M. 1993. *Internal Reflection Spectroscopy*, Marcel Dekker, ISBN 0-8247-8730-7, New York, New York.

Montalvo, J., Faught, S., & Buco, S. A 1991. Comparative Study of NIR Diffuse Reflectance of Cottons Grouped According to Fiber Cross-Sectional Dimensions, Part III. Experimental. *Applied Spectroscopy*, Vol. 45, No.5, pp.795-807, ISSN 0570-4928.

Moros, J., Garrigues, S. & de la Guardia, M. 2010. *Trends in Analytical Chemistry*, Vol.29, No.7, pp. 578-591, ISSN 0165-9936.

Pelletier, M., & Pelletier, C.C. 2010. Spectroscopic Theory for Chemical Imaging, In "*Raman, Infrared, and Near-Infrared Chemical Imaging*", Šašić, S., & Ozaki, Y., pp.1-20, John Wiley and Sons, ISBN 978-0-470-38204-2, Hoboken, New Jersey.

Perkampus, H.-H. 1995. *Encyclopedia of Spectroscopy*, VCH Verlagsgesellschaft mbH, ISBN 3-527-29281-0, Weinheim, Germany.

Rodgers, J. 2002. Influences of Carpet and Instrumental Parameters on the Identification of Carpet Face Fiber by NIR. *American Association of Textile Chemists and Colorists Review*, pp. 27-32, ISSN 1532-8813.

Rodgers, J. E. & Beck, K. 2005. Rapid Determination by NIR of the Cotton Content of Blend Fabrics after Dyeing, *Proceedings of the Beltwide Cotton Conference*, ISSN 10592644, New Orleans, LA,January 2005, pp. 2731-2736.

Rodgers, J. & Beck, K., NIR 2009. Characterization and Measurement of the Cotton Content of Dyed Blend Fabrics. *Textile. Research Journal*, Vol. 79, No.8, pp.675-686, ISSN 0040-5175.

Rodgers, J. & Ghosh, S., 2008. NIR Analysis of Textiles, In: *Handbook of Near-Infrared Analysis*, 3rd edition, Burns, D. & Ciurczak, E. pp 485-520.; CRC Press, ISBN 978-0-8493-7393-0, Boca Raton, Florida.

Siddaiah, M., Lieberman, M., Prasad, N., & Kreinovich, V. 1999. A Geometric Approach to Classification of Trash in Ginned Cotton. Departmental Technical Reports, Paper 581, pp. 77-82.

Siddaiah, M., Hughs, E., Lieberman, M., & Foulk, J. 2006. *Proceedings of the Beltwide Cotton Conference*, ISSN 10592644, San Antonio, TX, January 2006, pp.1926-1937.

Silverstein, R. & Webster, F. 1998. *Spectrometric identification of organic compounds*, John Wiley & Sons Ltd., ISBN 0-471-39362-2, Toronto, Canada.

Thibideaux, D., & Baril, A. 1981. Laboratory techniques for predicting cotton dust residues in open-end spinning. *Textile Research Journal*, Vol.51, No.11, pp.688-695, ISSN 0040-5175.

Thibodeaux, D. 1992. NIR as a Tool for Measuring Cotton Quality., *International Cotton Conference*, Bremen, Germany, March 1992, pp. 27-38.

Thomasson, J. & Shearer, S. 1995. Correlation of NIR Data with Cotton Quality Characteristics. *Transactions of the American Society of Agricultural Engineers*, Vol. 38, No. 4, pp. 1005-1010 ISSN 0001-2351.

Veit, D., Hormes, I., Bergmann, J., & Wulfhorst, B., 1996. Image Processing as a Tool to Improve Machine Performance and Process Control, *International Journal of Clothing Science and Technology*, Vol. 8, No. 1-2, pp. 66-72, ISSN 0955-6222.

Verschaege, L. 1989. Cotton Fibre Impurities: Neps, Motes, and Seed Coat Fragments, *Centre for Agricultural Bioscience International: International Conference on Analytical Chemistry: Review Articles on Cotton Production Research No. 1. ISBN 08-5198 633-1,* Wallingford, UK.

Wakelyn, P., Bertoniere, N., French, A., Thibodeaux, D., Triplett, B., Rousselle, M-A., Goynes, Jr., W., Edwards, J., Hunter, L., McAlister, D., & Gamble, G. 2007. *Cotton Fiber Chemistry and Technology*, CRC Press, ISBN -10: 1-4200-4587-3, Boca Raton, FL.

Jiskoot, W., Daan, J., & Crommelin, A. 2005. *Methods for structural analysis of protein pharmaceuticals*, Library of Congress, ISBN 0-9711767-2-8, Arlington, VA.

Xu, B., Fang, C. & Watson, M. 1999. Clustering Analysis for Cotton Trash Classification, *Textile Research Journal*, Vol.69, No.9, pp.656-662, ISSN 0040-5175.

9

Fourier Transform Infrared Spectroscopy for the Measurement of Spectral Line Profiles

Hassen Aroui[1], Johannes Orphal[2] and Fridolin Kwabia Tchana[3]
[1]*Laboratoire de Dynamique Moléculaire et Matériaux Photoniques,*
Université de Tunis, Ecole Supérieure des Sciences et Techniques de Tunis,
[2]*Institute for Meteorology and Climate Research (IMK),*
Karlsruhe Institute of Technology (KIT),
[3]*Laboratoire Interuniversitaire des Systèmes Atmosphériques (LISA),*
UMR CNRS 7583 Universités Paris Est Créteil et Paris- Diderot,
Institut Pierre Simon Laplace,
[1]*Tunisia*
[2]*Germany*
[3]*France*

Introduction

Our knowledge of the process of global change in Earth's atmosphere is primarily based on experimental observations made in situ or by remote-sensing. The data obtained reveal the state of the atmosphere and provide input data for validation of theoretical models of the atmosphere. In this context, high resolution molecular spectroscopy plays a key role as it is at the heart of optical remote sensing measurements. In fact, infrared spectroscopy is a powerful tool to identify and quantify atmospheric trace species, through the use of characteristic spectral signatures of the different molecular species and their associated vibration-rotation bands in the mid- or near-infrared. Different methods, based on quantitative spectroscopy, permit tropospheric or stratospheric measurements: in situ long path absorption, atmospheric absorption/emission by Fourier transform spectroscopy with high spectral resolution instruments on the ground, airborne, balloon-borne or satellite-borne measurements (Camy-Peyret et al., 2001). In all cases, the quality of the analysis and interpretation of atmospheric spectra requires reference spectroscopic information (positions, intensities, broadenings, profiles...) measured in the laboratory. Such information exists and is compiled in databases such as HITRAN, GEISA and ATMOS, available to the international scientific community.

Fourier Transform Infrared (FTIR) can be utilized to measure some components of an unknown mixture and is currently applied to the analysis of solids, liquids, and gases. The term FTIR refers to the manner in which the data is collected and converted from an interference pattern to a spectrum.

Fourier transform spectrometers have progressively replaced dispersive instruments for most applications due to their superior speed and sensitivity. They have greatly extended the capabilities of infrared spectroscopy and have been applied to many fields that are very difficult or nearly impossible to study by dispersive instruments. Instead of analyzing each spectral component sequentially, as in a dispersive IR spectrometer, all frequencies are examined simultaneously in FTIR spectroscopy.

By interpreting the infrared absorption spectrum, information about the structure and the nature of the chemical bonds in a molecule can be determined. For most common materials, the spectrum of an unknown mixture can be identified by comparison to a library of known compounds.

The FTIR technique is a powerful tool for identification of chemical species and their structures, for interpretation of the atmospheric spectra by producing an infrared absorption spectrum that is like a molecular fingerprint. This technique can be used to identify chemicals from spills, paints, polymers, coatings, drugs, and contaminants. It can measure more than 120 gaseous pollutants in the ambient air (U.S. Environmental Protection Agency, 2011), such as carbon monoxide, sulfur dioxide, ozone, and many others. This technique can also quantify toxic organic pollutants, such as toluene, benzene, methanol etc. This quantification is based on the fact that every gas has its own specific spectrum. The FTIR sensor monitors the entire infrared spectrum and reads the different fingerprints of the gases present in the atmosphere.

FTIR spectroscopy is also extensively used in the study of absorption line profiles to determine spectroscopic parameters related to line broadening.

This chapter describes laboratory measurements of line parameters for atmospheric trace species, using Fourier transform spectroscopy in the infrared spectral range. An illustration is given with the example of atmospheric molecules such as NH3, OCS and CH3Br, for which we have determined line intensities, self- and foreign-gas broadening coefficients, pressure-induced line shifts, as well as line-mixing coefficients for these molecules for more than 300 rovibrational lines located in the spectral range 1000-3000 cm^{-1}, at room and low temperatures.

A non-linear least-squares multi-spectrum fitting procedure, including Doppler, and line-mixing effects, has been used to retrieve line parameters from more than twenty experimental spectra, recorded at different pressures of the molecules. The accuracies of the results are also estimated and discussed as a function of the measured parameter.

The results, obtained for several branches of the molecules considered here, will be presented and analyzed as a function of gas pressure, temperature, rotational quantum numbers and vibrational bands.

From the intensity measurements, we have determined effective transition dipole moments, vibrational band strengths as well as Herman–Wallis parameters for some bands taking into account Coriolis and *l*-type interactions.

For the NH_3 molecule, the results concerning line-mixing demonstrate a large amount of coupling between the symmetric and asymmetric components of inversion doublets. The pressure shift coefficients and line-mixing parameters are both positive and negative. Some groups of lines illustrate a correlation between line-mixing and line shift phenomena, demonstrated by a nonlinear pressure dependence of the line position.

Excluding the Introduction and the Conclusion, this chapter is divided into three sections. The FTIR spectroscopy technique is described in detail in section. In particular, the model used to describe the instrument line shape function of the Fourier transform spectrometer is presented. Measurement of line profiles using Fourier transform spectroscopy is presented in section. In this section, the measured broadening and shift coefficients, as well as line intensities and line-mixing parameters are presented and discussed. The rotational dependence of the pressure broadening coefficients is fitted using empirical polynomial equations to provide empirical interpolation and extrapolation models.

Fourier transform spectroscopy

Quantitative absorption spectroscopy

Infrared spectroscopy is the absorption measurement of different IR frequencies by a sample located in the path of an IR beam. Using various sampling accessories, the spectrometers can accept a wide range of sample types such as gases, liquids, and solids.

The basic principle for the evaluation in IR spectroscopy is the Beer-Lambert law which was defined in 1852. This law relates the absorption of light to the properties of the material through which the light is traveling. Absorption spectroscopy consists of the application of this law which illustrates that when traversing a measurement cell, the light intensity decreases exponentially:

$$I = I_0(\sigma)\exp\{-\alpha(\sigma)\ell\}, \tag{1}$$

where $\alpha(\sigma)$ (in cm^{-1}) is the absorption coefficient of the material, σ is the wavenumber (in cm^{-1}), ℓ is the absorption path length (in cm). $I_0(\sigma)$ and $I(\sigma)$ are the intensity of a collimated light beam (in Wm^{-2}) respectively before and after absorption by the sample.

The transmission spectrum $T(\sigma)$ is the ratio of $I(\sigma)$ by $I_0(\sigma)$:

$$T(\sigma) = \frac{I(\sigma)}{I_0(\sigma)}. \tag{2}$$

Eq. (1) can be expressed as a function of the absorption cross section $\sigma_e(\sigma)$ (in cm^{-2} $molecule^{-1}$).

$$I = I_0(v)\exp\{-\sigma_e(\sigma)N\ell\}. \tag{3}$$

N is the particle density (in molecule cm^{-3}). At standard conditions of pressure (P = 1 atm) and the temperature (T = 273.15 K), N is equal to the Loschmidt number[1] (Auwera, 2004):

$$n_L = 2.6867661\ (47) \times 10^{19} \text{ molecule cm}^{-3} \text{ atm}^{-1}.$$

[1] Computed from the value of the molar gas constant *R* available in February 2003 on the website of the Physics Laboratory of the National Institute of Science and Technology (http://physics.nist.gov/).

Fourier transform spectroscopy

Introduction

Since the invention of the first spectrophotometers in the beginning of 20th century, a rapid technological progress has occurred. The first-generation spectrometers were all dispersive with a prism or grating as dispersive elements. In the mid 1960s infrared spectroscopy has seen the advent of the Fourier transform spectrometer. These second-generation infrared spectrometers have significant advantages compared to dispersive spectrometers.

General description

The basic experimental set up of the Fourier transform spectrometer is constituted by the Michelson interferometer. As shown in Fig. 1, light from a source (S) is collimated and then divided at a beam splitter (L) into two beams of equal amplitude. These beams are reflected back on themselves by two separate mirrors, one fixed (M_2) and the other movable (M_1). Each single beam strikes the beam splitter again, where they are recombined and directed to the detector (D). The two components in the recombined beam interfere with each other and form a spot whose intensity depends upon the different paths traversed by the two beams before recombination. As one mirror moves, the path length of one beam changes and the spot on the detector becomes brighter and dimmer successively, in synchronization with the mirror position.

It should be noted that the same types of radiation sources can be used for both dispersive and Fourier transform spectrometers.

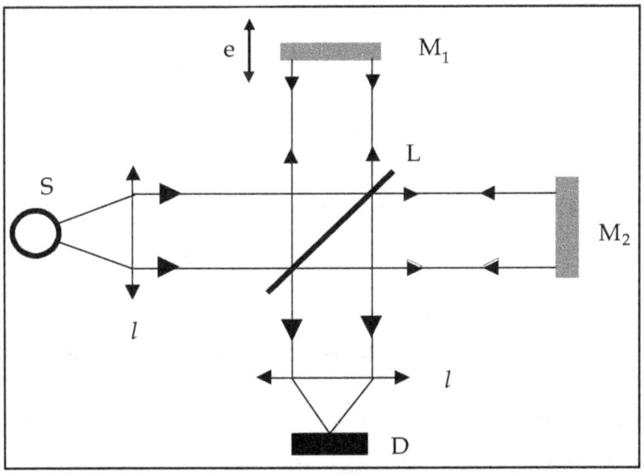

Fig. 1. Michelson interferometer showing three compartments: the radiation source (S), the detector (D), and the interferometer with its two mirrors M_1 and M_2 and its beam splitter L

Mathematic formulation

The interferogram expression

Fourier transform spectroscopy principles are extensively described in the literature (Bell, 1972; Davis et al., 2001; Griffiths & Haseth, 1986). Basically, a Fourier transform spectrometer generates the quantity $I'(\delta)$ which is the cosine Fourier transform of the incident light beam of spectral irradiance $B(\sigma)$ over a range of optical path differences δ defined by the positions of the moving mirror with respect to the fixed mirror (Bell, 1972):

$$I'(\delta) = \int_0^{\infty} B(\sigma)\left[1 + \cos(2\pi\sigma\delta)\right] d\sigma \,. \tag{4}$$

In FTIR spectroscopy, only the modulated part is retained:

$$I(\delta) = \int_0^{\infty} B(\sigma)\cos(2\pi\sigma\delta)\, d\sigma \,. \tag{5}$$

The real quantity $I(\delta)$ is called the interferogram. An example of such interferogram is shown in Fig. 2 for an infrared source, recorded from $-\delta_{max}$ to $+\delta_{max}$, where δ_{max} is the maximum optical path difference achieved.

The spectral distribution (spectrum) can be recorded by computation of the Fourier transform of $I(\delta)$:

$$\tilde{B}(\sigma) = \int_{-\infty}^{+\infty} I(\delta)\exp\{-i2\pi\sigma\delta\}\, d\delta \,. \tag{6}$$

This equation requires that the interferogram is known from $\delta_{max}=-\infty$ up to $\delta_{max}=\infty$. Since $I(\delta)$ can be split into the sum of even and odd functions (Brassewell, 1965):

$$I(\delta) = \frac{1}{2}\left[I(\delta) + I(-\delta)\right] + \frac{1}{2}\left[I(\delta) - I(-\delta)\right] = E(\delta) + O(\delta) \,. \tag{7}$$

Eq. (6) can be written as:

$$\tilde{B}(\sigma) = 2\int_0^{\infty} E(\delta)\cos(2\pi\sigma\delta)\, d\delta - 2i\int_0^{\infty} O(\delta)\sin(2\pi\sigma\delta)\, d\delta \,. \tag{8}$$

This equation shows that the observed spectrum $B(\sigma)$ is real and even, if the interferogram is an even function. Only the side of the interferogram corresponding to $\delta \geq 0$ is recorded; the spectrum is then computed using the cosine Fourier transformation only.

Instrumental line shape function

When the spectral analysis takes into account experimental parameters such as the truncation of the interferogram, the finite size of the source, phase errors, optical misalignments, etc., an instrumental line shape function, $F_{App}(\sigma)$, is introduced, and the recorded spectrum can be written as:

$$\tilde{B}(\sigma) = B(\sigma) \otimes F_{App}(\sigma) \tag{9}$$

The symbol $\otimes$ means the convolution of $B(\sigma)$ with $F_{App}(\sigma)$. In the following, we discuss the effect of $F_{App}(\sigma)$ on the recorded spectra.

Truncation of the interferogram

Mathematically, the truncation of the interferogram recorded up to a finite maximum optical path difference δ_{max} yields:

$$F_{App}(\sigma) = 2\delta_{max}\,\text{sinc}(2\pi\sigma\delta_{max}), \tag{10}$$

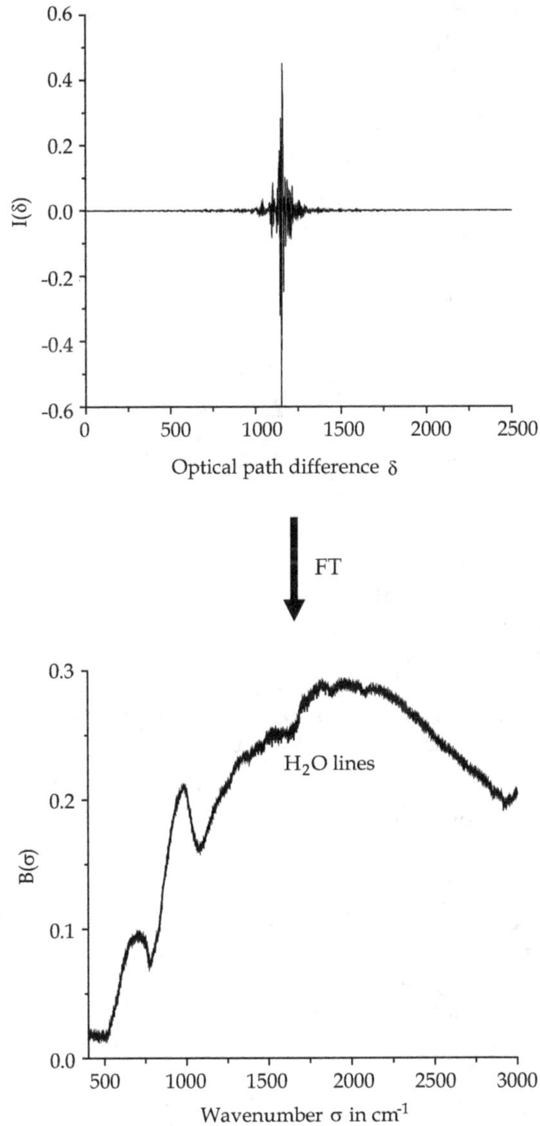

Fig. 2. Example of an interferogram and its spectrum calculated using a Fourier Transform algorithm

where sinc(x)=sin(x)/x. The full width at half maximum (FWHM) of this function defines the theoretical maximum spectral resolution R_{th} that can be achieved with a Fourier transform spectrometer (Bell, 1972):

$$R_{th} = \frac{1.2067}{2\delta_{max}}. \tag{11}$$

The sinc(x) function exhibits large oscillations: the amplitude of the first lobe is equal to 22% of the maximum amplitude of the function. These oscillations can be rather annoying for the spectral analysis since they lead to distortion of spectral lines. They can be reduced or eliminated through multiplication of the interferogram I(δ) with a so-called apodization function which decrease with increasing optical path difference, thus reducing the truncation effect of the interferogram at δ_{max}.

A typical apodization function is a triangle function whose Fourier transform is the $sinc^2(2\pi\delta_{max}\sigma)$ function. However, because R_{th} is inversely proportional to δ_{max}, apodization has the disadvantage of decreasing the spectral resolution. Then the apodized maximum spectral resolution achievable by a Fourier transform spectrometer is equal to the full width at half maximum of the $sinc^2(x)$ function.

$$R_{th} = \frac{1.79}{2\delta_{max}} \tag{12}$$

This definition of the resolution is used for Bruker FTIR spectrometers. For the Bomem instruments, $R = 0.5/\delta_{max}$ is used.

Detector non-linearity

For line intensity measurements, the major distortion of spectra may be due to the nonlinearities of the detector (MCT detectors are nonlinear). To remove these nonlinearities, some studies have been performed (Abrams et al., 1994; Auwera, 2004; Guelachvili, 1986). Abram et al. showed that the recorded interferogram can be described as power series of the true interferogram (Abrams et al., 1994). Later Auwera developed an algorithm based on Abram's work to avoid nonlinearities (Auwera, 2004).

This phenomenon is manifested when the detector is illuminated by a relatively intense source with large spectral coverage. This nonlinearity affects the interferograms near the zero optical path difference where the signal is the largest. Such behaviour affects consequently the entire spectrum.

Note that, using the sensitivity curve of the detector, one should be sure to remain permanently in the interval of linearity of detection by reducing the most possible the explored spectral region and by limiting the intensity of the source to avoid the effects of saturation of the detector. Since the true physical measured quantity is the interferogram, for broad band measurement in absorption spectroscopy, these effects will affect more strongly the center burst of the interferogramm, reducing artificially its amplitude where the signal is largest. Thus, one of the experimental signs signaling the presence of gross non-linearity effects is the appearance, after Fourier Transform, of an artefactual signal below the detector or beam-splitter cut-off (see for instance the curve below of Figure 2, where no signal should be present below 500 cm^{-1}).

Extended size of the source

The collimated beam inside the Michelson interferometer is produced using an extended source at the focus of a lens. This beam passes through an iris occulting part of the image of the source. Because of the finite size of this iris, the collimated beam is slightly divergent in the interferometer. When this effect is taken into account, the interferogram $I(\delta) = B\cos(2\pi\sigma_0\delta)$ of a monochromatic source at wavenumber σ_0 becomes

$$I(\delta) = B\frac{\pi d^2}{4F^2}\,\mathrm{sinc}\left(\frac{\pi d^2\sigma_0\delta}{8F^2}\right)\cos\left(2\pi\sigma_0\delta\left[1-\frac{d^2}{16F^2}\right]\right). \tag{13}$$

In this expression, d is the diameter of the iris and F the focal length of the collimating optic. This equation shows that the beam divergence has two effects:

- a change on the optical path difference $\delta' = \delta\left(1-\frac{d^2}{16F^2}\right)$. The first effect is easy to account for by using δ' in all calculations.
- a multiplication of the interferogram by a sinc(x) function at the wavenumber σ_0 of the source. This leads to a convolution of the spectrum with a box of width $\sigma_0 d^2/(8F^2)$.

Sampled interferogram and spectrum

First the interferogram is sampled at regular intervals as the moving mirror traverses the optical path difference δ. To avoid the loss of the signal information, for a spectrum with a $\Delta\sigma$ extend, the sampling step $\Delta\sigma$ must satisfy the Nyquist condition:

$$\Delta\delta < \frac{1}{2\Delta\sigma}. \tag{14}$$

Thus the continuous interferogram is multiplied by a sampling comb, modelled theoretically by a comb of delta functions. Sampling the interferogram has obvious effects such as replacing the integrals of Eqs. (6) and (8) by sums over the samples. The number of sample points in the spectrum is equal to $2\delta_{max}\Delta\sigma+1$.

Practically in the spectrum the wavenumber interval between two points is equal to $\delta\sigma = \frac{1}{2\delta_{max}}$. Thus the spectral resolution is chosen such that there are at least two sample points per line width.

Zero filling

Zero filling is a data processing technique where zero points are added to the end of the interferogram before the digital Fourier transformation. It is the process of interpolating extra data points into a spectrum so that the spectral lines have a smoother shape with a better digital resolution, using the same Fourier coefficients. FTIR software automatically provides zero-filling by extending the length of an interferogram with a zero straight line. However there is no new information added to the spectrum.

Generally, the original interferogram size should always be at least doubled by zero filling, i.e. zero filling factor (ZFF) of two is chosen. Zero-filling is an interpolation that does not affect the instrument line-shape, and in most cases, is therefore superior to polynomial or spline interpolation methods that are applied in the spectral domain.

Phase errors

In Fourier transform spectroscopy interferograms are apodized prior to the transformation and the calculated spectrum is phase corrected. If the zero point for the Fourier transformation is displaced, a phase error occurs because the apodization remains fixed while the interferogram is shifted by the phase correction. To avoid distortions of the spectrum resulting from phase errors, double-sided interferograms are needed. In practice, single-sided interferograms are recorded double-sided on a small range of optical path difference, allowing determination of the phase error. In such a case, the phase error is determined at low resolution, typically about 1 cm^{-1} or less. This is not a problem because it varies only slowly with wavenumber.

FTIR advantages

Compared with the dispersive spectrometer, the FTIR instrument has several advantages:

- Better speed and sensitivity: A complete spectrum is obtained during a single scan of the moving mirror, while the detector observes all frequencies simultaneously. This is called the "Felgett advantage".
- The so-called "Jaquinot advantage" consists of an increase of the optical throughput. A circular optical aperture is used in FTIR systems. The beam area of a Fourier transform instrument is usually about 100 times larger than the small slit width of a dispersive spectrometer. Thus, more radiation energy is made available. This constitutes a major advantage for many samples that are energy-limited.
- The so-called "Connes advantage" consists of the use of a helium neon laser as the internal reference. This source radiation provides an automatic and stable calibration for all wavelengths. This eliminates, to some extent, the need for external calibration sources.
- FTIR spectrometers are usually equipped with a powerful, computerized data system. It can perform a wide variety of data processing tasks such as Fourier transformation, interactive spectral subtraction, baseline correction, smoothing, integration, spectral line fits and library searching. Development of the fast Fourier transform algorithm (Cooley and Tukey, 1965) has facilitated these tasks and reduced the computation time.

Measurement of line profiles using Fourier transform infrared spectroscopy

Introduction

Analysis of line profiles are useful for atmosphere monitoring and studying the evolution of its composition including chemical trace species related to the reduction in stratospheric ozone, starting from the analysis of observations obtained by infrared spectroscopy. The retrieved spectroscopic parameters are indispensable for the interpretation of high-resolution infrared spectra of gaseous species, and are of increasing importance for qualitative and quantitative applications of spectroscopy. Also the pressure effects of gases on spectral lines are of significant interest in modeling and interpretation of radiative transfer calculations in climate models. In this context, inversion of atmospheric spectra to determine column densities of trace atmospheric species requires accurate knowledge of spectroscopic parameters. The needed parameters are, among others, line positions, line intensities, lower state transition energies and line widths as a function of temperature and quantum numbers.

However, spectroscopic information contained in the databases such as HITRAN09 (Rothman et al., 2009), GEISA (Jacquinet-Husson et al., 2005), JPL (Pickett et al., 1998) and the more specialized MASTER (Perrin et al., 2005), are often still partly incomplete, e.g. for some species or transitions, experimental or theoretical line pressure shift and line-mixing parameters are missing. For this reason, Fourier transform spectroscopy is extensively used to generate these parameters. Compared to diode laser spectrometers, FT spectrometers have the advantage of recording spectra over a large frequency range.

In this context, we present in the following Fourier transform measurements of NH_3, CH_3Br and OCS line profile parameters (line intensity, line width, line-mixing and line shift).

When the gas pressure increases, inelastic collisions transfer populations from one radiative state to another. At low pressure, the lines are isolated; this effect gives a line broadening and shift proportional to the pressure. At high pressure the lines are overlapped, interference effects occur between adjacent lines. The resulting profile can no longer be described by the superposition of Lorentzian profiles. In this case, in addition to the previous parameters, one must determine the line-mixing coefficients.

A large amount of laboratory data using various spectroscopic techniques of increasing precision has been assembled over the last two decades. Among the various molecular species, ammonia (NH_3) is probably one of the most investigated species. This molecule has always played an important spectroscopic role because of its inversion spectrum and its large molecular dipole, which are important properties for theoretical models. Moreover, ammonia is present in many planetary atmospheres (Ho & Townes, 1983; Kunde et al., 1982). It is also commonly used as an interstellar thermometer and is considered as an industrial and biological pollutant (Brassington, 1988).

The methyl halides (i.e. CH_3Cl, CH_3Br, and CH_3I) have been the subject of very detailed studies of the Earth's atmosphere. They are atmospheric components which take part in atmospheric photochemical reactions. Methyl bromide (CH_3Br) is a halogen chemical compound used in agriculture as a plant fungicide, and automobile using leaded petrol (Thomas et al., 1977).

This molecule significantly contributes to ozone depletion since CH_3Br is dissociated by UV radiation, producing Br radicals that catalyze the destruction of ozone (McElroy et al., 1986). These bromide atoms are far more destructive of ozone than the chlorine atoms coming from the chlorofluorocarbons compounds (CFC) (Kurylo & Rodriguez, 1998). For this reason, since 2005 the use of CH_3Br has been prevented under the Montreal protocol. Despite the important role of CH_3Br in the atmosphere, no spectroscopic data on CH_3Br is available in the main databases (Jacquinet-Husson et al., 2005; Perrin et al., 2005; Pickett et al., 1998; Rothman et al., 2009).

Carbonyl sulfide (OCS) is one of the principal and most long-lived reservoirs of sulfur in the Earth's troposphere (Watts, 2000) and has been detected in different astrophysical objects (the atmosphere of Venus (Bezard et al., 1990), the Orion molecular clouds (Evans et al., 1991), the comets Hyakutake and Hale-Bopp (Dello Russo et al., 1998; Woodney et al., 1997), and the starburst galaxies (Martin et al., 2005; Mauersberger et al., 1995)).

Experimental set-up

The results presented and discussed in the following are obtained using spectra recorded with the high resolution Bruker IFS125HR Fourier transform spectrometer (an upgraded version of the previous version, IFS120HR) located at the LISA facility, in Créteil. A general view of the laboratory is presented in Fig. 3. It shows the Fourier transform spectrometer Bruker IFS125HR, two multiple-reflections (white-type) Pyrex absorption cells and a gas handling system. The Fourier transform spectrometer is characterized by a maximum optical path difference (MOPD) of up to 473.68 cm (maximum resolution = 0.0019 cm^{-1}).

The instrument has all the equipment and accessories required to operate from the far infrared (20 cm^{-1}, 500 microns) to the ultraviolet region (45000 cm^{-1}).

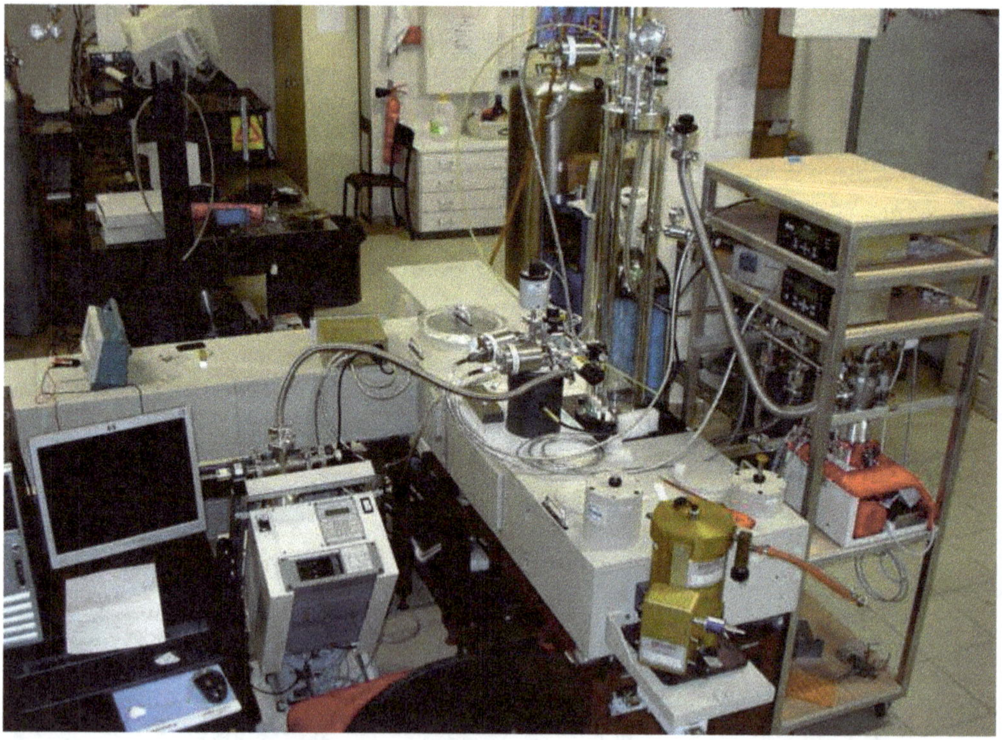

Fig. 3. View of the laboratory showing the Fourier transform spectrometer Bruker IFS125HR, two multiple-reflections Pyrex absorption cells and a gas handling system including a turbomolecular pump. In the back is an optical table with a difference frequency generation (DFG) laser system developed at LPPM in Orsay and LISA in Creteil

The two multiple reflections cells are of the so-called White-type. They consist of a Pyrex tube containing the mirrors separated by 20 and 80 cm, respectively, thus providing base lengths of 0.849 and 3.249 m (this value takes into account the distance between the surface of the field mirror and the windows of the cell (2 × 2.45 cm)). The largest path length that one can achieve is 7.249 m with the short cell and 32.049 m with the long one. The temperature of the sample in these cells is well defined by an air-conditioning system regulating the temperature of the room, with an uncertainty of ± 1 K.

The gas handling system is made of stainless steel tubing and can be evacuated to less than 10^{-6} mbar by a Pfeiffer Turbomolecular Pumps. Commercial gas lecture bottles with pure samples or buffer gases like N_2, He, Ne, Ar etc. can be connected to the system to allow transfer of the gases to be studied into the absorption cells.

Fig. 4 presents the optical design of the Fourier transform spectrometer Bruker IFS 125HR. This instrument is a fast-scanning, asymmetrical Michelson interferometer, characterized by a resolving power better than one million. Four different parts can be distinguished: the source compartment (bottom right), the interferometer itself (above the sources), the sample position (where short path absorption cells are placed) and detector compartment. The whole optical path is maintained under vacuum (pressure below 0.06 mbar). This instrument provides three different sources, a high-pressure Hg lamp for the far-infrared, a globar (Silicon Carbide, SiC) heated to about 1100 K for the mid-infrared, and a tungsten lamp for the near-infrared and visible regions. An external light beam can also enter the instrument through the input port. The light is focused onto an iris, whose size ranges from 0.5 to 12 mm (aperture changer).

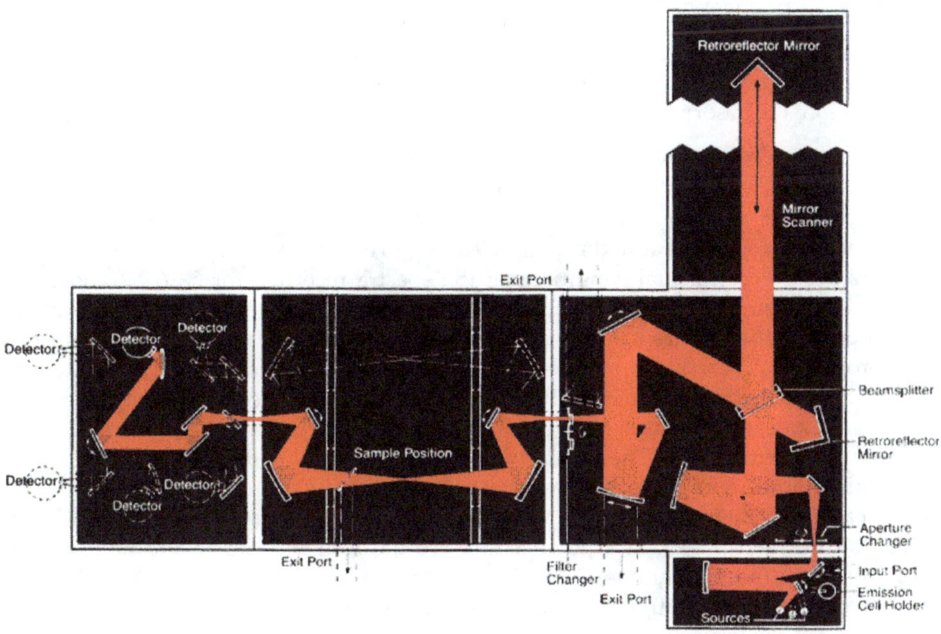

Fig. 4. Optical scheme of the Bruker IFS125HR Fourier transform spectrometer

The light is then collimated into a 7 cm diameter beam, sent into the Michelson interferometer. Various beam splitters, made of Mylar, KBr/Ge, CaF_2/Si and Quartz/TiO_2 are available, and used for different spectral regions. Exiting from the interferometer, the light is focused through an optical filter and then through the secondary iris. The beam then goes through the sample. Finally, the beam reaches the detectors compartment. We use four detectors, a Si bolometer operating at 4.2 K, a HgCdTe (MCT) detector, and an InSb detector, both cooled to 77 K, and a Si photodiode at room temperature.

For the studies of the OCS molecule, this spectrometer was equipped with a KBr/Ge beamsplitter, Globar source (Silicon Carbide, SiC), an InSb detector cooled at 77 K, and optical and electronic filters covering the spectral region 1850 - 2150 cm^{-1}.

For NH_3 and CH_3Br spectra in the region 1000-1800 cm^{-1}, we used also a liquid nitrogen-cooled HgCdTe detector. These spectra were recorded with an aperture diameter of 1.15 mm, 40 kHz scanner frequency, and a maximum optical path difference MOPD = 225 cm for OCS and NH_3, and 450 cm for CH_3Br. According to the Bruker definition this corresponds to a resolution of 0.004 cm^{-1} for OCS and NH_3 spectra and 0.002 cm^{-1} for CH_3Br.

A White-type multipass absorption cell, made of Pyrex glass and equipped with CsBr windows, was used to record OCS spectra broadened by O_2 and N_2. The cell is in thermic equilibrium with the air conditioned room with a temperature stabilized at 295 K. For NH_3 self-broadening experiments the gas was contained in a stainless steel gas cell equipped with CaF_2 windows with a path-length of 2.5 cm. For NH_3-H_2 collisions, the gas mixture was contained in a Pyrex cell, with a path length of 15 cm, equipped with ZnSe windows. For CH_3Br spectra, a multipass cell, of 1m base length equipped with KCl windows, was used for a total absorption path of 415 cm.

For self-broadening studies, the cell was filled with an increasing pressure of the active gas. For foreign gas broadenings, the cell was first filled with the active gas, then, after thermalization, the absorption cell was filled with an increasing pressure of perturber gas.

The following procedure was used for measurements: First a background spectrum was collected while the cell was being continuously evacuated. Next, the cell was filled with the considered gas at an increasing pressure. For OCS or NH_3 perturbed by N_2 or H_2, the pressure of these gases was added in stages leading to a series of at least 6 pressures. Table 1 summarizes the experimental conditions of the spectra recorded for OCS-N_2 and OCS-O_2 collisions. The number of spectra for each gas is chosen in order to have sufficient information on all line parameters when analyzing a transition using the multi-spectrum fitting procedure. The sample pressure in the cell was measured using a calibrated MKS Baratron capacitance manometers (2, 10, 100, 1000 Torr full scale) each of them characterized by its stated uncertainty according to the manufacturer. For all these manometers the uncertainty is less than 1 %. The spectra were recorded at a stabilized room temperature of 295 K with an uncertainty of ± 1 K. All spectra were ratioed against the empty cell, single-channel background spectrum which was taken in order to ensure the best possible signal-to-noise in the ratioed spectra. The spectra were the result of the co-addition of a large number of scans (interferograms). For CH_3Br spectra, the number of scans was 200. Every scan has been individually transformed to spectrum using the Fourier transform procedure included in the Bruker software OPUS package (Wartewig, 2003), selecting a Mertz phase error correction (Griffiths & Haseth, 1986; Mertz, 1965).

Optical path (m)		3.249	
Resolution (cm^{-1})		0.004	
Maximum optical path difference (cm)		225	
Collimator focal length (mm)		418	
Aperture diameter (mm)		1.15	
Useful spectral domain (cm^{-1})		1850-2150	

Spectrum number	**$OCS-N_2$**		**$OCS-O_2$**	
	OCS pressure (Torr)[a]	**N_2 pressure (Torr)[a]**	**OCS pressure (Torr)[a]**	**O_2 pressure (Torr)[a]**
1	0.0859 (1)	7.74 (2)	0.0754 (1)	5.80 (1)
2	0.0859 (1)	15.05 (4)	0.0754 (1)	13.98 (3)
3	0.0859 (1)	25.53 (6)	0.0754 (1)	22.09 (6)
4	0.0859 (1)	34.89 (9)	0.0754 (1)	32.41 (8)
5	0.0859 (1)	46.90 (1)	0.0754 (1)	51.80 (1)
6	0.0859 (1)	60.00 (2)	0.0754 (1)	83.20 (2)

[a] The values given in parentheses correspond to the estimated errors.

Table 1. Experimental data

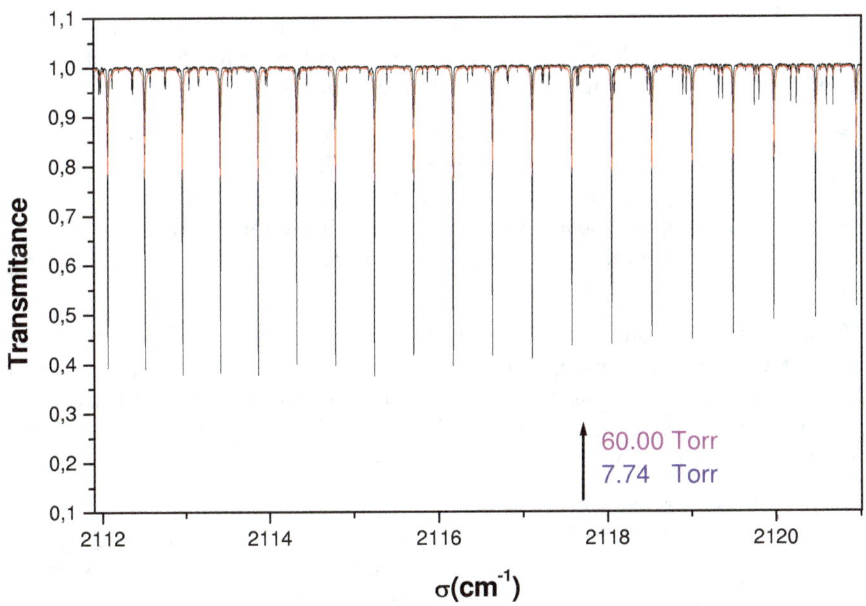

Fig. 5. Transmittance spectrum of OCS around 2116 cm^{-1} showing some lines of the R branch of $4\nu_2$ band broadened by N_2. The pressures are 0.086 Torr of OCS, and 7.74 and 60.00 Torr of N_2

The OCS spectra were calibrated with residual CO_2 and H_2O lines observed in the spectra. For CH_3Br the calibration was done using the low pressure ν_2 band spectrum of NH_3.

Fitting procedure

For OCS and NH_3 spectra, the analyses have been done taking into account interference effects. Thus within the impact theory of the spectral line shape and for moderately a overlapping line at low pressure considered for these molecules, the collisional absorption coefficient $\alpha(\sigma)$ could be written as (Gentry & Larrabee Strow, 1997; Lévy et al., 1992; Pine, 1997; Rosenkranz, 1975; Thibault et al., 1992):

$$\alpha(\sigma) = \frac{P_{NH_3}}{\pi} \sum_{\text{lines } k} S_k \frac{PY_k(\sigma - \sigma_k) + P\gamma_k}{(\sigma - \sigma_k)^2 + (P\gamma_k)^2}, \tag{15}$$

where k represents the line $v_iJ_iK_i \rightarrow v_fJ_fK_f$, S_k its intensity, σ_k its wavenumber, γ_k its halfwidth and Y_k its line mixing parameter related to the off diagonal element of the relaxation matrix.

Spectra were analyzed by means of nonlinear least squares fitting procedures, using the following theoretical expression $\tau^C(\sigma)$ for the transmission and the collisional absorption coefficient in Eq. (15):

$$\tau^C(\sigma) = \int_{-\infty}^{+\infty} F_{App}(\sigma - \sigma') \times \exp\left[-\ell \times \int_{-\infty}^{+\infty} \alpha_{Dop}(\sigma' - \sigma'') \times \alpha(\sigma'') \times d\sigma''\right] \times d\sigma', \tag{16}$$

where α_{Dop} is the Doppler profile, F_{App} is the Fourier transform instrument function with a gaussian shape and ℓ is the cell length. Eq. (16) is a convolution of the collisional absorption coefficient with the instrument function and the Doppler profile.

Otherwise the collisional parameters for a given temperature have been deduced by means of nonlinear least square multipressure fitting in which all spectra at various pressures are successively adjusted using Eq. (16). For a line k, the parameters deduced from the fits are $P\gamma_k$, σ_k, S_k and PY_k. These spectroscopic parameters have been iteratively varied by a fitting program to obtain the best agreement between the experimental and calculated absorption coefficients.

Because of the spectral density for the bands of the molecules considered here and the strong values of the self-broadening coefficient, a line by line study could not have been possible with the relatively high pressure, so the line parameters were derived using the multi-spectrum fitting method applied to the measured shapes of the lines, including the interference effects caused by the line overlaps.

An example of a multi-spectrum fit in the case of ${}^{P}{}_{a}P(8,7)$ and the two components of the ${}^{P}{}_{as}P(8,8)$ doublet all pertaining to the ${}^{P}P$ branch of the ν_4 band is given in Fig. 6 with the corresponding pressures. Plots (a) and (b) show that the overlapping increases with NH_3 pressure. They also demonstrate clearly the influence of line-mixing in the overlapped contours as illustrated by large discrepancies obtained when this process is disregarded in Fig. 6(a).

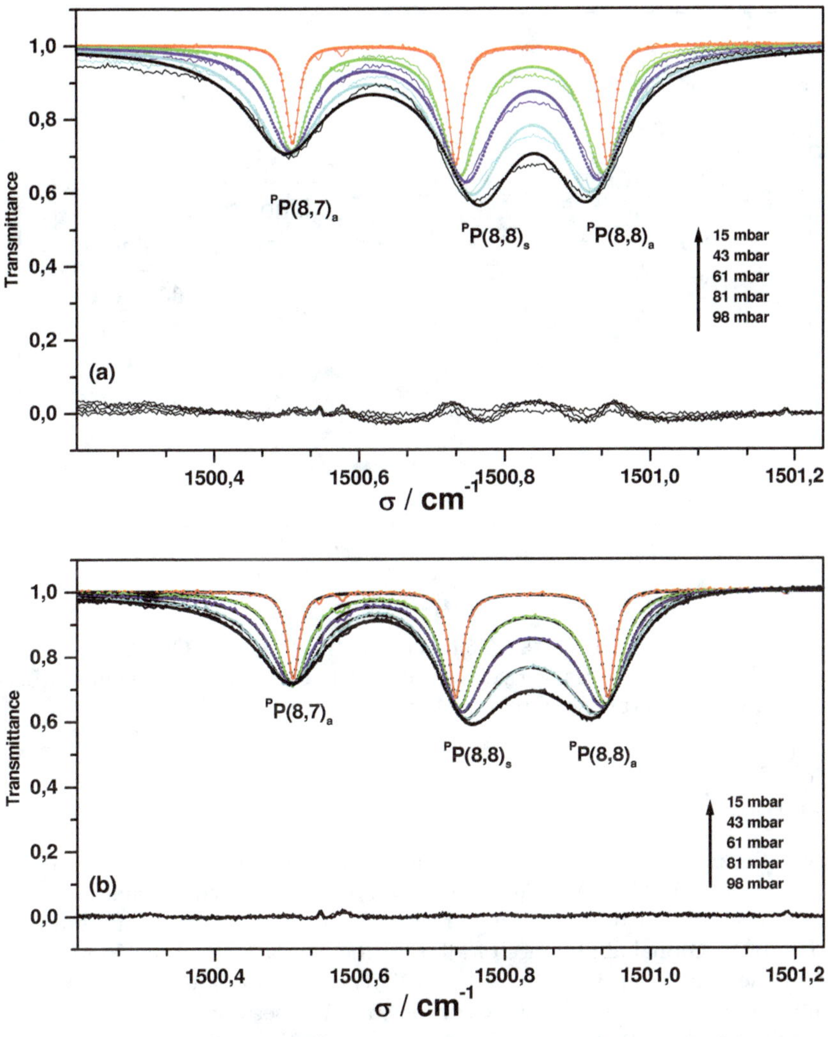

Fig. 6. Graphic demonstration of line-mixing obtained for pure NH_3 spectra for $^{P}{}_{a}P(8,7)$ line and $^{P}{}_{as}P(8,8)$ inversion doublet of the ν_4 band at various pressures. (—) and (•) are the measured and calculated values, respectively. Measured minus calculated deviations are shown in the lower part of graphs. Plots (a) and (b) have been obtained without (Y=0) and with (Y≠0) the line-mixing, respectively. These plots illustrate the shift of $^{P}{}_{as}P(8,8)$ components line center towards each other

Fig. 7 shows the results of multi-pressure spectrum fits for the $^{Q}P_{79}(26,E,5)$ line in the ν_2 band of CH_3Br, self-broadened at 296 K at different pressures. In this figure symbols and (–) are, respectively, the calculated and measured values. This line profile was calculated using a Voigt function. Indeed, no characteristic signature due to the presence of collisional narrowing or line-mixing has been observed in any residual.

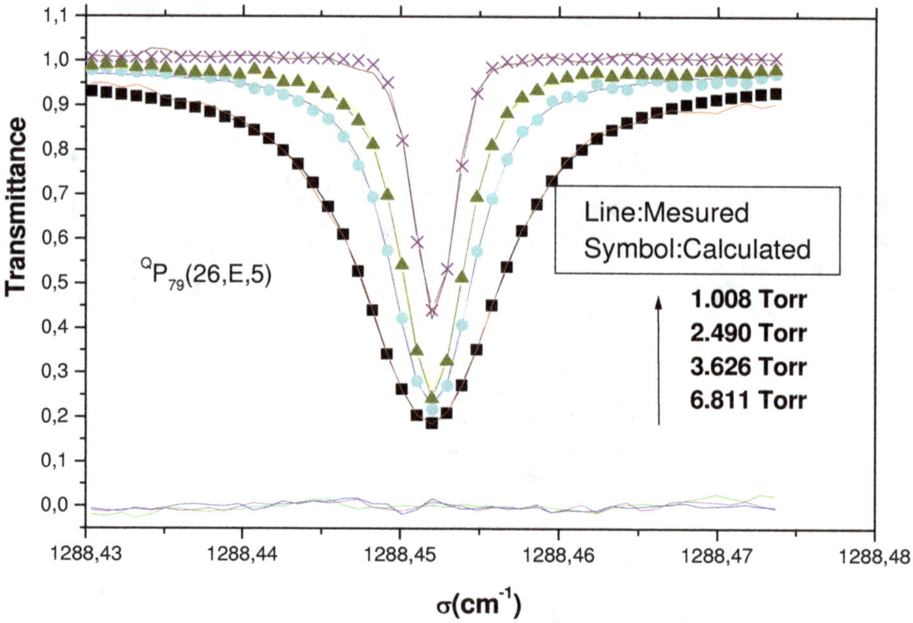

Fig. 7. Results of spectrum fits for the $^{Q}P_{79}(26,E,5)$ line in the ν_2 band of CH_3Br self-broadened at 296 K. (symbol) and (—) are, respectively, the calculated and measured values. Measured minus calculated deviations are shown in the lower part of the graph

Line intensities

Line intensity results

The line intensities presented here are for NH_3 in ν_2 and ν_4 band spectra. As shown in Fig. 8 for the $^{P}{}_{s}P(2,1)$, $^{P}{}_{s}P(3,1)$, $^{P}{}_{s}P(4,1)$, $^{P}{}_{s}P(6,1)$ and $^{P}{}_{s}P(7,1)$ lines of the ν_4 band, the intensity parameter S (cm^{-2}) derived from the fits is proportional to the real NH_3 pressure $P' = P \times 0.985$ (0.985 is the fractional abundance of NH_3 for the gas sample). The $^{P}{}_{s}P(5,1)$ line which has practically the same intensity value as the $^{P}{}_{s}P(3,1)$ line is not plotted. The line intensities S_0 ($cm^{-2}\ atm^{-1}$) are deduced from the slopes of the straight lines obtained from a linear least-squares procedure and the length $\ell = 2.5$ cm of the absorption cell. The results are listed in Refs. (Aroui et al., 2003; Hadded et al., 2001) for symmetric and asymmetric transitions for the ν_2 and ν_4 bands of NH_3.

For the two bands, the intensities of transitions with a nuclear spin statistical weight of the lower level $g_s = 2$ for $K = 3n$ ($n = 1, 2,...$) are larger than those with $g_s = 1$. S_0 increases with K for a given J when dividing by 2 the intensities of the lines with $g_s = 2$. However, for the ν_2 band when K is close or equal to J, line intensities exhibit a slight decrease. On the other hand, for a fixed value of K, S_0 decreases significantly as J increases. Otherwise the intensities of the asymmetric transitions are slightly larger than those of symmetric transitions. In the ^{P}P branch of the ν_4 band, line intensities also increase with K for a given J, but, as shown in Fig. 8, the evolution with J presents a maximum for J = 4 for the $^{P}P(J,K=1)$ lines and decrease monotonically for $K \geq 2$.

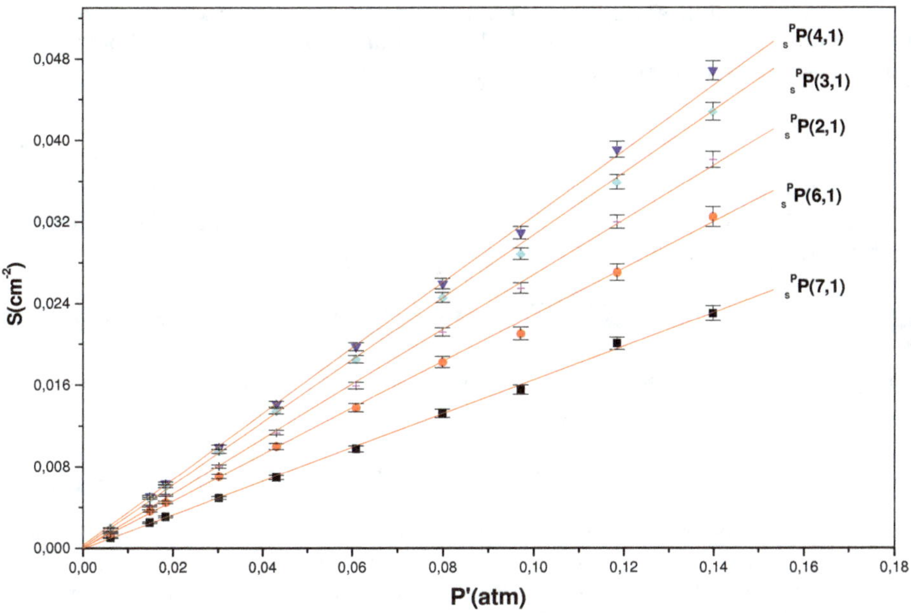

Fig. 8. Pressure dependence of the intensity parameter S for the ${}^P{}_sP(2,1)$, ${}^P{}_sP(3,1)$, ${}^P{}_sP(4,1)$, ${}^P{}_sP(6,1)$ and ${}^P{}_sP(7,1)$ lines in the ν_4 band of NH_3. The slopes of the best fit line represent the line intensity

As expected the line intensities in the ν_2 band (Aroui et al., 2003) are much larger than those reported previously (Aroui et al., 1998; Cottaz et al., 2001; Hadded et al., 2001) for the ν_4 band and also several times larger than the intensities in the ν_1 and ν_3 bands (Markov, 1993).

Vibrational transition moments and correction function

For NH_3 molecule, the effective transition dipole moments $R_{eff}(J,K)$ in the ν_2 band are derived from the line intensities S_0 such as (Aliev et al., 1987):

$$R_{eff}^2(J,K) = \frac{S_0 TQ \ \exp(E(J,K)/kT)}{3054.8 \ g_i\sigma \ \left[1-\exp(h\sigma/kT)\right] \ H_{JK}}, \tag{17}$$

where H_{JK} is the Höln-London factor (Aroui et al., 2003). σ is the wave number of the line under study. E(J,K) is the rotational energies of the lower level of the transitions. $Q = Q_rQ_v$ is the total partition function, with the rotational partition function Q_r = 563.56 at T = 295 K and the vibrational partition function Q_v = 1.011.

The effective transition dipole moments R_{eff}, determined from the observed line intensities using Eq. (17) show significant rotational dependencies (Aroui et al., 1998, 2003) which are mainly caused by vibration-rotation interactions (*l*-type interactions in the ν_4 band and Coriolis interaction between ν_4 and $2\nu_2$ bands): they decrease with J for a fixed value of K and increase with K for a given J.

For the interpretation of these dependencies, the zero-order theory, where the correction for vibration-rotation interactions is not included, cannot describe the transition dipole moment variations. By taking in to account the effects of these interactions, one can write

$$R_{eff}^2(J,K) = \left(R_0^v\right)^2 F_{HW}, \tag{18}$$

where R_0^v is the vibrational transition moment, and F_{HW} is the Herman-Wallis correction function describing the dependence of R_{eff} on J and K. For the ν_2 parallel band, for example, using the second-order development of Watson (Watson, 1992) we may write,

$$F_{HW} = \left\{1 + A^J m_J + A^K m_K + A^{JJ(Q)}\left[\overline{J(J+1)} - m_J^2\right] + A^{JJ(PR)} m_J^2 + A^{KK}\overline{K^2} + A^{JK} m_J m_K\right\}^2, \tag{19}$$

where the A's are correction parameters. For the R-transitions of this band, we have $m_J = J+1$, $m_K = 0$, $\overline{J(J+1)} = (J+1)^2$ and $\overline{K^2} = K^2$, thus we can write F_{HW} in the form:

$$F_{HW} = \left[1 + A^J(J+1) \ + \ A^{JJ(R)}(J+1)^2 \ + \ A^{KK}K^2\right]^2. \tag{20}$$

The correction parameters A^J, $A^{JJ(R)}$ and A^{KK} as well as the vibrational transition moment R_0^v, were determined, using linear least-squares fits, from the experimental values of effective transition moments. The resulting correction parameters and vibrational transition moments for the ν_2 band are listed in Table 2 along with the standard deviation of the fits (Aroui et al., 2003). In this table, the A^J and $A^{JJ(R)}$ parameters cannot be determined for ν_{2s}, since, for this partial band, the fits were performed only for 7 transitions pertaining to the $J = 10$ manifold, whereas for asymmetric ν_{2a} partial band the fits were performed for 62 transitions.

For the ν_4 band, the correction function, the vibrational transition moments and the correction parameters are reported previously (Aroui et al., 1998). For this band significant differences between the correction parameters obtained for the symmetric and asymmetric transitions, so one can determine the vibrational band strengths, S_{0S}^V and S_{0a}^V respectively, for the symmetric and asymmetric partial bands ν_{4s} and ν_{4a} of the ν_4 band. The values are reported in previous work (Aroui et al., 1998).

Parameter	s←a transitions	a←s transitions
R_0^v (D)	0.24604 ± 0.00209	0.23868 ± 0.00154
A^J	$(-3.88 \pm 1.47)\times10^{-3}$	—
$A^{JJ(R)}$	$(-1.57 \pm 0.72)\times10^{-4}$	—
A^{KK}	$(7.66 \pm 0.62)\times10^{-4}$	$(5.03 \pm 0.37)\times10^{-4}$

Table 2. Vibrational transition moments and Herman-Wallis correction parameters for the symmetric and asymmetric transitions of the ν_2 band of NH_3 (Aroui et al., 2003)

Pressure broadenings

Pressure broadening coefficients

For NH_3-H_2 collisions, a typical plot of $P\gamma$ versus the pressure P of hydrogen is shown in Fig. 9 for the ${}^P{}_sP(3,1)$ line of NH_3 in the ν_4 band at the temperatures 235 and 296 K. The gases were contained in a 15 cm path-length Pyrex cell. One can observe that straight line fits go through the measured points very well. The pressure broadening coefficients γ for each temperature are determined as the slopes of the best-fit lines. This figure illustrates an increase of line width with H_2 pressure but a decrease of broadening coefficient as function of temperature (Nouri, 2004). This dependence is usually well represented by the simple power law

$$\gamma = \gamma(T_0)\left(\frac{T_0}{T}\right)^n, \tag{21}$$

where $\gamma(T_0)$ is the broadening coefficient at the reference temperature T_0 = 296 K. From the measured values of γ at the considered temperatures, one can determine the values of exponent n as the slope of the graphs of $ln\gamma(T)$ versus lnT. The straight lines obtained for all transitions validate Eq. (21) within the indicated range of T. The derived values of n are given in Ref. (Nouri, 2004) for the ^PP, ^RP, ^PR, ^RQ, ^RR, and ^PQ branches in the ν_4 band and for R branch in the $2\nu_2$ band of NH_3.

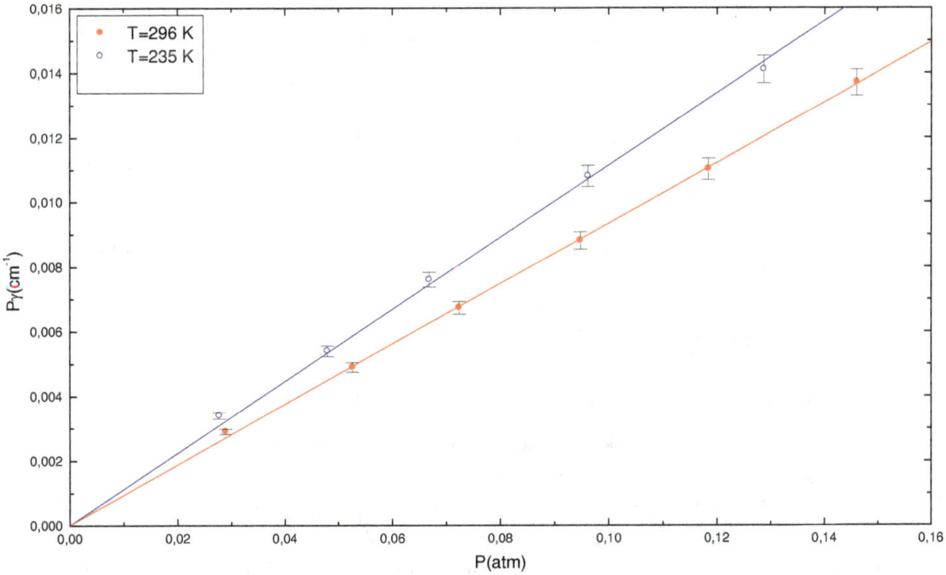

Fig. 9. Pressure dependence of the broadening line width $P\gamma$ for the $^P{}_sP(3,1)$ line of the ν_4 band perturbed by H_2; (o) T=235 K ; (•) T=296 K. The slope of the best fit line represents the broadening coefficient

As shown previously, distinctive dependence on the rotational quantum number m can be observed for the molecules considered here (Aroui et al., 2003, 2009; Bouanich et al., 1986; Cottaz et al., 2001; Domenech et al., 2000; Hadded et al., 2001; Koshelev et al., 2006, 2009; Markov et al., 1993; Mouchet et al., 1985). For OCS-N_2 and OCS-O_2 systems, Fig. 10 is a plot of the measured broadening coefficients versus $|m|$ ($m = -J$ for $P(J)$ lines and $m = J+1$ for $R(J)$ lines) for the P and R branches of the $4\nu_2$ band (Galalou et al., 2011). It shows the decrease of these coefficients with increasing $|m|$.

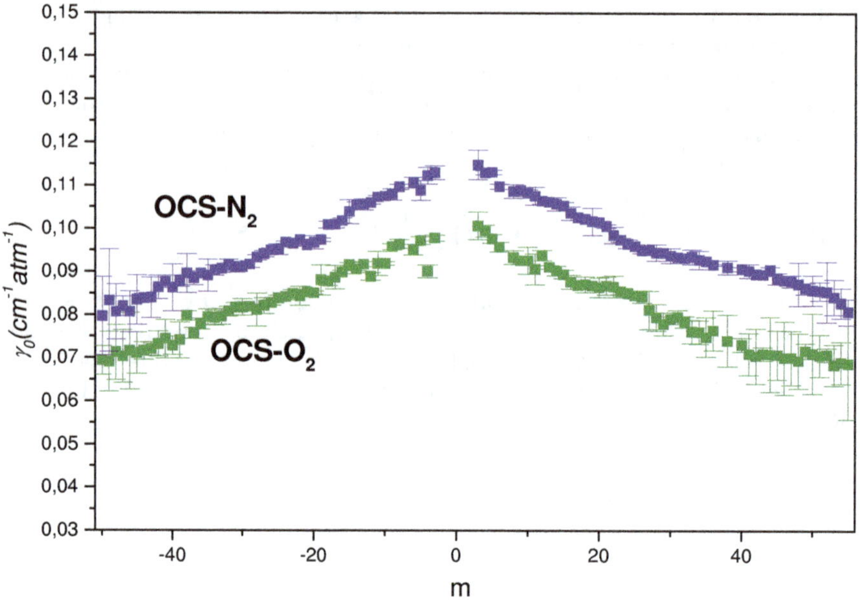

Fig. 10. Quantum number | m | dependence of the measured N_2, and O_2 broadening coefficients for P and R branches of the $4\nu_2$ band of OCS. Uncertainties become larger for high values of J (above 30) because of weaker line intensities

Contrary to the other symmetric top molecules, the broadening coefficients of NH_3 increase with K for a given J (Markov et al., 1993; Pine & Markov, 2004; Aroui et al., 2003, 2009). This behavior is more pronounced for the small J values. This increase could be related to the rotational energy separation between NH_3 levels which decrease with K. Note that the J = K lines have almost a constant value of broadening coefficient. These J and K dependencies have previously been observed in other bands (Hadded et al., 2001; Markov et al., 1993).

For CH_3Br molecule, previous works (Gomez et al., 2010, Jacquemart et al., 2007) report the observed and predicted self broadening coefficients of the ^PP, ^RQ and ^RR branches of the ν_6 band of CH_3Br. Fig. 11 shows that for a give K, the broadening coefficients of the ν_2 band increase with J in the interval $J \leq J_{max} = 20$ to nearly $\gamma = 0.45$ cm^{-1} atm^{-1} at $J = J_{max}$. Then these coefficients decrease at higher J values. The same trend is observed for the ν_6 band for K = 6 (Jacquemart et al., 2007). This pattern may be easily understood by considering the resonance condition that exists between the energy gaps of the two partners during collision. For the self broadening corresponding to the maximum contribution of the electrostatic interactions, J_{max} is given by the resonance condition (Giraud et al., 1971; Pourcin et al., 1981):

$$J_{max} \approx \frac{B_1}{B_2}\frac{\ell_1}{\ell_2} J_{2p} \tag{22}$$

J_{2P} is the most populated level of the perturbing molecule at the temperature considered (for CH_3Br at room temperature J_{2p} = 18). B_1 and B_2 are the rotational constants of the two partners ($B_1 = B_2$ for CH_3Br auto-perturbed). The spherical harmonic orders ℓ_i depend on the nature of the interaction. Therefore for the self- broadening dominated by the dipole-dipole interaction ($\ell_1 = \ell_2 = 1$), the maximum of collision effectiveness is around $J_{max} = 18$ in reasonable agreement with the experimental observation in Fig. 11.

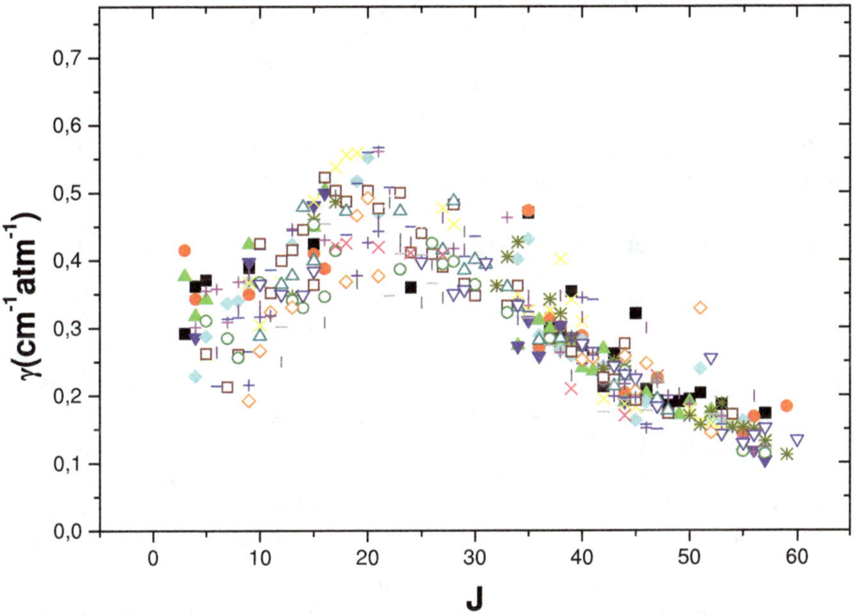

Fig. 11. Experimental self-broadening coefficients γ as function of the rotational quantum number J for the ν_2 band ^PQ branch of $^{12}CH_3{}^{79}Br$ and $^{12}CH_3{}^{81}Br$ for K=0 to 9. For a given J value, the symbols indicate the different K-lines for which measurements are performed

Polynomial fit of broadening coefficients

A constant need exists to examine the accuracy of the spectral line parameters listed in databases (Jacquinet-Husson et al., 2005; Perrin et al., 2005; Pickett et al., 1998; Rothman et al., 2009). For example, the NH_3 dataset in these databases has been revised by the inclusion of recent results (Brown & Peterson, 1994; Kleiner et al., 1999, 2003; Nemtchinov et al., 2004) but still needs to be improved and purged from possible errors in line parameters. However, generating the spectroscopic parameters represents a lot of work. This is why sometimes, due to the absence of the necessary information, the spectroscopic parameters obtained for lines of a given vibrational band are attributed to the corresponding lines of another band. For example, in the Hitran database, the air-broadening coefficients of OCS lines of the excited vibrational states are based on the ν_1 band measurement.

In this context, in order to provide empirical interpolation and extrapolation models for all spectral lines, some authors used empirical polynomial equations (Brown & Peterson, 1994; Aroui et al., 2009; Devis et al., 2002; Fabian & Yamada, 1999; Jacquemart et al., 2007) to fit the J and K dependencies of their pressure broadening coefficients.

Some least squares regressions fail to fit the observed values for particular quantum numbers. Other fits were satisfactory for nonpolar buffers, but fit poorly for self-broadening at low J quantum number.

In a recent paper (Aroui et al., 2009), the measured broadening coefficients of nine branches in the ν_2, $2\nu_2$ and ν_4 bands of NH_3 are fitted using the following equation.

$$\gamma(m,K) = \beta_0 + \beta_1 m + \beta_2 K + \beta_3 m^2 + \beta_4 K^2 + \beta_5 mK \tag{23}$$

The fitted parameters β_0, β_1, β_2, β_3, β_4 and β_5 are in unit of cm^{-1} atm^{-1}. They are given in Ref. (Aroui et al., 2009), along with the data points included in the fits. Most of the parameters are statistically well-determined and have remarkable regularities.

Vibrational dependence of pressure broadening

Some authors have discussed (Markov et al., 1993; Pine & Markov, 2004; Nemtchinov et al., 2004, Brown & Peterson, 1994; Cottaz et al., 2001; Aroui et al., 2009) the vibrational dependence of broadening coefficients. In most cases the comparison of the measured results for different bands exhibits little vibrational dependence. Within experimental error, the broadening coefficients of ammonia, for example, are only slightly greater in the ν_4 band than in the ν_2 and $2\nu_2$ bands. This is explained partially by the smaller splitting of the inversion doublets in the ν_4 band than in the ν_2 and $2\nu_2$ ones. For example, when the ν_2 and $2\nu_2$ bands are compared, the differences between line broadening coefficients with similar quantum numbers do not exceed 4% with a weighted average ratio $\langle \gamma(\nu_2)/\gamma(2\nu_2) \rangle = 1.031$.

However, for this molecule, Cottaz et al. (Cottaz et al., 2001) have highlighted a vibrational broadening effect when they compared their results for the R(3,1) line measured in six different vibrational bands. The present result for this line is slightly greater (3%) than that obtained by these authors.

For OCS-N_2 and OCS-O_2, pressure broadening coefficients in previous works for different bands (Bouanich et al., 1986; Domenech et al., 2000; Koshelev et al., 2006, 2009; Mouchet et al., 1985; Galalou et al., 2011), illustrate that no vibrational effect can be clearly identified. To clarify this effect, Ref. (Galalou et al., 2011) presents data of pressure broadening coefficients reported in earlier works. The observation of these results does not illustrate any clear differences between the bands.

Line-mixing effect

Line-mixing effects have received much less interest than line intensities and line broadening. This is mainly because they are rather small and their measurements are therefore difficult. However these spectroscopic parameters are necessary to detect molecules in tropospheric and planetary spectra, and to determine vertical concentration profiles. Moreover, even when laboratory measurements of these parameters are available, effects such as line-mixing may still require a sophisticated theoretical model in order to unravel observed spectra. When these effects are not negligible, the observed lineshapes exhibit significant deviations from the conventional Voigt profile, which may be attributed to line-mixing.

In the past decade, line-mixing effects have been examined for diverse molecules such as CO_2 (Ozanne et al., 1999), CH_4 (Pieroni et al., 1999), and OCS (Broquier & Picard-Bersellini, 1984). However for C_{3v} molecules the number of works is less. Measurements of these effects have been reported for CH_3F (Thibault et al., 1999), CH_3Cl (Hartmann et al., 1995) and CH_3Br (Tran et al., 2008). Otherwise, for NH_3, which presents line doubling due to its inversion motion, line-mixing effects were the subject of numerous works (Aroui et al., 1998, 2009; Dhib et al., 2007; Hadded et al., 2001). These works have clearly demonstrated the necessity of taking into account line-mixing in the molecular spectra.

We present below some line-mixing results for the NH_3 molecule. Plots of PY deduced from least squares fits (Eq. (16)) as a function of the NH_3 pressure P are shown in Fig. 12 for the components of the doublet, pertaining to the ν_4 band. The line-mixing parameter Y (in atm^{-1}) are determined as the slopes of the best-fit lines and are presented in Table 3 for the $^PP(5,K)$ and $^PP(6,K)$ manifolds of the ^PP branch of the ν_4 band of NH_3 at 295 K. The values given in parentheses correspond to the estimated errors corresponding to one standard deviation value.

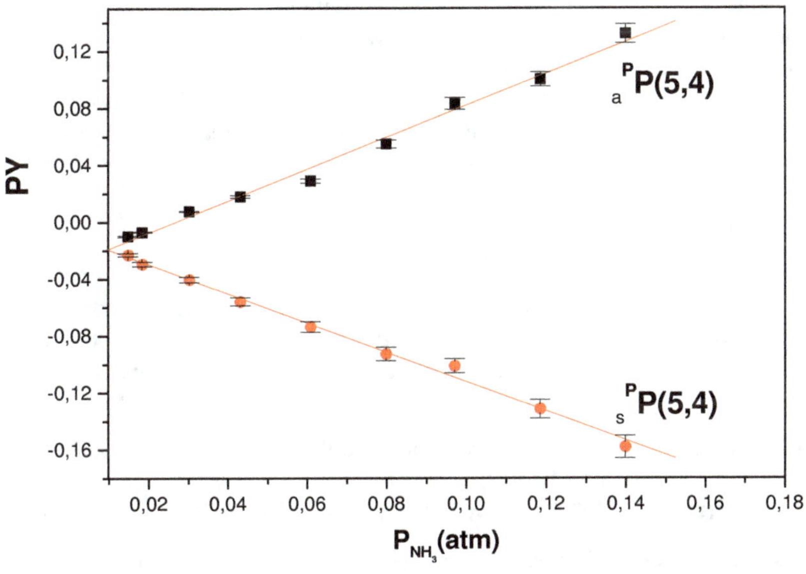

Fig. 12. Variation of PY as function of pressure P of NH_3 for the components of $^P_{as}P(5,4)$ doublet of the ν_4 band of NH_3 self-perturbed

This table shows that symmetric (s) and asymmetric (a) components of inversion doublets have opposite sign of Y. For numerous lines, Y appears to increase with K quantum number for a given value of J. This behavior could be explained in part by the decrease of the frequency difference between the manifold transitions with K, and particularly by the decrease of the $\Delta\sigma_0$ splitting inversion doublets components. This may be easily understood, since line-mixing parameter Y is inversely proportional to $\Delta\sigma_0$. Indeed, for a line k, Y_k is related to $\Delta\sigma_0$ and the off-diagonal element W_{lk} (line-coupling coefficient) of the relaxation matrix by (Gentry & Larrabee Strow, 1997; Pine, 1997; Pieroni et al., 1999):

$$Y_k = 2\sum_{l \neq k} \frac{d_l}{d_k} \frac{W_{lk}}{\Delta\sigma_0} . \tag{24}$$

$\Delta\sigma_0 = \sigma_{0l} - \sigma_{0k}$ where σ_{0k} and σ_{0l} are the unperturbed positions of the *k* and *l* lines respectively; d_k and d_l are their reduced matrix elements of the dipole moment.

However, Table 1 of Ref. (Aroui et al., 2009) presents exceptions to this behavior: the parameter Y of ^PQ branch lines decreases with quantum number K. The situation is different for ^RQ and ^RR branches where no systematic dependence of interference parameter with J and K is observed.

Otherwise the $\Delta\sigma_0$ splitting could be large enough to contain lines other than those pertaining to the doublet of interest. So the line-mixing effects can occur not only between the two components of the doublets but also between more than two transitions, provided that the selection rules are valid.

On the other hand, the ν_2 band has a large inversion splitting, 35.7 cm^{-1} for the vibrational quantum number $v_2 = 1$. Thus the line-mixing is negligible and the inversion contribution to this effect should be smaller than for the ν_3 and ν_4 bands. In these bands (with very small inversion splitting, about 0.5 cm^{-1}) the line-mixing is essentially dominated by collisional transitions across the inversion doublets with the selection rules $\Delta J = 0$, $\Delta K = 0$ and a$\rightarrow$s. These two bands are thus the more appropriate to study line-mixing effects in the infrared bands of NH_3.

Pressure induced shift

Measurements of line-shifts have been investigated by numerous authors for various molecules. For NH_3, line shifts were measured by Dhib et al. in the ν_4 and $2\nu_2$ bands (Dhib et al., 2007), by Raynaud et al. in the ν_2 and $(2\nu_2-\nu_2)$ bands (Raynaud et al., 1994), by Baldacchini et al. in the ν_2 band, (Baldacchini et al., 1982), and by Aroui et al. in ν_2, $2\nu_2$ and ν_4 bands (Aroui et al., 2009).

For OCS, line shifts were measured by Domenech et al. (Domenech et al., 2000) for the $2\nu_3$ band, by Koshelev et al. in the ν_3 band (Koshelev et al., 2006) and in the ground vibrational state (Koshelev & Tretyakov, 2009), and by Babay (Babay, 1997) in the ν_2 band.

Fig. 13 shows typical plots of line shift $\sigma_k-\sigma_{0k}$ as function of NH_3 pressure for the $^R{}_{as}R(2,1)$ and $^P{}_{as}P(4,4)$ lines of the ν_4 band of NH_3. The self-shift coefficients δ_0 are determined as the slopes of the best-fit lines.

As may be seen by Fig. 13 and as illustrated by the results listed in Ref. (Aroui et al., 2009), shift coefficients are both positive and negative. A sample of these results is given by Table 3 for the ^PP branch of the ν_4 band. About two thirds of the lines have positive shifts, and the more bipolar shifts are obtained for the ν_4 band. Several symmetric and asymmetric components of the isolated inversion doublets in this band have opposite sign for the shift coefficients, so they appear to be shifting towards each other. As expected this mutual line-attraction is negligible in the ν_2 band, for which the splitting is very large compared to that of the ν_3 and ν_4 bands.

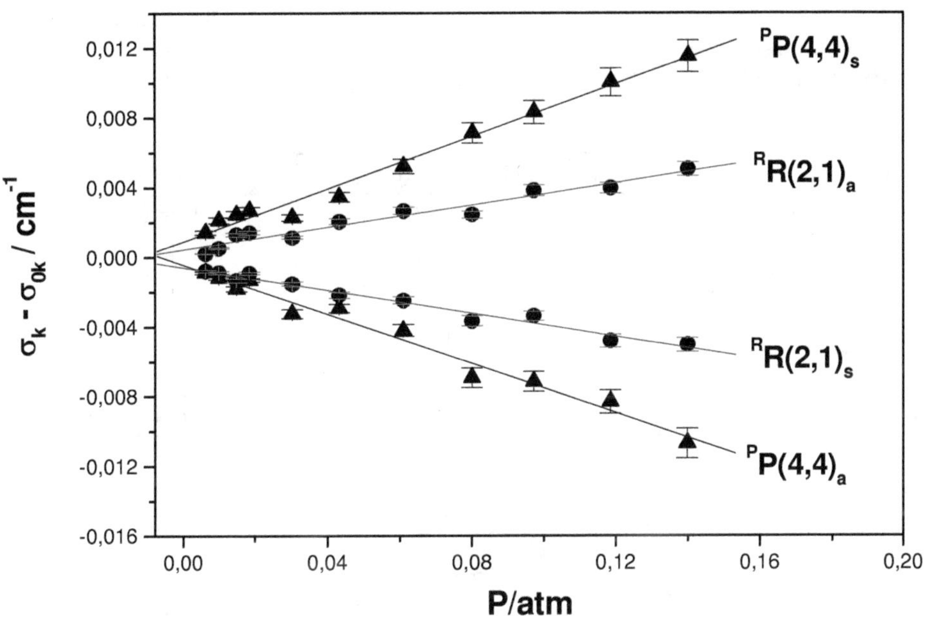

Fig. 13. Plots of the pressure-induced variation of σ-σ_0 versus NH_3 pressure P for the components of the doublets ${}^{R}{}_{as}R(2,1)$ and ${}^{P}{}_{as}P(4,4)$ of the ν_4 band of NH_3 self-perturbed

The self-shift coefficients of ammonia are larger than those obtained for other perturbers like He, H_2 and CO_2 (Hadded et al., 2001, Dhib et al., 2006) since the collisions in NH_3 are long range dipole-dipole collisions while the foreign gases collisions are shorter range collisions.

For all rotational lines of the molecules considered in this work, the shift coefficients do not reveal any systematic dependencies on the J and K quantum numbers with the exception seen in Table 3 for the NH_3 lines with K approaching J, for which the shift seems to increase with increasing K. For most of the lines of this molecule, the shifts are large, ranging from about -0.140 to +150 cm^{-1} atm^{-1}. The smallest one is $\delta_0 = 0.003 \pm 0.002$ cm^{-1} atm^{-1}. These shifts are in excellent agreement with the submillimeter measurement (Belov et al., 1982) of the ground state of NH_3 where $\delta_0 = 0.153 \pm 0.002$ cm^{-1} atm^{-1}, and somewhat larger than the ν_2 band diode laser measurement (Baldacchini et al., 1982) with $\delta_0 = 0.113 \pm 0.025$ cm^{-1} atm^{-1}. Similar large values of shifts are reported for other molecules such as CH_3CN studied by Rinsland et al. (Rinsland et al., 2008) using a multi-spectrum fitting technique.

For the R and P branches of the $4\nu_2$ band of OCS molecule perturbed by O_2 and N_2, the shift coefficients are mostly positive. The largest one is $\delta_0 = (17.91 \pm 3.02)\times10^{-3}$ $cm^{-1}atm^{-1}$, obtained for the R(51) line. These shifts are small compared with those of NH_3 and CH_3CN; this makes them difficult to measure as evidenced from their relatively large errors (Galalou et al., 2011).

In the $2\nu_3$ (Domenech et al., 2000), ν_3 (Koshelev et al., 2006) and ν_2 (Babay, 1997) bands of OCS molecule the shift coefficients were found negative for all studied lines. In the ground vibrational state (Koshelev & Tretyakov, 2009), the upper limits of the pressure shift coefficient is $\delta_0 = \pm\ 0.38\ 10^{-3}cm^{-1}atm^{-1}$. This value is about 10 times smaller than those of the $4\nu_2$ band. These reasons, and the fact that the line shift coefficients do not illustrate any systematic variation with m, leads one to conclude that they depend on the vibrational band but are rotationally independent.

Lines	$\sigma_0(cm^{-1})$	$\delta_0(10^{-3}\ cm^{-1}\ atm^{-1})$	$Y(atm^{-1})$
$^{P}{}_{s}P(5,1)$	1536.2088	8.9 (5.3)	0.18 (0.02)
$^{P}{}_{a}P(5,1)$	1538.0102	-10.7 (5.7)	0.44 (0.10)
$^{P}{}_{s}P(5,2)$	1539.7596	10.7 (5.8)	0.43 (0.07)
$^{P}{}_{a}P(5,2)$	1541.0041	9.0 (5.0)	-0.25 (0.04)
$^{P}{}_{s}P(5,3)$	1542.9798	12.3 (4.1)	0.45 (0.02)
$^{P}{}_{a}P(5,3)$	1543.8550	-16.8 (6.6)	-0.42 (0.04)
$^{P}{}_{s}P(5,4)$	1545.8043	43.6 (5.6)	1.22 (0.06)
$^{P}{}_{a}P(5,4)$	1546.3315	-35.6 (6.6)	-1.10 (0.06)
$^{P}{}_{s}P(5,5)$	1548.1839	55.0 (10.0)	4.00 (0.09)
$^{P}{}_{a}P(5,5)$	1548.4290	-53.8 (10.9)	-4.16 (0.16)
$^{P}{}_{s}P(6,1)$	1519.6627	38.0 (6.3)	-0.62 (0.03)
$^{P}{}_{a}P(6,1)$	1522.3847	-13.5 (6.3)	0.74 (0.07)
$^{P}{}_{s}P(6,2)$	1522.7763	40.8 (6.0)	-0.81 (0.19)
$^{P}{}_{a}P(6,2)$	1524.7088	21.9 (4.7)	0.32 (0.05)
$^{P}{}_{s}P(6,3)$	1525.7625	47.3 (6.4)	-0.33 (0.03)
$^{P}{}_{a}P(6,3)$	1527.0615	18.4 (6.0)	-0.41 (0.03)
$^{P}{}_{s}P(6,4)$	1528.3774	18.6 (6.2)	0.74 (0.05)
$^{P}{}_{a}P(6,4)$	1529.2897	8.6 (4.7)	-0.67 (0.01)
$^{P}{}_{s}P(6,5)$	1530.6141	53.4 (5.5)	1.11 (0.04)
$^{P}{}_{a}P(6,5)$	1531.1591	-26.4 (7.3)	-1.28 (0.02)
$^{P}{}_{s}P(6,6)$	1532.4503	111.0 (9.4)	3.83 (0.11)
$^{P}{}_{a}P(6,6)$	1532.6830	-100.8 (13.4)	-4.03 (0.09)

Table 3. Self shift coefficients δ_0 (in 10^{-3} cm^{-1} atm^{-1}) and self mixing parameters Y (in atm^{-1}) for the ^{P}P branch of ν_4 band of NH_3 at 295 K. The values given in parentheses correspond to the estimated errors expressed as one time standard deviation

It should be noted that one can determine the air-induced line shape parameters G (broadening coefficients, shift coefficient, or line-mixing parameter) for the collision between an active molecule A and air using the measured A-N_2 and A-O_2 parameters assuming binary collisions and a standard atmospheric composition of air (Tejwani & Varnassi, 1971):

$$G(A-Air) = 0.79G(A-N_2) + 0.21G(A-O_2) \tag{25}$$

Line-mixing and shift correlation

It should be noted that for the NH_3 molecule, within the measurement errors, almost all line-mixing and -shifts vary linearly with the considered pressures. However for some lines with small rotational energy gaps, the observed shift- and mixing-coefficients exhibit quadratic dependencies on pressure. This situation is more remarkable for strongly overlapped inversion doublets of the ν_4 band which have small inversion splitting $\Delta\sigma_0$ compared to the rotational energy separation. This is illustrated by Fig. 14 for ${}^{P}{}_{as}P(8,8)$ doublet with a splitting $\Delta\sigma_0$=0.210 cm^{-1}.

This non linear dependence of line-mixing or line shift on pressure can be expressed as (Smith, 1981):

$$\sigma_{s/a} = \sigma_{0s/a} + P\delta_{0s/a} \pm \frac{S_s}{S_a}\frac{\Delta\sigma_0}{4}P^2Y_{s/a}^2, \tag{26}$$

where $S_{s/a}$, $\sigma_{0s/a}$, $\delta_{0s/a}$ and $Y_{s/a}$ are, respectively, the intensity, the unperturbed position, the shift coefficient and line-mixing parameter of the (s) or (a) components of the doublets. The sign (±) is considered for the (s) or (a) transitions of NH_3. This equation illustrates a correlation between line-mixing and shift phenomena. This quadratic evolution of frequency shift with pressure has been also observed previously by Thibault et al. (Thibault et al., 1992) for the 0-1 and 0-2 bands of CO perturbed by He. Also, nonlinear NO-He and NO-Ar shifts have been observed by Vyrodov et al. using laser-induced fluorescence (Vyrodov et al., 1995).

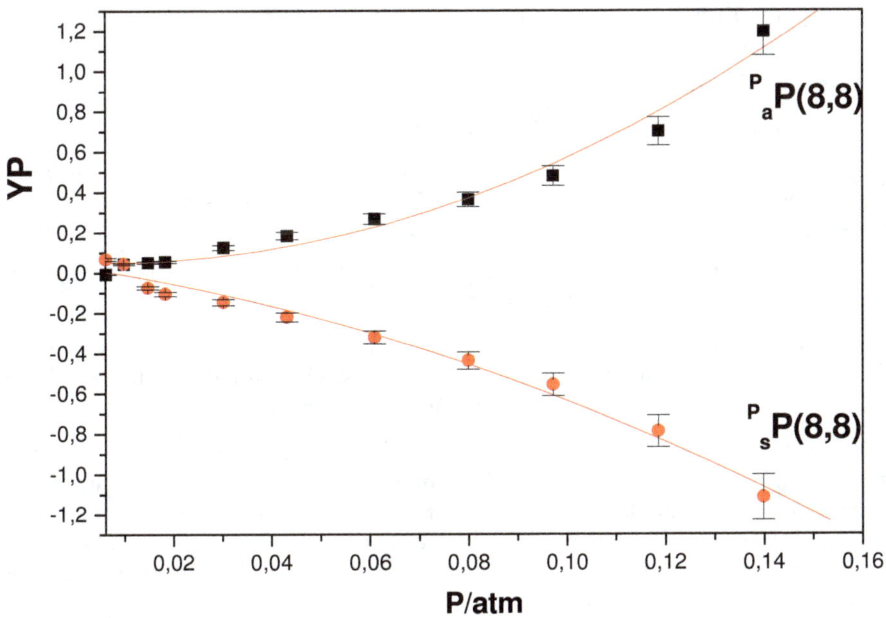

Fig. 14. Variation of PY as function of pressure P for the two components of ${}^{P}{}_{as}P(8,8)$ doublet of the ν_4 band of NH_3 self-perturbed

Line parameter accuracies

Considering the unknown systematic uncertainties in sample pressure and temperature, and the absorption path length, the errors are often estimated as one or two standard deviations (1 or 2σ) derived from the linear fits, and vary widely depending on the quality of spectral lines, which is related to the overlapping of neighboring lines and the base line location. The experimental errors reported in Table 3 are estimated as one standard deviation as derived from the linear fit of line mixing and line shifts parameters.

This table also reveals that the line-mixing parameters are more accurate than the shifts. The mean value of accuracy is about 15% for line-mixing parameters and 25% for the shift coefficients that can attain more than 50% or are of the same order of magnitude as the parameter themselves. This makes them negligible and difficult to measure reliably.

Today the achievable accuracies with FTIR are about 3% for line intensities, 5% for pressure broadening coefficients, and 15-20% for pressure-induced line shifts and line-mixing parameters.

It should be noted that, in some instances, the broadening coefficient accuracies of some lines tends to have greater values, which can attain 10-15%. These errors can be explained by the relatively low intensities and the overlaps with neighboring lines.

For the temperature dependence of the broadening coefficients, since the values of temperature coefficient n are determined from a power-law relation, the estimated errors in the derived values of n are larger than the corresponding errors in the measured broadening coefficients. Assuming an error of 5% for broadening measurements, the uncertainty of n values could be about 15%.

Conclusion

The purpose of this paper was to highlight and demonstrate the ability of the Fourier Transform Infrared Spectrometer to measure spectroscopic parameters such as line intensities, line broadenings, line shifts and line-mixing parameters for a select number of small molecules.

Compared to the dispersive spectrometers, this technique has the advantage to measure with high precision complete sets of line parameters over large spectral regions.

Fourier transform spectroscopy was used here to study infrared spectra of three molecules: one linear, OCS, and two symmetric tops: NH_3 and CH_3Br which are of considerable interest for quantitative and qualitative studies of planetary and Earth's atmospheres.

These spectra were fitted using a multi-spectrum fitting technique and the first order model absorption coefficient taking into account line-mixing effects. In the considered pressure range, this model is adequate to extract with high enough accuracy line parameters for the studied molecules. For CH_3Br a Voigt profile is used to extract the line parameters, indeed in this case no significant deviations are observed when the measured and calculated spectra are compared.

The estimated errors vary widely depending on the quality of spectral lines with, in general, improving accuracies in the order: line intensities, line broadenings, line-mixing and line shifts. Among these spectroscopic parameters, the line shifts are the smallest, so are the most difficult to measure with a high accuracy.

More measurements were performed in the ν_2, $2\nu_2$ and ν_4 bands of NH_3 at room temperature and at low temperatures, as well as in the ν_2 and ν_6 bands of CH_3Br and in the $4\nu_2$ band of OCS. However, in the ^PP branch of the ν_4 band of NH_3, the shifts are correlated to the line-mixing parameters and exhibited significant non-linearity. The broadening coefficients do not show any pronounced vibrational effect. For NH_3, the self-shift coefficients are rather large and more bipolar than foreign gas shifts with no significant branch or rotational dependencies.

Acknowledgments

The authors are grateful to the "Ministry of Education and Scientific Research in Tunisia" who has supported this work.

References

Abrams, M.C., Toons, G.C. & Schindler, R.A., 1994. A practical example of the correction of Fourier transform spectra for detector nonlinearity. *Appl. Opt.*, 33, 6307-6314.

Aliev, M.R., Papousek, D. & Urban, S., 1987. Third-order theory of the line intensities in the allowed and forbidden vibrational-rotational bands of C_{3v} molecules. *J. Mol. Spectrosc.*, 124, 285-305.

Aroui, H. Broquier, M. Picard-Bersellini, A. Bouanich, J.P. Chevalier, M. & Gherissi, S., 1998. Absorption intensities, pressure-broadening and line-mixing parameters of some lines of NH_3 in the ν_4 Band. *J. Quant. Spectrosc. Radiat. Transfer.*, 60, 1011-1023.

Aroui, H., Laribi, H., Orphal, J. & Chelin, P., 2009. Self-broadening, self-shift and self-mixing in the ν_2, $2\nu_2$ and ν_4 bands of NH_3. *J. Quant. Spectrosc. Radiat. Transfer.*, 110, 2037-2059.

Aroui, H., Nouri, S. & Bouanich, J.P., 2003. NH_3 self-broadening coefficients in the ν_2 and ν_4 bands and line intensities in the ν_2 band. *J. Mol. Spectrosc.*, 220, 248-58.

Auwera, J.V., 2004. Agrégation de l'Enseignement Supérieur (HDR thesis), Quantitative high resolution Fourier transform infrared spectroscopy. May, Université libre de Bruxelle.

Babay, A., 1997. Thèse de doctorat. Université des Sciences et Technologies de Lille.

Baldacchini, G., Marchetti, S., Montelatici, V., Buffa, G. & Tarrini, O., 1982. Experimental and theoretical investigations of self-broadening and self-shifting of ammonia transitions in the ν_2 band. *J. Chem. Phys.*, 76, 5271-77.

Bell, R.J., 1972. Introductory Fourier Transform Spectroscopy. *Academic press Inc. New York and London.*

Belov, S.P., Krupnov, A.F., Markov, V.N., Mel'nikov, A.A., Skvortsov, V.A. & Tret'yakov, M.Y., 1982. The study of microwave pressure lineshifts. *J. Mol. Spectrosc.*, 94, 264-82.

Bezard, B., DeBergh, C., Crisp, D. & Maillard, J.P., 1990. The deep atmosphere of Venus revealed by high-resolution nightside spectra. *Nature.*, 345, 508-11.

Bouanich, J.P., Walrand, J., Alberty, S. & Blanquet, G., 1987. Diode-laser measurements of oxygen-broadened line widths in the ν_1 band of OCS. *J Mol Spectrosc.*, 123, 37-47.

Brassewell, R., 1965. The Fourier Transform and its applications. *McGraw-Hill Book Company, New York.*

Brassington, D.J., 1988. I*n Proceedings of International Symposium on Monitoring of gaseous Pollutant by Diode Laser*, Freiburg, Germany, 17 October.

Broquier, M. & Picard-Bersellini, A., 1984. Overlapping effects and collisional narrowing in rotational doublets of OCS in the infrared. *Chem. Phys. Lett.*, 111, 602-06.

Brown, L.R. & Peterson, D.B., 1994. An empirical expression for linewidths of ammonia from far-infrared measurements. *J. Mol. Spectrosc.*, 168, 593–606.

Camy-Peyret, C., Payan, S., Jeseck, P. & Té, Y., 2001. Spectroscopie ultra-sensible des gaz; détection de polluants. C. R. Acad. Sci. Paris., t. 2, Série IV, 905-922.

Cooley, J.W. & Tukey, J.W., 1965. An algorithm for the machine calculation of complex Fourier series, Maths. Comput. 19, 297-301.

Cottaz, C., Tarrago, G., Kleiner, I. & Brown, L.R., 2001. Assignments and intensities of $^{14}NH_3$ hot bands in the 5- to 8-μm ($3\nu_2$-ν_2, ν_2+ν_4-ν_2) and 4μm ($4\nu_2$-ν_2, ν_1-ν_2, ν_3-ν_2 and $4\nu_4$-ν_2) regions. *J. Mol. Spectrosc.*, 209, 30–49.

Davis, S.P., Abrams, M.C. & Brault, J.W., 2001. Fourier Transform Spectroscopy. *Academic press Inc.*

Dello Russo, N., DiSanti, M.A., Mumma, M.J., Magee-Sauer, K. & Rettig, T.W., 1998. Carbonyl sulfide in comets C/1996 B2 (Hyakutake) and C/1995 O1 (Hale-Bopp): evidence for an extended source in Hale-Bopp. *Icarus.*, 135, 377-88.

Devi, V.M., Benner, D.C., Brown, L.R., Smith, M.A.H., Rinsland, C.P., Sams, R.L. & Sharpe, S.W., 2002. Multi-spectrum analysis of self- and N_2-broadening, shifting and line-mixing. coefficients in the ν_6 band of $^{12}CH_3D$. *J. Quant. Spectrosc. Radiat. Transfer.*, 72, 139-91.

Dhib, M., Aroui, H. & Orphal, J., 2007. Experimental and theoretical study of line shift and mixing in the ν_4 band of NH_3 perturbed by N_2. *J. Quant. Spectrosc. Radiat. Transfer.*, 107, 372-84.

Dhib, M., Echargui, M.A., Aroui, H. & Orphal, J., 2006. Shifting and line-mixing parameters in the ν_4 band of NH_3 perturbed by CO_2 and He: experimental results and theoretical calculations. *J. Mol. Spectrosc.*, 238, 168-77.

Domenech, J.L., Bermejo, D., & Bouanich, J.P., 2000. Pressure lineshift and broadening coefficients in the $2\nu_3$ band of $^{16}O^{12}C^{32}S$. *J Mol Spectrosc.*, 200, 266-76.

Evans, N.J., Lacy, J.H. & Carr, J.S., 1991. Infrared molecular-spectroscopy toward the Orion IRC2 and IRC7 sources: a new probe of physical conditions and abundances in molecular clouds. *Astrophys J.*, 383, 674-92.

Fabian, M. & Yamada, K.M.T., 1999. Absolute Intensity of the NH_3 ν_2 Band. *J. Mol. Spectrosc.*, 198, 102-109.

Galalou, S., Ben Mabrouk, K., Aroui, H., Kwabia Tchana, F., Willaert, F. & Flaud, J-M., 2011. N_2 and O_2 Pressure Broadening and Pressure Shift in the $4\nu_2$ Band of $^{16}O^{12}C^{32}S$. *J. Quant. Spectrosc. Radiat. Transfer.*, 112 , 2750-2761.

Gentry, G. & Larrabee Strow, L., 1987. Line-mixing in a N_2-broadened CO_2 Q branch observed with a tunable diode laser. *J. Chem. Phys.*, 86, 5722–30.

Giraud, M., Robert, D. & Galatry, L., 1971. Sur la détermination du moment quadrupolaire de molecules linéaires à partir de l'élargissement des raies spectrales. *C.R. Acad. Sci. Paris.*, 272, 1252–55.

Gomez, L., Jacquemart, D., Bouanich, J.P., Boussetta, Z. & Aroui, H., 2010. Theoretical calculations of self-broadening coefficients in the ν_6 band of CH_3Br. *J. Quant. Spectrosc. Radiat. Transfer.*, 111, 1252-1261.

Griffiths, P.R. & de Haseth, J.A., 1986. Fourier Transform Infrared Spectroscopy. *Wiley, New York.*

Guelachvili, C., 1986. Distortion-free interferograms in Fourier transform spectroscopy with nonlinear detectors. *Appl. Opt.* 25, 4644-48.

Hadded, S., Aroui, H., Orphal, J., Bouanich, J.P. & Hartmann, J.M., 2001. Line broadening and mixing in NH_3 inversion doublets perturbed by NH_3, He, Ar, and H_2. *J Mol Spectrosc.*, 210, 275-83.

Hartmann, J.M., Bouanich, J.P., Blanquet, G., Walrand, J. & Lacome, N., 1995. Simple modeling of Q branch absorption-II: application to molecules of atmospheric interest. *J. Quant. Spectrosc. Radiat. Transfer.*, 54, 723-35.

Ho, P.T.P. & Townes, C.H., 1983. Interstellar Ammonia. *Ann. Rev. Astr. Astrophys.*, 21, 239-270.

Jacquemart, D., Kwabia Tchana, F., Lacome, N. & Kleiner, I., 2007. A complete set of line parameters for CH_3Br in the 10 µm spectral region. *J. Quant. Spectrosc. Radiat. Transfer.*, 105, 264–302.

Jacquinet-Husson, N., Scott, N.A., Chédin, A., Garceran, K., Armante, R., Chursin, A.A. et al., 2005. The 2003 edition of the GEISA/IASI spectroscopic database. *J. Quant. Spectrosc. Radiat. Transfer.* 95, 429-67.

Kleiner, I., Brown, L.R., Tarrago, G., Kou, Q.L., Picqué, N., Guelachvili, G. Dana, V. & Mandin, J.Y., 1999. Positions and intensities in the $2\nu_4/\nu_1/\nu_3$ vibrational system of $^{14}NH_3$ near 3 µm. *J. Mol. Spectrosc.*, 193, 46-71.

Kleiner, I., Tarrago, G., Cottaz, C., Sagui, L., Brown, L.R., Poynter, R.L., Pickett, H.M., Chen, P., Pearson, J.C., Sams, R.L., Blake, G.A., Matsuura, S., Nemtchinov, V., Varanasi, P., Fusina, L. & Lonardo, G. Di., 2003. NH_3 and PH_3 line parameters: the 2000 HITRAN update and new results. *J. Quant. Spectrosc. Radiat. Transfer.*, 82, 293-312.

Koshelev, M.A. & Tretyakov, M.Y., 2009. Collisional broadening and shifting of OCS rotational spectrum lines. *J. Quant. Spectrosc. Radiat. Transfer.*, 110, 118-28.

Koshelev, M.A., Tretyakov, M.Y., Lees, R.M. & Xu, L.H., 2006. Tunable diode laser measurements of N_2- and O_2-pressure broadening and pressure-induced shifts for $^{16}O^{12}C^{32}S$ transitions in the ν_3 band. *J Mol Struc.*, 780-781, 7-16.

Kunde, V., Hanel, R., Maguire, W., Gautier, D., Baluteau, J.P., Marten, A., Chedin, A., Husson, N. & Scott, N., 1982. The tropospheric gas composition of Jupiter's north equatorial belt /NH_3, PH_3, CH_3D, GeH_4, H_2O/ and the Jovian D/H isotopic ratio. Astrophysical J., 263, 443-467.

Kurylo, M.J. & Rodriguez, J.M., 1999. Chapter 2, Short-lived Ozone-Related Compounds. In: Scientific Assessment of Ozone Depletion: 1998. Global Ozone Research and Monitoring. *Project - Report 44, WMO, Geneva, Switzerland*, 21-56.

Lévy, A., Lacome, N. & Chackerian, C.Jr., 1992. Spectroscopy of the Earth's Atmosphere and Interstellar Medium. *Academic Press, New York*, 261-337.

Markov, V.N., Pine, A.S., Buffa, G. & Tarrini, O., 1993. Self broadening in the ν_1 band of NH_3. *J. Quant. Spectrosc. Radiat. Transfer.*, 50, 167-78.

Martin, S., Martin-Pintado, J., Mauersberger, R., Henkel C. & Garcia-Burillo, S., 2005. Sulfur chemistry and isotopic ratios in the Starburst Galaxy NGC 253. *Astrophys J.*, 620, 210-216.

Mauersberger, R., Henkel, C. & Chin, Y.N., 1995. Dense gas in nearby galaxies viii. The detection of OCS. *Astron Astrophys.*, 294, 23-32.

McElroy, M.B., Salawitch, R.J., Wosfy, S.C., & Logan, J.A., 1986. Antarctic ozone: Reductions due to synergistic interactions of chlorine and bromine. *Lett. Nat.*, 321, 759-762.

Mertz, L., 1965. Transformations in optics. New York: Wiley.

Mouchet, A., Blanquet, G., Herbin, P., Walrand, J., Courtoy, C.P. & Bouanich, J.P., 1985. Diode laser measurements of N_2-broadened line widths in the ν_1 band of OCS. *Can J Phys.*, 63, 527-31.

Nemtchinov, V., Sung, K. & Varanasi, P., 2004. Measurements of line intensities and half-widths in the 10 μm bands of NH_3. *J. Quant. Spectrosc. Radiat. Transfer.*, 83, 243-65.

Nouri, S., Aroui, H., Bouanich, J.P., Orphal, J. & Hartmann, J.M., 2004. Temperature dependence of pressure broadening of NH_3 perturbed by H_2 and N_2. *J. Mol. Spectrosc.*, 227, 60-66.

Ozanne, L., Bouanich, J.P., Rodrigues, R., Hartmann J.M., Blanquet G. & Walrand J., 1999. Diode laser measurements of He and N_2 broadening coefficients and line-mixing effects in the Q branch of the ν_1-ν_2 band of CO_2. *J. Quant. Spectrosc. Radiat. Transfer.*, 61, 153–84.

Perrin, A., Puzzarini, C., Colmont, J-M., Verdes, C., Wlodarczak, G., Cazzoli, G. & al., 2005. Molecular line parameters for the "MASTER" (millimeter wave acquisitions for stratosphere/troposphere exchange research) database. *J Atmos Chem.* 51, 161-205.

Pickett, H.M., Poynter, R.L., Cohen, E.A., Delitsky, M.L., Pearson, J.C. & Muller, H.S.P., 1998. Submillimeter, millimeter, and microwave spectral line catalog. *J. Quant. Spectrosc. Radiat. Transfer.* 60, 883-90.

Pieroni, D., Nguyen, V.T., Brodbeck, C., Claveau, C., Valentin, A., Hartmann, J.M., Gabard, T., Champion, J.P., Bermejo, D. & Domenech, J.L., 1999. Experimental and theoretical study of line-mixing methane spectra. I. The N_2- broadened ν_3 band at room temperature. *J. Chem. Phys.*, 110, 7717–32.

Pine, A.S. & Markov, V.N., 2004. Self- and foreign-gas-broadened lineshapes in the ν_1 band of NH_3. *J. Mol. Spectrosc.*, 228, 121–42.

Pine, A.S., 1997. N_2 and Ar broadening and line-mixing in the P and R branches of the ν_3 band of CH_4. *J. Quant. Spectrosc. Radiat. Transfer.*, 57, 157-176.

Pourcin, J., Jacquemoz, A., Fournel, A. & Sielmann, H., 1981. Pressure broadening of HCl pure rotational lines with a far-infrared optically pumped laser. *J. Mol. Spectrosc.*, 90, 43–50.

Raynaud, F., Lemoine, B. & Rohart, F., 1994. High precision pressure-induced lineshifts measured with a frequency-stabilized diode laser: application to ν_2 and ($2\nu_2$-ν_2) bands of NH_3. *J. Mol. Spectrosc.*, 168, 584-92.

Rinsland, C.P., Devi, V.M., Chris Benner, D., Blake, T.A., Sams, R.L., Brown, L.R., Kleiner, I., Dehayem-Kamadjeu, A., Müller, H.S.P., Gamache, R.R., Niles, D.L. & Masiello, T., 2008. Multi-spectrum analysis of the ν_4 band of CH_3CN: Positions, intensities, self and N_2-broadening, and pressure induced shifts. *J. Quant. Spectrosc. Radiat. Transfer.*, 109, 974-94.

Rosenkranz, P.W., 1975. Shape of the 5mm oxygen band in the atmosphere. *IEEE Trans. Antennas Propag.*, 23, 498-506.

Rothman, L.S., Gordon, I.E., Barbe, A., Benner, D.C., Bernath, P.F., Birk, M., Boudon, V., Brown, L.R., Campargue, A., Champion, J.P., Chance, K., Coudert, L.H., Dana, V., Devi, V.M., Fally, S., Flaud, J-M., Gamache, R. R., Goldman, A., Jacquemart, D., Kleiner, I. et al., 2009. The HITRAN 2008 molecular spectroscopic database. *J. Quant. Spectrosc. Radiat. Transfer.* 110, 533-72.

Smith, E.W., 1981. Absorption and dispersion in the O_2 microwave spectrum at atmospheric pressures. *J. Chem. Phys.*, 74, 6658–6673.

Tejwani, G.D.T. & Varnassi, P., 1971. Theoretical line widths in N_2O-N_2O and N_2O-air collisions. *J. Quant. Spectrosc. Radiat. Transfer.*, 11, 1659-64.

Thibault, F., Boissoles, B., Grigoriev, I.M., Filippov, N.N. & Tonkov, M.V., 1999. Line-mixing effects in the ν_3 band of CH_3F in helium: experimental band shapes and ECS analysis. *Eur. Phys. J. D.*, 6, 343-53.

Thibault, F., Boissoles, J., Le Doucen, R., Farrenq, R., Morillon-Chapey, M. & Boulet, C., 1992. Line by line measurements of interference parameters for the 0-1 and 0-2 bands of CO in He, and comparison with coupled-states calculations. *J. Chem. Phys.*, 97, 4623-32.

Thomas, V.M., Bedford, J.A. & Cicerone, R.J., 1997. Bromine emissions from leaded gasoline. *Geophys. Res. Lett.*, 24, 1371–1374.

Tran, H. Jacquemart, D., Mandin, J.Y. & Lacome, N., 2008. Line mixing in the ν_6 Q branches of self-and nitrogen-broadened methyl bromide. *J. Quant. Spectrosc. Radiat. Transfer.*, 109, 119-31.

U.S. Environmental Protection Agency, 2011. Gaseous Pollutants - Fourier Transform Infrared Spectroscopy (FTIR)

Vyrodov, A.O., Heinze, J. & Meier U.E., 1995. Collisional broadening of spectral lines in the A-X system of NO by N_2, Ar and He at elevated pressure measured by laser-induced fluorescence. *J. Quant. Spectrosc. Radiat. Transfer.*, 53, 277-87.

Wartewig, S., 2003. IR and Raman spectroscopy: fundamental processing. *Weinheim: Wiley-VCH.*

Watson, J.K.G., 1992. Quadratic Herman-Wallis factors for symmetric- and asymmetric-top molecules. *J. Mol. Spectrosc.*, 153, 211-24.

Watts, S.F., 2000. The mass budgets of carbonyl sulfide, dimethyl sulfide, carbon disulfide and hydrogen sulfide. *Atmos Environ.*, 34, 761-79.

Woodney, L.M., McMullin, J. & Ahearn, M.F., 1997. Detection of OCS in comet Hyakutake, (C/1996 B2). *Planet Space Sci.*, 45, 717-9.

10

Fourier Transform Infrared Spectroscopy for Natural Fibres

Mizi Fan[1,2], Dasong Dai[1,2] and Biao Huang[2]
[1]*Department of Civil Engineering, Brunel University, London, UB8 3PH,*
[2]*School of Material and Engineering, Fujian Agricultural and Forestry University,*
[1]*UK*
[2]*P. R. China*

Introduction

Infrared spectroscopy is nowadays one of the most important analytical techniques available to scientists. One of the greatest advantages of the infrared spectroscopy is that virtually any sample in any state may be analyzed. For example, liquids, solutions, pastes, powders, films, fibres, gases and surfaces can all be examined with a judicious choice of sampling technique. The review by Annette, Sudhakar, Ursula and Andrea [1-2] also demonstrates the applicability of dispersion infrared spectroscopy for natural fibres studies.

Fourier transform infrared spectroscopy (FTIR) has facilitated many different IR sampling techniques, including attenuated total reflection and diffuses reflectance infrared Fourier transform (DRIFT) spectroscopy. It has dramatically improved the quality of infrared spectra and minimized the time required to obtain data. The increased speed and higher ratio of signal-to-noise of FTIR relative to dispersion infrared has lead to a substantially greater number of applications of infrared in natural fibres research. In addition, the constant advancing of computer and computing science has made infrared spectroscopy techniques striding further: The availability of a dedicated computer, which is required for the FTIR instrumentation, has allowed the digitized spectra to be treated by sophisticated data processing techniques and increased the utility of the infrared spectra for qualitative and quantitative purposes. With interferometric techniques, the infrared spectroscopy is being launched into a new era and interest in this technique is at an all time high.

Cellulose, which acts as the reinforcing material in the cell wall, is the main constitute in natural fibres. The cellulose molecules are laid down in microfibrils in which there is extensive hydrogen bonding between cellulose chains, producing a strong crystalline structure. Much work has been published on the characterization of the hydrogen bonds in cellulose by using various techniques, among which FTIR has proved to be one of the most useful methods [3-6]. Furthermore, FTIR can provide researchers with further information on the super-molecular structure. FTIR can also be used to determine the chemical compositions of native natural fibres and the modified natural fibres.

This chapter of the book describes the application of FTIR in the hydrogen bonds analysis, determination of structures and chemical compositions, and the morphology characterization for natural fibres.

Hydrogen bonds analysis of natural celluloses by using FTIR

A hydrogen bond is the attractive interaction of a hydrogen atom with an electronegative atom, such as nitrogen, oxygen or fluorine, that comes from another molecule or chemical group. Cellulose occurs in the form of long, slender chains, polymer of 1-4 linked β-D-glucose (Figure 1). Hydroxyl groups in C2, C3 and C6 contribute to the formation of various kinds of inter- and intra-molecular hydrogen bonds. The formation of inter- and intra- molecular hydrogen bonds in the cellulose not only has a strong influence on the physical properties of cellulose, including solubility [7-8], hydroxyl reactivity [9-10] and crystallinity [11-12], but also plays an important role in the mechanical properties of the cellulose [13]. Calculated by Tashiro and Kobayashi [14] showed that hydrogen bonds contribute about 20% the strain energy to the cellulose. It is apparent that the investigation of hydrogen bonds on cellulosic fibres and other materials gives rise to great benefits for the research on all other aspects of natural fibres and related materials.

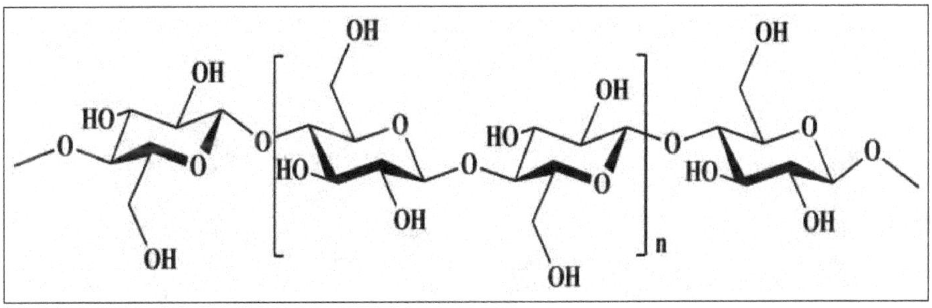

Fig. 1. Chemical structure of cellulose

X-ray diffraction has been a powerful tool [15-19] to investigate hydrogen bonds visualization, lengths and angles. FTIR is even a more advanced tool to study hydrogen bonds in cellulose. IR was firstly used to investigate hydrogen bonds in cellulose in the 1950s [e.g. 20-22] and then the whole area of OH stretching wave-number in IR spectra for cellulose I and cellulose II [23-24]. The OH stretching region always covers 3-4 sub-peaks and these sub-peaks cannot be determined in the original data set. Some mathematical methods (e.g. deconvolution [25-27] and second-derivative [28-30]) were used to identify the exact peak for hydrogen bonds. Hinterstoisser and Salmén [3, 31] recently used DMA-FTIR to investigate OH stretching vibration regions between 3700 and 3000 cm^{-1} in the cellulose. In their experiments, cellulose sheets were stretched sinusoidally at low strains while being irradiated with polarized infrared light. For the obtained dynamic IR signals (the in-phase and the out-of-phase responses of the sample), the dynamic IR cross-correlation can be defined. The responses of the OH-groups to an external perturbation can be recorded as in-phase and out-of phase spectra. The cross correlation of these spectra gave the 2D synchronous (Figure 2) and asynchronous (Figure 3) plots, clearly showing the separated bands in the OH-vibration range and the relation of the OH-groups among them. It is apparent that most of the researchers have focused on the establishment of cellulose structure by investigating hydrogen bonds with FTIR. These (the structure of nature fibres) will be discussed in the next section. Few reports have described the correlation of hydrogen bonds with other characteristics of cellulose by using FTIR technologies.

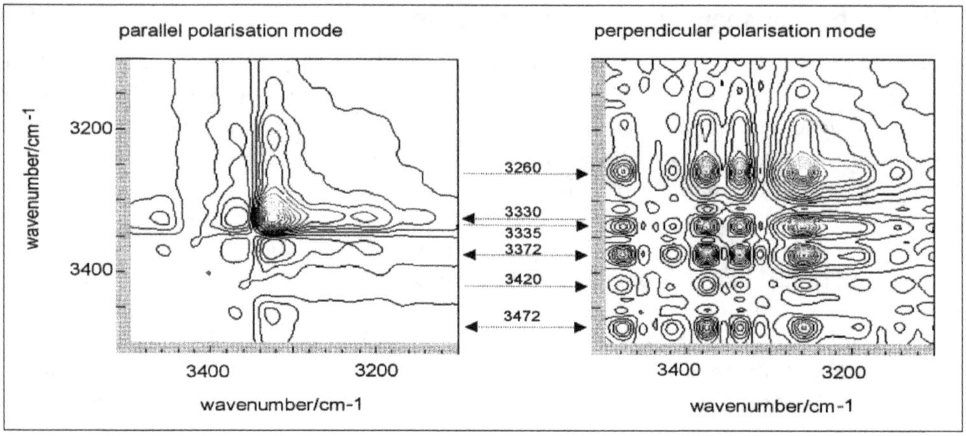

Fig. 2. Synchronous 2D plot-cross-correlation of in- and out-of-phase spectra [3]

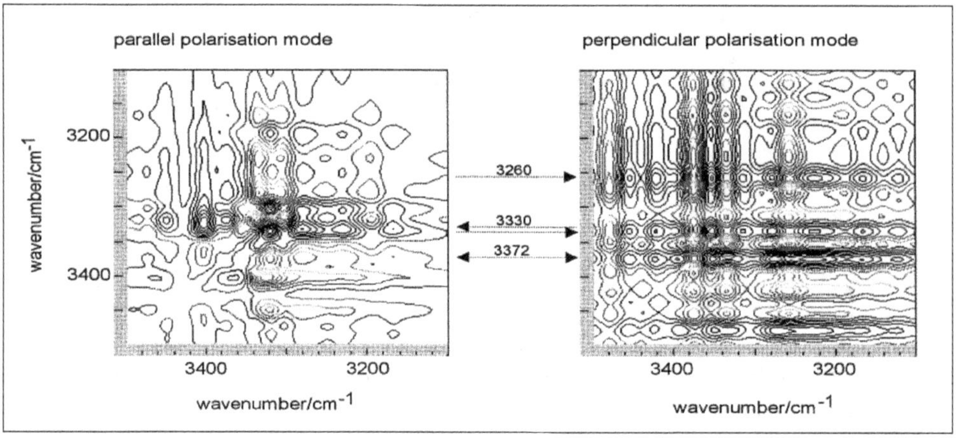

Fig. 3. Asynchronous 2D plot-cross-correlation of in- and out-of-phase spectra [3]

FTIR is very useful for examining the variation of hydrogen-bonds due to various defects [32]. The Nano-cellulose and Bio-composite Research Centre at Brunel University has investigated dislocations in natural fibres (hemp fibres) by using hydrogen-bonding characteristics under FTIR procedure. The test pieces were made from dislocation cluster (region) with the size of a single dislocation from a few microns to 100μm. The test pieces were then processed and examined by using FTIR measurement by using a Perkin-Elmer spectrometer and the standard KBr pellet technique. The recorded FTIR spectra (e.g. 3660–3000cm^{-1}) were deconvolved using Peak Fit V.4.12 software (Figure 4) and the peak positions of the major IR bands can be summarized and compared (Table 1). It can be found that the absorbance of hemp fibres without dislocations in the X–H (O–H and C–H) stretching region is much higher than that with dislocations. The peak positions of the four bands for hemp fibres with and without dislocations are 3450cm^{-1}, 3346cm^{-1}, 3262cm^{-1} and 3161cm^{-1} for the hemp without dislocations, and 3451cm^{-1}, 3350cm^{-1}, 3264cm^{-1} and 3167cm^{-1} for the dislocation regions. These bands are related to the valence vibration of hydrogen bonded OH groups [26]: i.e. band 1 to the intra-molecular hydrogen bond of O(2)H---O, band 2 to the intra-molecular hydrogen bond of O(3)H---O, band 3 to the intermolecular hydrogen bond of O(6)H---O and band 4 to the O---H stretching respectively.

Peak wavenumber (without dislocation) (cm^{-1})	Peak wavenumber (with dislocation) (cm^{-1})	$\Delta v(cm^{-1})$	Bonds
3327	3332	5	OH stretching
2883	2882	-1	C–H symmetrical stretching
1724	1724	0	C=O stretching vibration
1623	1624	1	OH bending of absorbed water
1506	disappear	-	C=C aromatic symmetrical stretching
1423	1423	0	HCH and OCH in-plane bending vibration
1368, 1363	1367,1363	-1/0	In-the-plane CH bending
1325	1325	0	S ring stretching
1314	1313	-1	CH_2 rocking vibration at C6
1259	1261	1	G ring stretching
1245	1244	-1	C-C plus C-O plus C=O stretch; G condensed > G etherfied
1232	1231	-1	COH bending at C6
1204	1199	-5	C-O-C symmetric stretching, OH plane deformation
1152	1156	4	C-O-C asymmetrical stretching
1046	1043	-3	C-C, C-OH, C-H ring and side group vibrations
1020	1018	-2	C-C, C-OH, C-H ring and side group vibrations
994	996	2	C-C, C-OH, C-H ring and side group vibrations
895	894	-1	COC,CCO and CCH deformation and stretching
662	663	1	C-OH out-of-plane bending

Table 1. Bonds wavenumber related to regions without and with dislocations

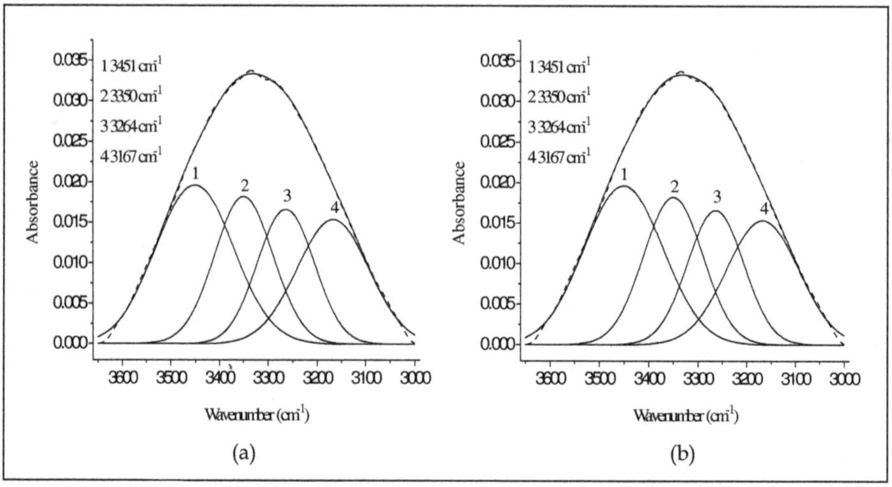

Fig. 4. Deconvolved FTIR spectra of the υ_{OH} region of hemp without dislocation (a) and dislocation regions (b). (Solid curves=calculated data; dotted curves=experimental data) [32]

It can be seen that the wave-numbers of peak position of dislocations are higher than those of hemp fibre without dislocation. This indicates that the degree of hydrogen bonding in dislocation regions is lesser than that in without dislocation regions. Furthermore, the absorbance of these bands in the dislocation regions is much lower than that in the regions without dislocations: for dislocation regions being about 79.3% for band 1, 64.4% for band 2, 64.9% for band 3 and 75.7% for band 4 those without dislocations respectively. These mean that the number of hydrogen bonds in dislocations is lower than without dislocation regions according to Beer–Lambert law.

Structure of natural fibres determined by using FTIR

The structure of cellulose has a remarkable and complex influence on the course of chemical reactions of the polymer (cellulosic materials). Generally, the structure of cellulose consists of three structural levels: namely (i) the molecular level of the single macromolecule; (ii) the supramolecular level of packing and mutual ordering of the macromolecules; (iii) the morphological level concerning the architecture of already rather complex structural entities, as well as the corresponding pore system [33]. This section only focuses on the molecular level and supramolecular level, and the morphological level will be discussed in the final section of this chapter.

Molecular orientation is one of the most important parameters, affecting the physical properties of macromolecular systems. It is often introduced in natural macromolecules by the mechanical deformation incurred during their processing. By using FTIR equipped with a microscopic accessory, Kataoka and Kondo [34] determined the molecular orientation of cellulose during the formation of wood cell wall by virtue of the C-O-C stretching mode parallel to molecular chains [23] (Figure 5). It was found that the molecular orientation of cellulose in the primary cell wall coincided with the direction of enlarging cellular growth. It is therefore that the cellulose in the (nascent) primary cell wall might be oriented during crystallization and subsequent formation of microfibrils due to the drawing stress/effect exerted during cellular enlargement. This force, distributed along molecular chains, can cause β-glucose chains in the nascent cellulose to crystallize in the I_α phase with a higher crystallinity, making the molecules orientated in the enlarging direction.

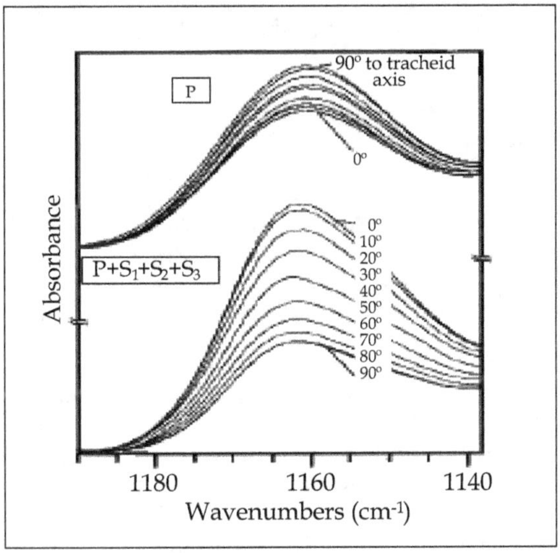

Fig. 5. Changes in FTIR spectra with a rotation of IR polarizer to the tracheid cell axis due to the C-O-C stretching mode: the primary (P) and the mature (P + S1 + S2 + S3) [34]

In order to better understand wood and wood fibres for their potential utilization in advanced materials, some researchers have employed FTIR in conjunction with mechanical loading to study the molecular responses to the stress/load, such as for spruce wood and cellulose paper materials [35] (Figure 6), illustrating the shift of the absorption peak at 1,160 cm-1, C-O-C vibration when the materials successively loaded from 0 up to 24 MPa at 0% RH. The decrease of the shift of absorption peak as the stress increased can be observed (6 wavenumbers in Figure 6). This decrease in wavenumber signifies an increase in the length of the covalent bonds involved in the vibration absorption, i.e. a decrease in the force constant of the bond. This demonstrates that FTIR-spectroscopy may be used to monitor molecular straining of cellulosic material under load and the molecular deformation is linearly related to the macroscopic load of the material. Using FTIR technologies, it was found that spectral deformations occurred in cellulose related groups, but no molecular deformation detected for the lignin or hemicelluloses of wood constituents. The molecular straining of the cellulose molecule resulted in greater macroscopic force under moist conditions compared to dry conditions, but an equal macroscopic strain under both conditions. This may be interpreted that moisture accessible regions are arranged tending in parallel with the cellulose load bearing entities, suggesting that the cellulose disordered regions may not exist as large regions across the cellulose aggregate structure, rather that are spread out. In addition, the moisture absorbing area of the cellulose structure is probably related to the surface areas of the cellulose.

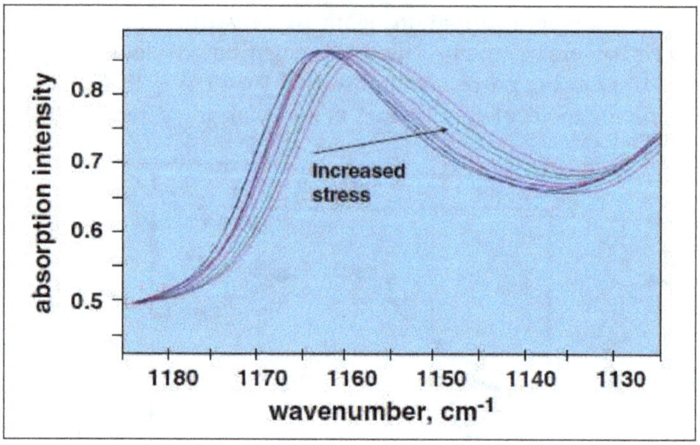

Fig. 6. Absorption spectra of the C–O–C vibration peak with increasing stress levels [35]

Polarized FTIR accompanied with a vapor-phase deuteration has been used to characterize orientation of the main chains and hence to study the molecular orientation of Nematic Ordered Cellulose (NOC) [36]. A ratio (R) of the absorbance of the band due to the particular molecular moiety for radiation polarized perpendicular to to parallel to the stretching direction was introduced to evaluate the orientation behaviour of the main chains and OH groups. Computation of the FTIR spectra (e.g. Figure 7) shows that R values for the main chain are 0.32, and OH group 0.81 for Intramolecular and 0.91 for intermolecular H.B.

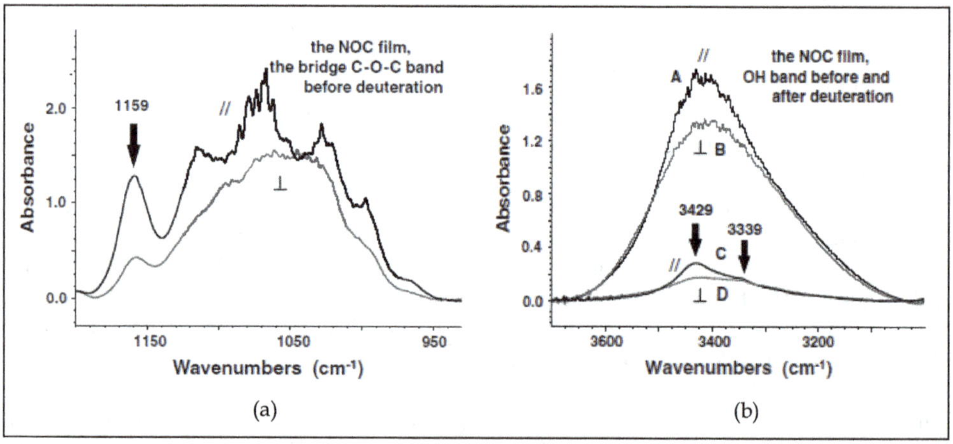

Fig. 7. The bridge C–O–C (a) and OH (b) stretching band for the NOC film before deuteration [for (b) (A, B)=before, (C, D)=after, (//)=electric vector parallel to and (⊥)=perpendicular to the stretching direction]

It is apparent that: (1) the R value for the β-glucan main chains of cellulose molecules is not necessarily in agreement with that for the side chains of OH groups; (2) the uniaxial drawing process to prepare the NOC film gave rise to the oriented main chains toward the stretching direction; (3) the nonoriented OH groups in the noncrystalline regions which occupy more than 80% of the drawn film samples could be the key for discouragement of the crystallization.

Supramolecular level investigated with FTIR mainly focuses on the crystal structure, which includes: 1) hydrogen bonding, 2) crystallinity measurement and 3) cellulose I_α and I_β determination. Kondo, Togawa and Brown [37] proposed a concept to describe how various states of molecular association can be categorized in cellulose. Figure 8 demonstrates the schematic representation of their concept.

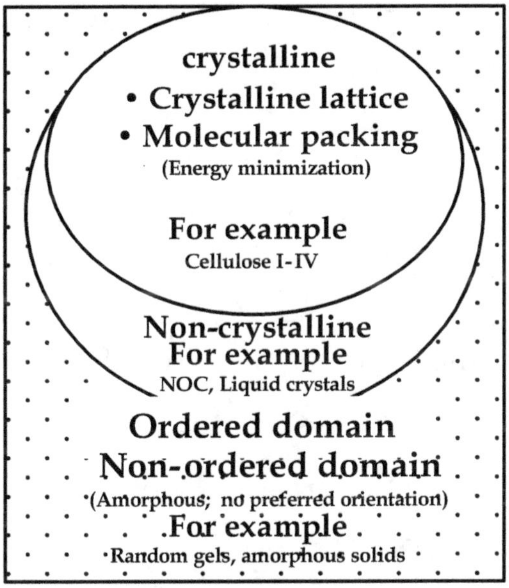

Fig. 8. Concept of glucan chain association for cellulose

According to two-phase model theory [38], there exist two regions in cellulose chain, namely amorphous and crystalline regions. Crystalline region in cellulose is an idealistic assembly of cellulose molecules in the biological system. There exist four different crystalline forms in cellulose. Researchers have developed various techniques to characterize the crystalline structure of cellulose, e.g. XRD, FTIR, Raman spectroscopy, and ^{13}C CP/MAS NMR. Among them FTIR is a more advanced tool for investigating the structure of cellulose. As mentioned above, since 1950s, some important work had been carried out by researchers and there are a number of literatures reporting on the IR/FTIR data of natural fibres [39].

The hydrogen bonds in cellulose mainly distribute in crystal domains and amorphous domains. It is possible to establish relation between the OH-bands and the cellulose structure. In 1913, Nishikawa and Ono [40] firstly revealed the crystalline nature of cellulose with X-ray diffraction. Cellulose has four polymorphic crystalline structures from cellulose I to cellulose IV. However, cellulose I and cellulose II have been most extensively studied. The other crystalline structures are still in question and yet to be studied further. According to Gardner-Blackwell model [41], hydrogen bonds for cellulose I include two intramolecular bonding, namely, O(2)H---O(6) bonding and O(3)H---O(5) bonding and one intermolecular bonding, O(6)H---O(3) (Figure 9a). Based on the Kolpak-Blackwell model [42], hydrogen bonds in cellulose II contains three intramolecular bonds: O(2)H---O(6) bonding, O(3)H---O(5) bonding and O(2)H---O(2) bonding, and two intermolecular bonding: O(6)H---O(2) and O(6)H---O(3) (Figure 9b). The IR assignments for OH regions in cellulose I and II are summarized in Table 2.

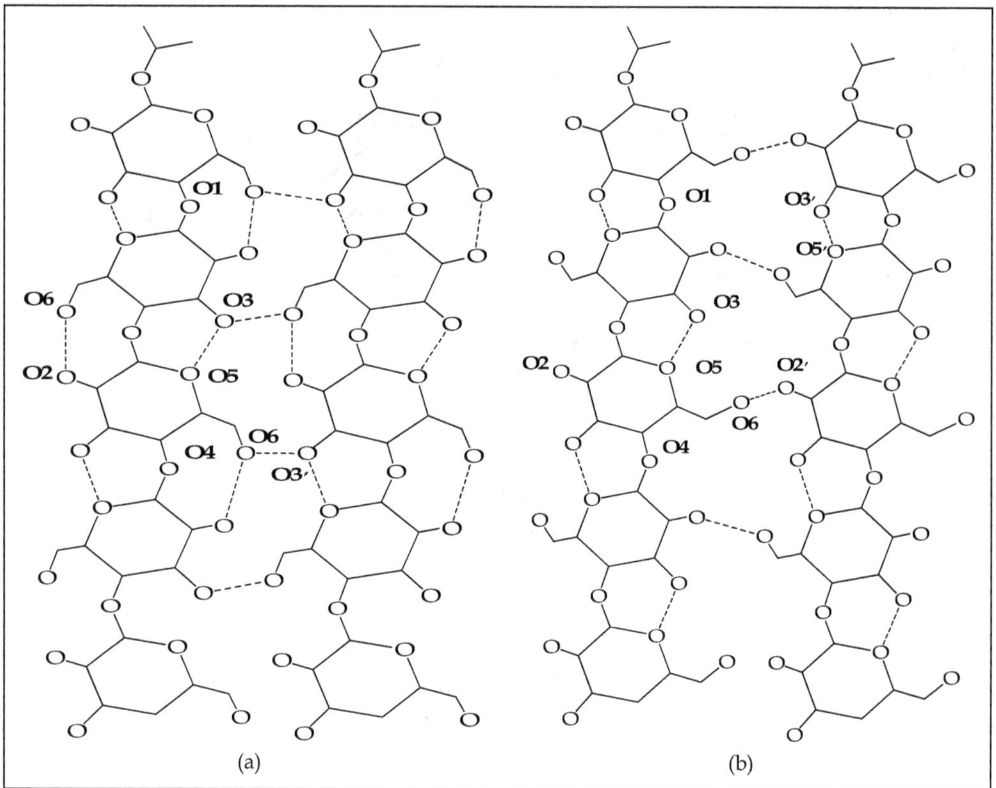

Fig. 9. Hydrogen-bonding network: (a) parallel to the *bc* plane (cellulose I); (b) in the centre chains (Cellulose II)

Peak wavenumber (cellulose I) (cm^{-1})	Peak wavenumber (cellulose II) (cm^{-1})	Bonds
	3175	OH stretching
3230-3310		O(6)H---O(3)
	3308	OH Inter H-bond
	3309	OH Inter H-bond
	3315	OH Intra H-bond
3340-3375		O(3)H---O(5)
	3374	OH Intra H-bond
3405-3460		O(2)H---O(6)
	3486	OH Intra H-bond

Table 2. Correlation of bonds and celluloses (structure) [43], [44]

Hatakeyama and his coworkers firstly studied the hydrogen bond in the amorphous regions of cellulose. These studies focus on investigating the effect of temperature on the formation of interchain hydrogen bonds [45], and the effect of hydrogen bonds on the temperature dependence of the dynamic modulus and the mechanical loss tangent [46]. In 1996, Kondo and Sawatari systematically examined the formation of hydrogen bonds in amorphous cellulose. The substituted amorphous cellulose derivatives, 6-O-, 2,3-di-O-, and tri-O-substituted methylcellulose, were used to model the components of amorphous cellulose. An artificial spectrum for amorphous cellulose was then quantitatively constructed by using compound IR spectra in order to investigate hydrogen bond formation in cellulose. The typical absorption wavenumber for the real and artificial spectra were summarized in Table 3.

Peak wavenumber (real) (cm^{-1})	Peak wavenumber (artificial) (cm^{-1})	Absorbance	Bond stretching
669	671	W	OH out-of-phase bending
899	892	M	Nonsymmetric out-phase ring
1040	1040	S	C-O
1070	1075	S	Skeletal vibrations C-O
1108	1108	S	Nonsymmetric in-phase ring
1159	1154	S	Nonsymmtric bridge C-O-C
1374	1375	M	CH bending
1420	1425	W	CH_2 symmetric bending
2892	2903	M	CH
3420	3457	S	OH

Table 3. Absorption wavenumber between the real and synthesized IR spectra of amorphous cellulose [43, 47]

The traditional two-phase cellulose model describes cellulose chains as containing both crystalline (ordered) and amorphous (less ordered) regions. A parameter termed the crystallinity index (CI) has been used to describe the relative amount of crystalline material in cellulose. The CI of celluloses has been measured using several different techniques including XRD, solid-state ^{13}C NMR, infrared (IR) spectroscopy and Raman spectroscopy. The determination of CI using FTIR spectroscopy is the simplest method. It should be noted that this method gives only relative values, because the spectrum always contains contributions from both crystalline and amorphous regions. In 1958, O'Connor [49] proposed Lateral Order Index (LOI, A_{1420}/A_{893}) to calculate the CI for cellulose. Later, Nelson and O'Connor [49-50] introduced Total Crystallinity Index (TCI, A_{1375}/A_{2900}) to evaluate the CI of cellulose. The absorbance ratio A_{1420}/A_{893} was defined as an empirical CI. The absorbance at 1420 and 894 cm^{-1} are sensitive to the amount of crystalline versus amorphous structure in the cellulose, that is, broadening of these bands reflects more disordered structure. As for TCI, various reports seem not to show a coherent result [51-52].

Chemical composition of natural fibres by using FTIR

Compositional variation and physical organization at the microscopic level determine the ability to perform a desired function for most materials. Lignocellulosic fibres from different lignocellulosic materials appear quite different, but the chemical composition is fairly similar although with different magnitudes of constituents. The major compositions of lignocellulosic fibres are cellulose, hemicellulose and lignin (see Figure 10), while the minor constituents include minerals, pectin, waxes and water-soluble components. The application of infrared spectroscopy in lignocellulosic fibres has a long history: The infrared spectroscopy was used to investigate the hydroxyl groups of cellulose in the 1930's [53] and significant efforts were made in the 1950's to assign the different absorption maxima in the IR spectrum of cellulose [54-59]; The absorption maxima in the IR spectra of lignin were investigated from 1940's [60-61] through 1950's [62-64]; The characteristic absorption maxima of hemicellulose were studied during the 50's [65-66].

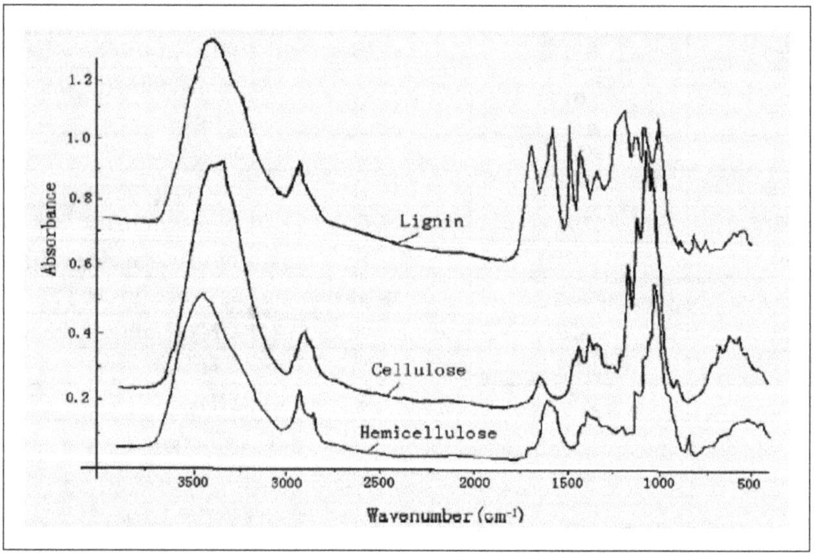

Fig. 10. IR spectra of cellulose, hemicellulose and lignin of natural fibres [60]

FTIR has been commonly used to characterize natural fibres with various treatments, e.g. grafting [67-68], coupling [69-71], mercerization [72-74]. With the aid of FTIR, researchers are able to obtain much more in-depth information of natural fibres after various modifications. FTIR is also an efficacy technique for the surface and interface characterizations of lignocellulosic fibres [75]. This allows further interpretation of the nature of adhesion between lignocellulosic with other substances. For example, Felix and Gatenholm [76] modified the lignocellulosic fibers with polypropylene–maleic anhydride copolymer. The spectrum of untreated fibres from the spectrum of treated fibres showed two peaks: one located at 1739 cm^{-1} and one at 1746 cm^{-1}, and the FTIR analysis indicated that the reaction between fibres and copolymer can be divided into two main steps: the copolymer is firstly converted into the more reactive anhydride form and then esterification takes place on the surface of cellulose fibres.

FTIR has recently been found most promising to examine the change of the chemical compositions of natural fibres (hemp fibres) due to inherent defects. An example of the results is given in Figure 11. A scrutiny of the IR spectra from 1370cm^{-1} to 1330cm^{-1} shows that the band at 1368cm^{-1} and 1363cm^{-1} almost disappears in dislocation regions (Figure 11a). These two bands, assigned as the in-plane CH bending, may be from hemicelluloses or cellulose, the near disappearance of these may be due to the removal of the hemicelluloses in dislocation regions. Hemicelluloses can form a linkage between cellulose and lignin, and lignin-carbohydrate complex with lignin by ether bonds [77]. The removal of hemicelluloses in dislocation regions may cause the decrease of transfer of shear stress under tensile loading and loss of lignin as well.

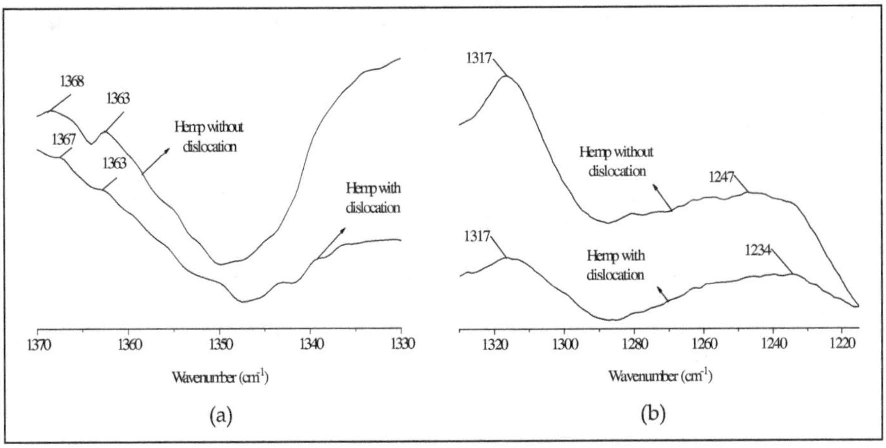

Fig. 11. FTIR spectra of hemp fibres from 1370 cm^{-1} to 1330 cm^{-1}(a) and from 1330 cm^{-1} to 1215 cm^{-1} (b) with and without dislocation [32]

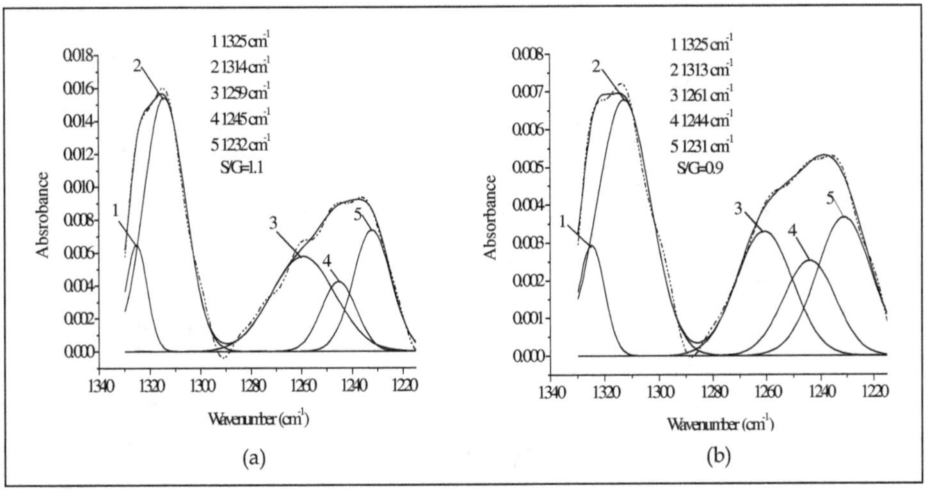

Fig. 12. Deconvolved FTIR spectra without dislocation (a) and dislocation regions (b). (Solid curves=calculated data; dotted curves=experimental data) [32]

The S ring (CH2 rocking at C6 in cellulose) and G ring stretching (C–C plus C–O plus C O stretch and COH bending at C6 in cellulose) could normally be observed in bands at 1325, 1314, 1259, 1245 and 1232cm^{-1} respectively for the hemp fibres without dislocation. Due to the overlapping of bands, only two peaks can be seen in Figure 11b. Lignin is composed of three basic units, namely p-hydroxyphenyl (H), guaiacyl (G) and syringyl (S) [78]. Guaiacyl (G) and syringyl (S) are the main units of lignin, but the ratio of S/G varies from one to another plant. It was reported recently by del Río et al. [79] that S/G values calculated upon FTIR were in agreement with those calculated upon Py-GC/MS at the bands of 1271cm^{-1} and 1327cm^{-1} respectively. However, the study on hemp fibre showed that the bands at 1271cm^{-1} and 1327cm^{-1}, assigned as G-ring stretching and S ring stretching respectively, were shifted to lower wavenumbers: for the hemp fibres without dislocations (Figure 12a), the G ring and S ring stretching appear at the bands of 1259cm^{-1} and 1325cm^{-1} and for the dislocation regions at 1261cm^{-1} and 1325cm^{-1} (Figure 12b).

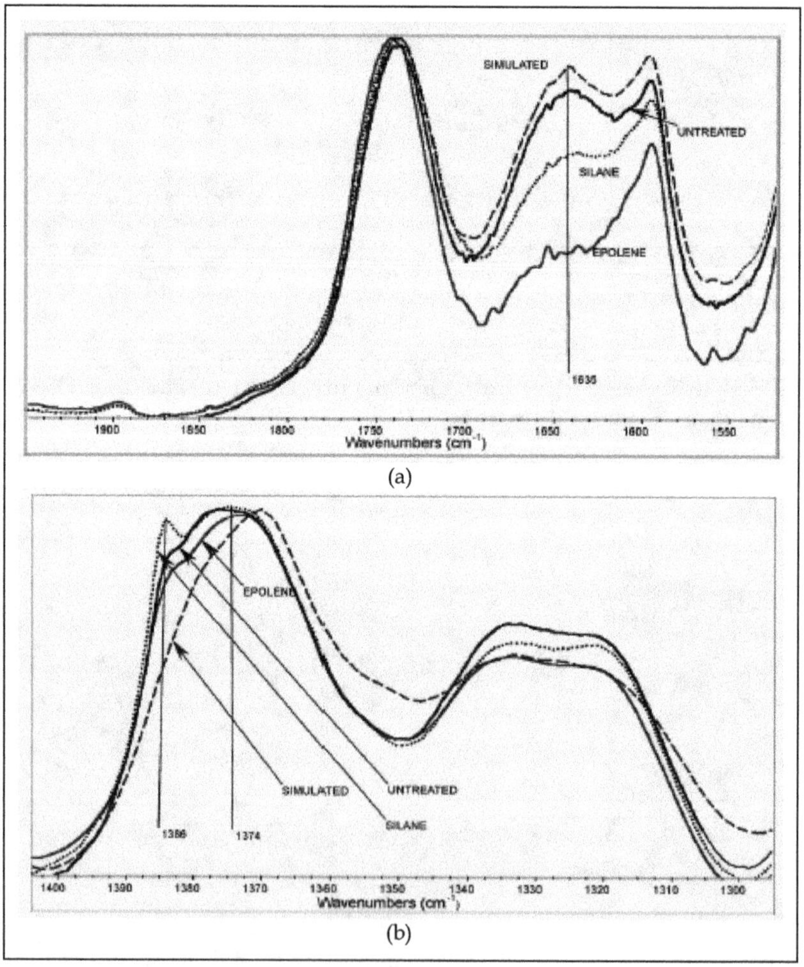

Fig. 13. FTIR spectra of various types of composites

The different molar contents of G-lignin and S-lignin of the hemp with and without dislocations gave rise to the ratio of S/G 0.9 for the former and 1.1 for the latter fibres. The lignin network in the parts without dislocations would be more rigid than that in dislocation regions. The lower absorbance in dislocations means that the lignin was removed from dislocation regions, and such the cellulose content in dislocations would be higher than that without dislocations.

FTIR can further be used to investigate the interfacial properties of natural fibre composites [80]. For example, Figures 13a and b exhibits the spectra for different types of composites containing 40% aspen fibres. The highest absorbance value corresponds to the untreated composites and the lowest value to that of composites modified with maleated polypropylene. The FTIR examination on the interface of wood fibre-reinforced polypropylene composites has also confirmed the efficacy of the technique [81]. The spectra are able to illustrate that the coupling agent was located around the wood fibers rather than randomly distributed in the polypropylene matrix, and the compatabilizer was attached to the wood fibers either by ester or hydrogen bonds.

Morphologies of natural fibres by using FTIR

FTIR spectroscopic imaging is the complete synthesis of FTIR spectroscopy with sample visualization and greatly extends the capabilities of conventional FTIR spectroscopy. Figure 14 illustrates a general configuration of an FTIR imaging micro- spectrometer. Spectral data can be represented as a picture, showing chemical information simultaneously from thousands of pixels. The main advantages of FTIR imaging are noninvasiveness, fast data collection and the ability to create visually appealing display. FTIR imaging not only provides new scientific capabilities, but it is also a compact and informative way to present results. It can collect more than 10,000 spectra in a few minutes. FTIR imaging has been

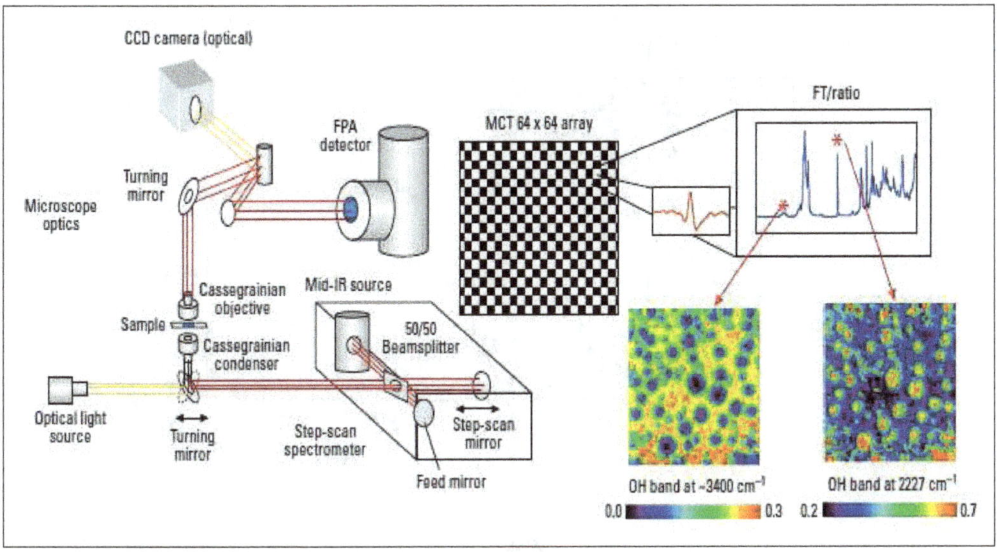

Fig. 14. Schematic of a typical FTIR imaging spectrometer [82]

found to be a remarkable tool for biological and materials analysis. It can be used extensively to investigate the chemical composition of stem [83-85] and cell wall structure [86] of natural fibres, and natural fibre composites [87].

FTIR imaging in conjunction with pyrolysis molecular beam mass spectrometry (py- MBMS) can work as a rapid analysis tool to evaluate difference in the chemical composition, for example, from the bark to the pith of wood stern (Figure 15) [85], and the data can statistically be processed to establish the correlation of the change in chemical features and the distance across the xylem (Figure 16).

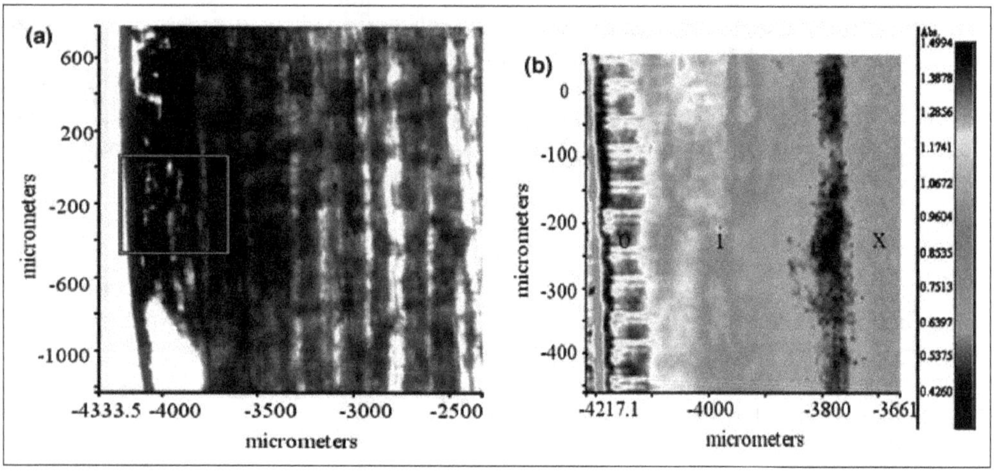

Fig. 15. (a) Visible image of the bark, cambium, and xylem of the control aspen stem. The area in the box was selected for FT-IR spectral analysis. (b) Spectral image of a portion of the outer bark [o], inner bark [I], cambium [c], xylem [x] showing the relative concentration of phenolic in these anatomical features [85]

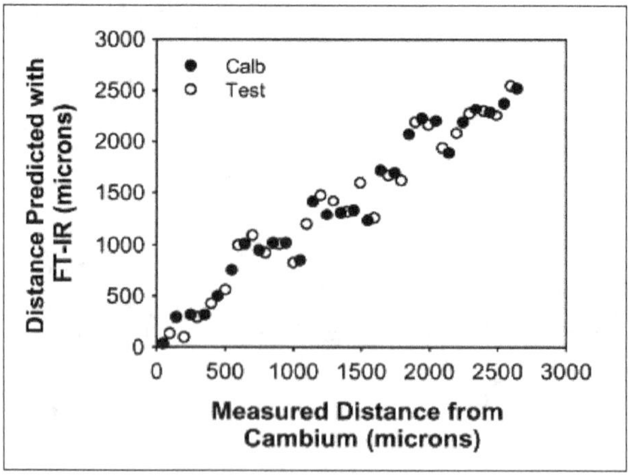

Fig. 16. PLS model predicting the distance from the bark to pith based on changes in the chemical composition. (Filled circles=calibration and open circles =test set) [85]

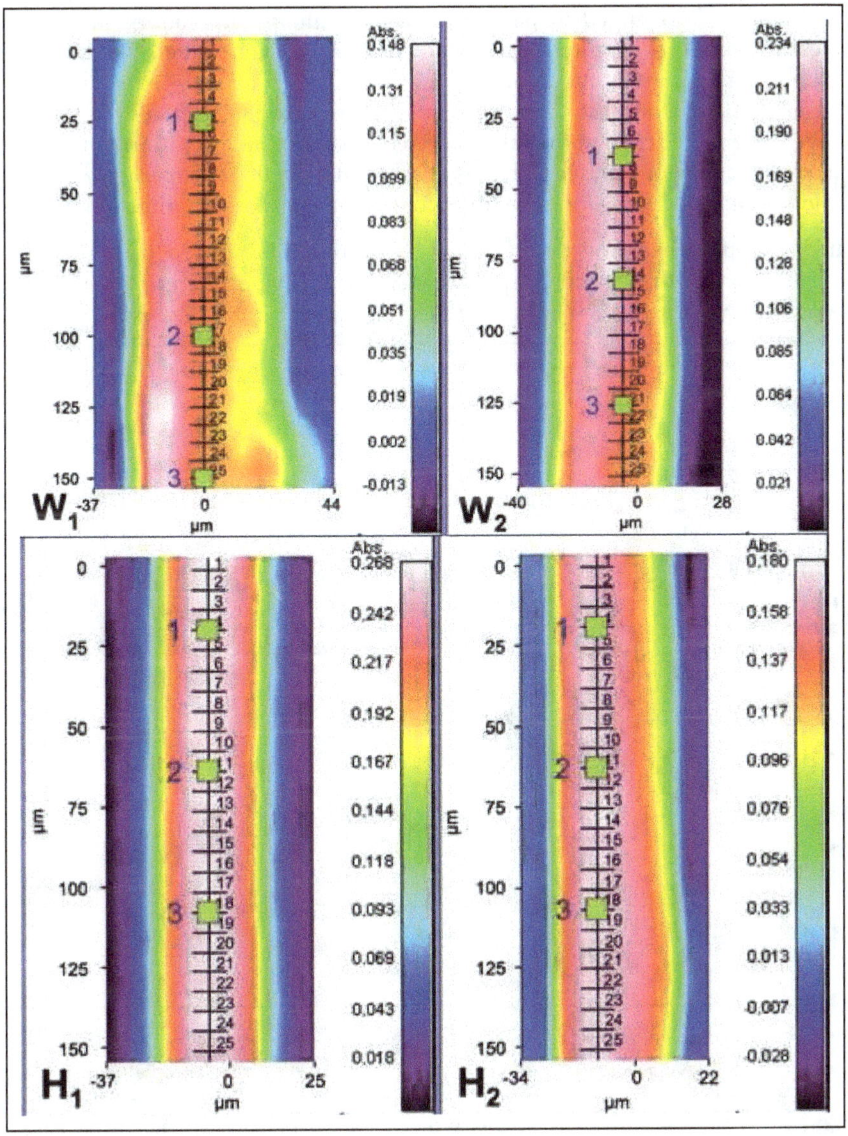

Fig. 17. Total IR absorbance full-spectral images of the two *W fibres* (*W*1 and *W*2) and the two *H fibres* (*H*1 and *H*2), showing the 25 pixel positions for each fibre used for evaluating the average orientation spectra as well as the three pixel positions for each fibre selected for evaluating the orientation of the different wood polymers in the fibres [86]

FTIR spectroscopy imaging has also been used to examine the orientation of the main wood compositions in transverse and longitudinal directions of wood fibres. For example, the examination by using FTIR on spruce fibres (Figures 17 and 18) [86] is able to illustrate that 1) glucomannan and xylan show a predominant orientation in the S_2 layer of cell wall,

2) hemicelluloses are arranged in parallel with the cellulose microfibrils and accordingly more or less in parallel with the longitudinal axis (the S_2 layer of the cell wall) of fibres, 3) only a little degree of orientation can be observed for lignin and 4) the variation in the molecular orientation along the fibres seems to be uniform in the pore-free regions. These results gave rise to a conclusion that all of three main components within fibres may have a clear anisotropic behaviour under mechanical stress, that is, their properties will be different in the longitudinal direction (along the fibre axis) and the transverse direction.

FTIR can be used to examine the structure of natural fibre based composites, such as, examining the surface distribution of polyacrylamide (PAM) or the in-plane distribution of cellulose within a paper sheet [87].

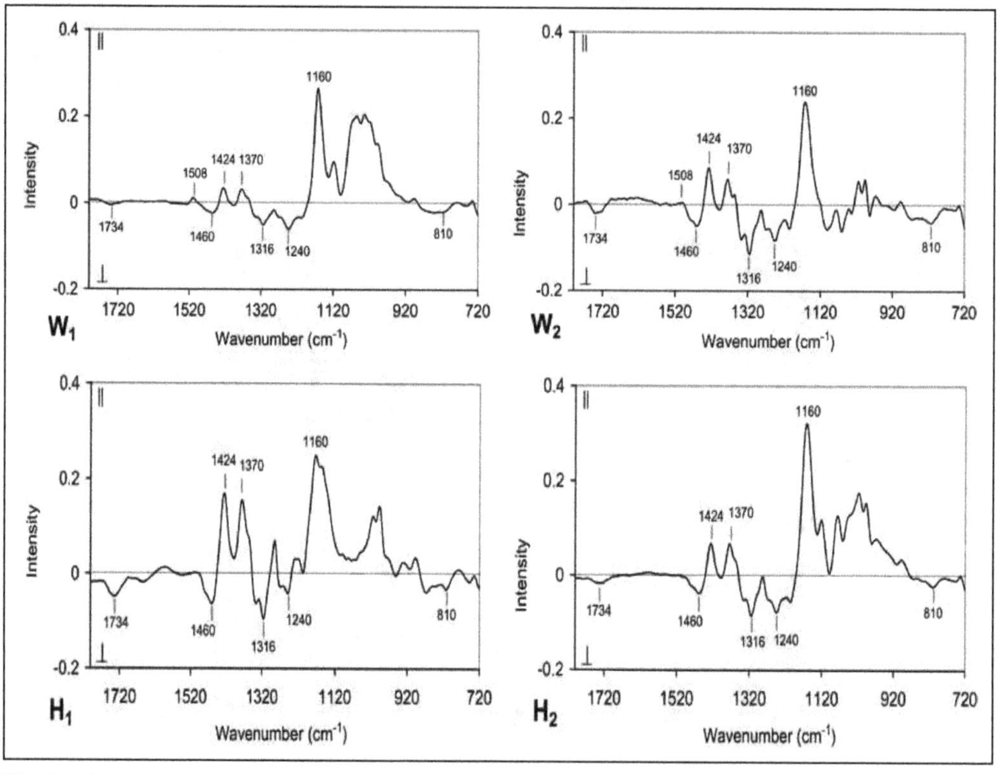

Fig. 18. Average orientation spectra of the two *W fibres* (*W1* and *W2*) and the two *H fibres* (*H1* and *H2*): *cellulose* 1160 cm^{-1}, 1316 cm^{-1}, 1370 cm^{-1} and 1424 cm^{-1}, *glucomannan* 810 cm^{-1}, *xylan* 1734 cm^{-1}, 1460 cm^{-1} and 1240 cm^{-1} and *lignin* 1508 cm^{-1} [86]

Conclusions

FTIR offers scientists an excellent range of solutions for understanding natural fibres and their related modification technologies and products, such as chemical compositions, microstructures, fibre architectures, characterisation of interface, and properties of both natural fibres and related composites.

FTIR is a powerful technique to examine the formation of inter- and intra- molecular hydrogen bonds in cellulose. The detailed database allows the establishment of strong correlation between the nature of hydrogen bonds and physical (e.g. solubility, hydroxyl reactivity, crystallinity) and mechanical properties of cellulose. The capability of accurate examination of hydrogen bonds has lead to an ever increasing uses of FTIR for investigating the defects (e.g. dislocation of hemp fibre) or deterioration (e.g. perturbation) of natural fibres and change of materials after modification.

The structure of cellulose has a profound influence on the course of chemical reactions of cellulose materials and the resulted properties. The molecular orientation and crystallization and formation of microfibrils not only vary from one plant to another, but could also change due to various environmental or other physical effects. FTIR is able to examine the nature of molecular chains, crystallinity and their correlations with various bonds.

In commons with other materials, the chemical composition at microscopic level determines the ability to perform various functions for the usefulness of natural fibres. FTIR has been mostly successful in accurate analysis of both major (cellulose, hemicellulose and lignin) and minor (mineral, pectin, waxes) constituents of natural fibres. Change in chemical compositions, interface and hence properties of natural fibres and composites could also be effectively identified by using FTIR.

FTIR is the most interesting and versatile of all analytical techniques and are well placed to become the technology of the century.

References

[1] Urban, M. W. (1993). Fourier Transform Infrared and Fourier Transform Raman Spectroscopy of Polymers, In: Urban, M. W. & Craver, C. D. (Ed.),3-40, ISBN 0841225257, the University of Virginia, American Chemical Society.

[2] Annette, N., Sudhakar, P., Ursula, K. & Andrea, P. Fourier Transform Infrared Microscopy in Wood Analysis. In: Ursula Kües(Ed.) Wood production, wood technology, and biotechnological impacts. Universitätsverlag Göttingen, 2007. 179-ISBN 978-3-940344-11-3.

[3] Hinterstoisser, B. & Salmén, L. (2000). Application of dynamic 2D FTIR to cellulose. Vibrational Spectroscopy, Vol.22, No.1-2, pp. 111-118, ISSN 0924-2031.

[4] Kokot, S., Czarnik-Matusewicz, B. & Ozaki, Y. (2002). Two-dimensional correlation spectroscopy and principal component analysis studies of temperature-dependent IR spectra of cotton–cellulose. Journal, Vol.67, No.6, pp. 456-469, ISSN 1097-0282.

[5] Guo, Y. & Wu, P. (2008). Investigation of the hydrogen-bond structure of cellulose diacetate by two-dimensional infrared correlation spectroscopy. Biopolymers, Vol.74, No.3, pp. 509-513, ISSN 0144-8617.

[6] http://en.wikipedia.org/wiki/Hydrogen_bond

[7] Kondo, T. (1994). Hydrogen bonds in regioselectively substituted cellulose derivatives. Journal of Polymer Science Part B: Polymer Physics, Vol.32, No.7, pp. 1229-1236, ISSN 1099-0488.

[8] Kondo, T. (1997). The assignment of IR absorption bands due to free hydroxyl groups in cellulose. Cellulose, Vol.4, No.4, pp. 281-292, ISSN 0969-0239.

[9] Kondo, T. (1997a). The relationship between intramolecular hydrogen bonds and certain physical properties of regioselectively substituted cellulose derivatives. Journal of Polymer Science Part B: Polymer Physics, Vol.35, No.4, pp. 717-723, ISSN 1099-0488.

[10] Kondo, T. (1997b). The relationship between intramolecular hydrogen bonds and certain physical properties of regioselectively substituted cellulose derivatives. Journal of Polymer Science Part B: Polymer Physics, Vol.35, No.4, pp. 717-723, ISSN 1099-0488.

[11] Itagaki, H., Tokai, M. & Kondo, T. (1997). Physical gelation process for cellulose whose hydroxyl groups are regioselectively substituted by fluorescent groups. Polymer, Vol.38, No.16, pp. 4201-4205, ISSN 0032-3861.

[12] Kondo, T., Sawatari, C., Manley, R. S. J. & Gray, D. G. (1994). Characterization of hydrogen bonding in cellulose-synthetic polymer blend systems with regioselectively substituted methylcellulose. Macromolecules, Vol.27, No.1, pp. 210-215, ISSN 0024-9297.

[13] Kondo, T. & Sawatari, C. (1994). Intermolecular hydrogen bonding in cellulose/ poly(ethylene oxide) blends: thermodynamic examination using 2,3-di-O- and 6-O-methylcelluloses as cellulose model compounds. Polymer, Vol.35, No.20, pp. 4423-4428, ISSN 0032-3861.

[14] O'Sullivan, A. (1997). Cellulose: the structure slowly unravels. Cellulose, Vol.4, No.3, pp. 173-207, ISSN 0969-0239.

[15] Tashiro, K. & Kobayashi, M. (1991). Theoretical evaluation of three-dimensional elastic constants of native and regenerated celluloses: role of hydrogen bonds. Polymer, Vol.32, No.8, pp. 1516-1526, ISSN 0032-3861.

[16] Gardner, K. H. & Blackwell, J. (1974). The hydrogen bonding in native cellulose. Biochimica et Biophysica Acta (BBA) - General Subjects, Vol.343, No.1, pp. 232-237, ISSN 0304-4165.

[17] Stipanovic, A. J. & Sarko, A. (1976). Packing Analysis of Carbohydrates and Polysaccharides. 6. Molecular and Crystal Structure of Regenerated Cellulose II. Macromolecules, Vol.9, No.5, pp. 851-857, ISSN 0024-9297.

[18] Woodcock, C. & Sarko, A. (1980). Packing Analysis of Carbohydrates and Polysaccharides. 11. Molecular and Crystal Structure of Native Ramie Cellulose. Macromolecules, Vol.13, No.5, pp. 1183-1187, ISSN 0024-9297.

[19] Langan, P., Nishiyama, Y. & Chanzy, H. (1999). A Revised Structure and Hydrogen-Bonding System in Cellulose II from a Neutron Fiber Diffraction Analysis. Journal of the American Chemical Society, Vol.121, No.43, pp. 9940-9946, ISSN 0002-7863.

[20] Nishiyama, Y., Sugiyama, J., Chanzy, H. & Langan, P. (2003). Crystal Structure and Hydrogen Bonding System in Cellulose Iα from Synchrotron X-ray and Neutron Fiber Diffraction. Journal of the American Chemical Society, Vol.125, No.47, pp. 14300-14306, ISSN 0002-7863.

[21] Marrinan, H. J. & Mann, J. (1954). A study by infra-red spectroscopy of hydrogen bonding in cellulose. Journal of Applied Chemistry, Vol.4, No.4, pp. 204-211, ISSN 1934-998X.

[22] Marrinan, H. J. & Mann, J. (1956). Infrared spectra of the crystalline modifications of cellulose. Journal of Polymer Science, Vol.21, No.98, pp. 301-311, ISSN 1542-6238.

[23] Mann, J. & Marrinan, H. J. (1958). Crystalline modifications of cellulose. Part II. A study with plane-polarized infrared radiation. Journal of Polymer Science, Vol.32, No.125, pp. 357-370, ISSN 1542-6238.

[24] Liang, C. Y. & Marchessault, R. H. (1959). Infrared spectra of crystalline polysaccharides. I. Hydrogen bonds in native celluloses. Journal of Polymer Science, Vol.37, No.132, pp. 385-395, ISSN 1542-6238.

[25] Marchessault, R. H. & Liang, C. Y. (1960). Infrared spectra of crystalline polysaccharides. III. Mercerized cellulose. Journal of Polymer Science, Vol.43, No.141, pp. 71-84, ISSN 1542-6238.

[26] Fengel, D. (1992). Characterization of Cellulose by Deconvoluting the OH Valency Range in FTIR Spectra. Holzforschung, Vol.46, No.4, pp. 283-288, ISSN 0018-3830.

[27] Fengel, D. (1993). Influence of Water on the OH Valency Range in Deconvoluted FTIR Spectra of Cellulose. Holzforschung, Vol.47, No.2, pp. 103-108, ISSN 0018-3830.

[28] Fengel, D. Jakob, H. & Strobel, C. (1995). Influence of the Alkali Concentration on the Formation of Cellulose II. Study by X-Ray Diffraction and FTIR Spectroscopy. Holzforschung, Vol.49, No.6, pp. 505-511, ISSN 0018-3830.

[29] Anthony J, M. (1988). Second derivative F.t.-i.r. spectra of celluloses I and II and related mono- and oligo-saccharides. Carbohydrate Research, Vol.173, No.2, pp. 185-195, ISSN 0008-6215.

[30] Anthony J, M. (1990). Second-derivative F.t.-i.r. spectra of native celluloses. Carbohydrate Research, Vol.197, No.0, pp. 53-60, ISSN 0008-6215.

[31] Anthony J, M. (1993). Second-derivative FTIR spectra of native celluloses from Valonia and tunicin. Carbohydrate Research, Vol.241, No.0, pp. 47-54, ISSN 0008-6215.

[32] Hinterstoisser, B. & Salmén, L. (1999). Two-dimensional step-scan FTIR: a tool to unravel the OH-valency-range of the spectrum of Cellulose I. Cellulose, Vol.6, No.3, pp. 251-263, ISSN 0969-0239.

[33] Dai, D. & Fan, M. (2011). Investigation of the dislocation of natural fibres by Fourier-transform infrared spectroscopy. Spectroscopy, Vol.55, No.2, pp. 300-306, ISSN 0924-2031.

[34] Klemm, D., Philipp, B., Heinze, T. & Heinze, U. (1998). Comprehensive cellulose chemistry Volume 1: Fundamentals and analytical methods. ISBN 3-527-29413-9, Weinheim, Germany.

[35] Kataoka, Y. & Kondo, T. (1998). FT-IR Microscopic Analysis of Changing Cellulose Crystalline Structure during Wood Cell Wall Formation. Journal, Vol.31, No.3, pp. 760-764, ISSN 0024-9297.

[36] Salmén, L. & Bergström, E. (2009). Cellulose structural arrangement in relation to spectral changes in tensile loading FTIR. Cellulose, Vol.16, No.6, pp. 975-982, ISSN 0969-0239.

[37] Hishikawa, Y., Togawa, E. & Kondo, T. (2010). Molecular orientation in the Nematic Ordered Cellulose film using polarized FTIR accompanied with a vapor-phase deuteration method. Cellulose, Vol.17, No.3, pp. 539-545, ISSN 0969-0239.

[38] Kondo, T.; Togawa, E. & Brown, R. M. (2001). "Nematic Ordered Cellulose": A Concept of Glucan Chain Association. Biomacromolecules, Vol.2, No.4, pp. 1324-1330, ISSN 1525-7797.

[39] Hearle, J. W. S. (1958). A fringed fibril theory of structure in crystalline polymers. Journal of Polymer Science, Vol.28, No.117, pp. 432-435, ISSN 1542-6238.

[40] Salmén, L., Åkerholm, M. (2005). Two-Dimensional Fourier Transform Infrared Spectroscopy Applied to Cellulose and Paper, In: Dumitriu S (Ed.), ISBN 3-540-37102-8, New York, USA

[41] Nishikawa, S. & Ono, S. (1913). Transmission of X-rays through fibrous, lamellar and granular substances. Proceedings of the Tokyo Mathematico-Physical, Vol.7, pp. 131

[42] Sugiyama, J., Vuong, R. & Chanzy, H. (1991). Electron diffraction study on the two crystalline phases occurring in native cellulose from an algal cell wall. Macromolecules, Vol.24, No.14, pp. 4168-4175, ISSN 0024-9297.

[43] Kolpak, F. J. & Blackwell, J. (1976). Determination of the Structure of Cellulose II. Macromolecules, Vol.9, No.2, pp. 273-278, ISSN 0024-9297.

[44] Kondo, T. (2005). Hydrogen Bonds in Cellulose and Cellulose Derivatives. In: Dumitriu S (Ed.), ISBN 3-540-37102-8, New York, USA

[45] Popescu, C.-M., Popescu, M.-C., Singurel, G., Vasile, C., Argyropoulos, D. S. & Willfor, S. (2007). Spectral Characterization of Eucalyptus Wood. Applied Spectroscopy, Vol.61, No.11, pp. 1168-1177, ISSN

[46] Hatakeyama, H., Hatakeyama, T. & Nakano, J. (1976). The effect of hydrogen-bond formation on the structure of amorphous cellulose. Applied Polymer Symposium, Vol. 28, pp.743-750, ISSN 0271-9460.

[47] Yano, S., Hatakeyama, H. & Hatakeyama, T. (1976). Effect of hydrogen bond formation on dynamic mechanical properties of amorphous cellulose. Journal of Applied Polymer Science, Vol.20, No.12, pp. 3221-3231, ISSN 1097-4628.

[48] Kondo, T. & Sawatari, C. (1996). A Fourier transform infra-red spectroscopic analysis of the character of hydrogen bonds in amorphous cellulose. Polymer, Vol.37, No.3, pp. 393-399, ISSN 0032-3861.

[49] O'Connor, R. T., DuPré, E. F. & Mitcham, D. (1958). Applications of Infrared Absorption Spectroscopy to Investigations of Cotton and Modified Cottons. Textile Research Journal, Vol.28, No.5, pp. 382-392, ISSN 0040-5175

[50] Nelson, M. L. & O'Connor, R. T. (1964). Relation of certain infrared bands to cellulose crystallinity and crystal latticed type. Part I. Spectra of lattice types I, II, III and of amorphous cellulose. Journal of Applied Polymer Science, Vol.8, No.3, pp. 1311-1324, ISSN 1097-4628.

[51] Nelson, M. L & O'Connor, R. T. Relation of certain infrared bands to cellulose crystallinity and crystal lattice type. Part II. A new infrared ratio for estimation of crystallinity in cellulose I and II. Journal of Applied Polymer Science, Vol.8, No.3, pp. 1325-1341, ISSN 1097-4628.

[52] Colom, X., Carrillo, F., Nogués, F. & Garriga, P. (2003). Structural analysis of photodegraded wood by means of FTIR spectroscopy. Journal, Vol.80, No.3, pp. 543-549, ISSN 0141-3910.

[53] Oh, S. Y., Yoo, D. I., Shin, Y. & Seo, G. (2005). FTIR analysis of cellulose treated with sodium hydroxide and carbon dioxide. Carbohydrate Research, Vol.340, No.3, pp. 417-428, ISSN 0008-6215.

[54] Ellis, J. W & Bath, J. (1938). The Near Infra-Red Absorption Spectrum of Sucrose Crystals in Polarized Light. Journal of Chemical Physics. Vol. 6, pp. 221-222,ISSN 0021-9606.

[55] Forziati, F.H., Stone, W.K., Rowen, J.W. & Appel, W.D. (1950). Cotton powder for infrared transmission measurements. Journal of Research of the National Bureau of Standards, Vol.45, No.(2, pp. 109-113.

[56] Rowen, J. W., Forziati, F. H. & Reeves, R. E. (1951). Spectrophotometric Evidence for the Absence of Free Aldehyde Groups in Periodate-oxidized Cellulose1. Journal of the American Chemical Society, Vol.73, No.9, pp. 4484-4487, ISSN 0002-7863.

[57] Shirk, H. G. & Greathouse, G. A. (1952). Infrared Spectra of Bacterial Cellulose. Analytical Chemistry, Vol.24, No.11, pp. 1774-1775, ISSN 0003-2700.

[58] Ermolenko, I. N., Zhbankov, R. G., Ivanov, V. I., Lenshina, N. I. & Ivanova, V. S. (1958). Investigation of some oxidation reactions of cellulose by infrared spectroscopy. Russian Chemical Bulletin, Vol.7, No.2, pp. 241-243, ISSN 1066-5285.

[59] Klein, E. & Snowden, J. (1958). Replacing Hydroxyl Groups in Cotton Cellulose. Industrial & Engineering Chemistry, Vol.50, No.1, pp. 80-82, ISSN 0019-7866.

[60] Liang, C. Y. & Marchessault, R. H. (1959). Infrared spectra of crystalline polysaccharides. II. Native celluloses in the region from 640 to 1700 cm^{-1}. Journal of Polymer Science, Vol.39, No.135, pp. 269-278, ISSN 1542-6238.

[61] Jones, E. J. (1948). The infrared spectrum of spruce native lignin. Journal of the American Chemical Society, Vol.70, No.5, pp. 1984-1985, ISSN 0002-7863.

[62] Buchanan, M. A., Brauns, F. E. & Leaf, R. L. (1949). Native Lignin. II. Native Aspen Lignin. Journal of the American Chemical Society, Vol.71, No.4, pp. 1297-1299, ISSN 0002-7863.

[63] Stevens, G. D. & Nord, F. F. (1951). Investigations on Lignin and Lignification. VIII.1 Isolation and Characterization of Bagasse Native Lignin. Journal of the American Chemical Society, Vol.73, No.10, pp. 4622-4625, ISSN 0002-7863.

[64] Kudzin, S. F. & Nord, F. F. (1951). Investigations on Lignin and Lignification. IV. Studies on Hardwood Lignin. Journal of the American Chemical Society, Vol.73, No.2, pp. 690-693, ISSN 0002-7863.

[65] Hergert, H. L. & Kurth, E. F. (1953). The Infrared Spectra of Lignin and Related Compounds. I. Characteristic Carbonyl and Hydroxyl Frequencies of Some Flavanones, Flavones, Chalcones and Acetophenones1. Journal of the American Chemical Society, Vol.75, No.7, pp. 1622-1625, ISSN 0002-7863.

[66] Srivastava, H. C. & Adams, G. A. (1959). Uronic Acid Components of Jute Fiber Hemicellulose1,2. Journal of the American Chemical Society, Vol.81, No.10, pp. 2409-2412, ISSN 0002-7863.

[67] Timell, T. E. (1959). The constitution of a hemicellulose ffrom sugar maple (acer saccharum). Canadian Journal of Chemistry, Vol.37, No.5, pp. 893-898, ISSN 0008-4042

[68] Castelvetro, V., Fatarella, E., Corsi, L., Giaiacopi, S. & Ciardelli, G. (2006). Graft Polymerisation of Functional Acrylic Monomers onto Cotton Fibres Activated by Continuous Ar Plasma. Plasma Processes and Polymers, Vol.3, No.1, pp. 48-57, ISSN 1612-8869

[69] Krouit, M., Bras, J. & Belgacem, M. N. (2008). Cellulose surface grafting with polycaprolactone by heterogeneous click-chemistry. European Polymer Journal, Vol.44, No.12, pp. 4074-4081, ISSN 0014-3057

[70] Singh, B., Gupta, M., Verma, A. & Tyagi, O. S. (2000). FT-IR microscopic studies on coupling agents: treated natural fibres. Polymer International, Vol.49, No.11, pp. 1444-1451, ISSN 1097-0126

[71] Abdelmouleh, M., Boufi, S., Belgacem, M. N., Duarte, A. P., Ben Salah, A. & Gandini, A. (2004). Modification of cellulosic fibres with functionalised silanes: development of surface properties. International Journal of Adhesion and Adhesives, Vol.24, No.1, pp. 43-54, ISSN 0143-7496

[72] Sgriccia, N., Hawley, M. C. & Misra, M. (2008). Characterization of natural fiber surfaces and natural fiber composites. Composites Part A: Applied Science and Manufacturing, Vol.39, No.10, pp. 1632-1637, ISSN 1359-835X

[73] Mwaikambo, L. Y. & Ansell, M. P. (1999). The effect of chemical treatment on the properties of hemp, sisal, jute and kapok for composite reinforcement. Die Angewandte Makromolekulare Chemie, Vol.272, No.1, pp. 108-116, ISSN 1522-9505

[74] Mwaikambo, L. Y. & Ansell, M. P. (2002). Chemical modification of hemp, sisal, jute, and kapok fibers by alkalization. Journal of Applied Polymer Science, Vol.84, No.12, pp. 2222-2234, ISSN 1097-4628

[75] Gwon, J. G., Lee, S. Y., Doh, G. H. & Kim, J. H. (2010). Characterization of chemically modified wood fibers using FTIR spectroscopy for biocomposites. Journal of Applied Polymer Science, Vol.116, No.6, pp. 3212-3219, 1097-4628

[76] Jacob, M., Joseph, S., Pothan, L. A. & Thomas, S. (2005). A study of advances in characterization of interfaces and fiber surfaces in lignocellulosic fiber-reinforced composites. Composite Interfaces, Vol.12, No.1/2, pp. 95-124, ISSN 0927-6440

[77] Felix, J. M. & Gatenholm, P. (1991). The nature of adhesion in composites of modified cellulose fibers and polypropylene. Journal of Applied Polymer Science, Vol.42, No.3, pp. 609-620, ISSN 1097-4628

[78] Sjöström, E. (1981). Wood Chemistry: Fundamentals and Applications. ISBN 0-12-647481-18, San Diego, USA.

[79] Ruan, D., Zhang, L., Mao, Y., Zeng, M. & Li, X. (2004). Microporous membranes prepared from cellulose in NaOH/thiourea aqueous solution. Journal of Membrane Science, Vol.241, No.2, pp. 265-274, ISBN 0376-7388

[80] del Río, J. C., Gutiérrez, A., Rodríguez, I. M.; Ibarra, D. & Martínez, Á. T. (2007). Composition of non-woody plant lignins and cinnamic acids by Py-GC/MS, Py/TMAH and FT-IR. Journal of Analytical and Applied Pyrolysis, Vol.79, No.1-2, pp. 39-46, ISSN 0165-2370

[81] Colom, X., Carrasco, F., Pagès, P. & Cañavate, J. (2003). Effects of different treatments on the interface of HDPE/lignocellulosic fiber composites. Composites Science and Technology, Vol.63, No.2, pp. 161-169, ISSN 0266-3538

[82] Hristov, V. & Vasileva, S. (2003). Dynamic Mechanical and Thermal Properties of Modified Poly(propylene) Wood Fiber Composites. Macromolecular Materials and Engineering, Vol.288, No.10, pp. 798-806, ISSN 1439-2054

[83] Koenig, J. L. (2001). FTIR images: a new technique produces images worth a thousand spectra. Analytical chemistry (Washington), Vol.73, No.13, pp. 360A, ISSN 0003-2700

[84] Himmelsbach, D. S., Khalili, S. & Akin, D. E. (2002). The use of FT-IR microspectroscopic mapping to study the effects of enzymatic retting of flax (Linum usitatissimum L)

stems. Journal of the Science of Food and Agriculture, Vol.82, No.7, pp. 685-696, ISSN 1097-0010

[85] Himmelsbach, D.S.; Khalili, S. & Akin, D.E. (1998) FT-IR microspectroscopic imaging of flax (Linum usitatissimum L.) stems. Molecular and Cellular Biology, Vol. 44, No.1, pp.99-108, ISSN 0270-7306

[86] Labbé, N., Rials, T. G., Kelley, S. S., Cheng, Z.-M., Kim, J.-Y. & Li, Y. (2005). FT-IR imaging and pyrolysis-molecular beam mass spectrometry: new tools to investigate wood tissues. Wood Science and Technology, Vol.39, No.1, pp. 61-76, ISSN 0043-7719

[87] Stevanic, J. S. & Salmén, L. (2009). Orientation of the wood polymers in the cell wall of spruce wood fibres. Holzforschung, Vol.63, No.5, pp. 497-503, ISSN 0018-3830

[88] Sakaemura, T., Mihara, I. & Yamauchi, T. (2009). Microscopic Attenuated Total Reflection/Fourier Transform Infrared Imaging of Paper Containing a Polyacrylamide Dry Strength Resin. Sen'i Gakkaishi, Vol.65, No.9, pp. 252-255, ISSN OO37-9875

11

High Resolution Mass Spectrometry using FTICR and *Orbitrap* Instruments

Paulo J. Amorim Madeira, Pedro A. Alves and Carlos M. Borges
Centro de Química e Bioquímica, Departamento de Química e Bioquímica, Faculdade de Ciências da Universidade de Lisboa
Portugal

Introduction

From the 1950s to the present, mass spectrometry has evolved tremendously. The pioneering mass spectrometrist had a home-built naked instrument, typically a magnetic sector instrument with electron ionisation. Nowadays, highly automated commercial systems, able to produce thousands of spectra per day, are now concealed in a "black box", a nicely designed and beautifully coloured unit resembling more an espresso machine or a tumble dryer than a mass spectrometer.

Mass spectrometry (MS) is probably the most versatile and comprehensive analytical technique currently available in the chemists and biochemists' arsenal. Mass spectrometry precisely measures the molecular masses of individual compounds by converting them into charged ions and analysing them in what is called a mass analyser. This is the simplest, but somewhat reductionist, definition of mass spectrometry. The days of the simple determination of the m/z ratio of an organic compound are over, today mass spectrometry can be used to determine molecular structures, to study reaction dynamics and ion chemistry, provides thermochemical and physical properties such as ionisation and appearance energies, reaction enthalpies, proton and ion affinities, gas-phase acidities, and so on.

Mass spectrometry is so versatile that even several areas of physics, pharmaceutical sciences, archaeology, forensic and environmental sciences, just to state a few, have benefited from the advances in this instrumental technique.

The history of mass spectrometry starts in 1898 with the work of Wien, who demonstrated that canal rays could be deflected by passing them through superimposed parallel electric and magnetic fields. Nevertheless, its birth can be credited to Sir J. J. Thomson, Cavendish Laboratory of the University of Cambridge, through his work on the analysis of negatively and positively charged cathode rays with a parabola mass spectrograph, the great grand-father of the modern mass spectrometers. (Thomson 1897; Thomson 1907) In the next two decades, the developments of mass spectrometry continued by renowned physicists like Aston, (Aston 1919) Dempster, (Dempster 1918) Bainbridge, (Bainbridge 1932; Bainbridge and Jordan 1936) and Nier. (Nier 1940; Johnson and Nier 1953)

In the 1940s, chemists recognised the great potential of mass spectrometry as an analytical tool, and applied it to monitor petroleum refinement processes. The first commercial mass spectrometer became available in 1943 through the Consolidated Engineering Corporation. The principles of time-of-flight (TOF) and ion cyclotron resonance (ICR) were introduced in 1946 and 1949, respectively. (Sommer, Thomas et al. 1951; Wolff and Stephens 1953)

Applications to organic chemistry started to appear in the 1950s and exploded during the 1960s and 1970s. Double-focusing high-resolution mass spectrometers, which became available in the early 1950s, paved the way for accurate mass measurements. The quadrupole mass analyser and the ion trap were described by Wolfgang Paul and co-workers in 1953. (Paul 1990) The development of gas chromatography/mass spectrometry (GC/MS) in the 1960s marked the beginning of the analysis of seemingly complex mixtures by mass spectrometry. (Ryhage 2002; Watson and Biemann 2002) The 1960s also witnessed the development of tandem mass spectrometry and collision-induced decomposition, (Jennings 1968) being a breakthrough in structural and quantitative analysis, as well as in the development of soft ionisation techniques such as chemical ionisation. (Munson and Field 1966)

By the 1960s, mass spectrometry had become a standard analytical tool in the analysis of organic compounds. Its application to the biosciences, however, was lacking due to the inexistence of suitable methods to ionise fragile and non-volatile compounds of biological origin. During the 1980s the range of applications in the field of the biosciences increased "exponentially" with the development of softer ionisation methods. These included fast atom bombardment (FAB) in 1981, (Barber, Bordoli et al. 1981) electrospray ionisation (ESI) in 1984-1988, (Fenn, Mann et al. 1989) and matrix-assisted laser desorption/ionisation (MALDI) in 1988. (Karas and Hillenkamp 2002) With the development of the last two methods, ESI and MALDI, the upper mass range was extended beyond 100 kDa and had an enormous impact on the use of mass spectrometry in biology and life sciences. This impact was recognised in 2002 when John Fenn (for his work on ESI) and Koichi Tanaka (for demonstrating that high molecular mass proteins could be ionised using laser desorption) were awarded with the Nobel Prize in Chemistry.

Concurrent with the development of ionisation methods, several innovations in mass analyser technology, such as the introduction of high-field and superfast magnets, as well as the improvements in the TOF and Fourier transform ion cyclotron resonance (FTICR) enhanced the sensitivity and the upper mass range. The new millennium brought us two new types of ion traps, the orbitrap which was invented by Makarov (Makarov 2000) and the linear quadrupole ion trap (LIT) which was developed by Hager. (Hager 2002)

The coupling of high-performance liquid chromatography (HPLC) with mass spectrometry was first demonstrated in the 1970s (Dass 2007); nevertheless, it was with the development and commercialisation of atmospheric pressure ionisation sources (ESI, APCI) that for the first time the combination of liquid chromatography and mass spectrometry entered the realm of routine analysis. (Voyksner 1997; Covey, Huang et al. 2002; Whitehouse, Dreyer et al. 2002; Rodrigues, Taylor et al. 2007)

Generally, a mass spectrometer is composed of five components (Fig. 1): inlet system, ion source, mass analyser, ion detector, and data system.

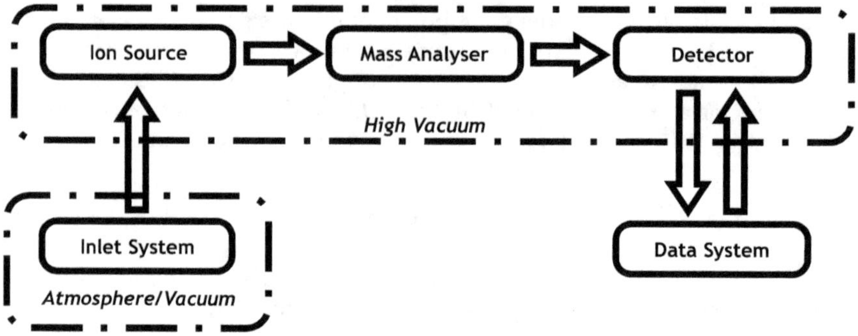

Fig. 1. Diagram of the major components common to all typical modern mass spectrometers

Samples are introduced in the mass spectrometer and transferred into the gas phase through the inlet system that could be at atmospheric pressure or under vacuum. In the ion source, the gas-phase analytes are ionised and transferred into the mass analyser where they are separated according to their mass-to-charge ratios (*m/z*). Ion detection can be accomplished by electron multiplier systems that enable *m/z* and abundance to be measured and displayed by means of an electric signal perceived by the data system, which also controls the equipment. All mass spectrometers are equipped with a vacuum system in order to maintain the low pressure (high vacuum) required for operation. This high vacuum is necessary to allow ions to reach the detector without undergoing collisions with other gaseous molecules. In fact, collisions would produce a deviation of the trajectory and the ion would lose its charge against the walls of the instrument. On the other hand, a relatively high pressure environment could facilitate the occurrence of ion-molecule reactions that would increase the complexity of the spectrum. In some experiments the pressure in the source region or in a part of the mass spectrometer is intentionally increased to study ion-molecule reactions or to perform collision-induced dissociations. The high vacuum is maintained using mechanical pumps in conjunction with turbomolecular, diffusion or cryogenic pumps. The mechanical pumps allow a vacuum of about 10^{-3} torr to be obtained. Once this vacuum is achieved the operation of the remainder of the vacuum system allows a vacuum as high as 10^{-10} torr to be reached.

Fourier transforms in mass spectrometry

In the following sections we will briefly describe two types of mass analysers that employ Fourier transforms to determine *m/z* ratios. We will describe the Fourier Transform Ion Cyclotron Resonance mass spectrometer (FTICR MS) and the *Orbitrap* in sections, respectively. The basic aspects of each mass analyser will be dealt with; nevertheless, the interested reader is encouraged to seek more information in the literature. For exam ple, in the case of FTICR mass spectrometry several reviews (Marshall, Hendrickson et al. 1998; Zhang, Rempel et al. 2005) and books are available. (Marshall and Verdun 1990; Gross 2004; Dass 2007; Hoffmann and Stroobant 2007) For the *Orbitrap*, the operation principles are well described in the papers published by Makarov, its inventor, (Makarov 2000; Hu, Noll et al. 2005; Makarov, Denisov et al. 2006) as well as by other authors, (Perry, Cooks et al. 2008) and in the more recent editions of some mass spectrometry textbooks. (Dass 2007; Hoffmann and Stroobant 2007)

Fourier transform ion cyclotron resonance mass spectrometry (FTICR MS)

The theory of cyclotron resonance was developed in the 1930s by Lawrence (1951 Nobel Prize in Physics). Lawrence built the first cyclotron accelerator to study the fundamental properties of the atom. Subsequently, Penning devised the first trap for charged particles by using a combination of static electric and magnetic fields to confine electrons. (Vartanian, Anderson et al. 1995) In the 1950s the principle of ion cyclotron resonance was first incorporated into a mass spectrometer, called the omegatron, by Sommer and co-workers, who successfully applied the concept of cyclotron resonance to determine the charge-to-mass ratio of the proton. (Sommer, Thomas et al. 1951) Major improvements in ICR awaited McIver's introduction of the trapped ion cell. Unlike the conventional drift cell, the trapped ion cell allowed for ion formation, manipulation and detection to occur within the same volume in space. The trapped ion cell differed from previous ICR cell designs by the inclusion of "trapping" electrodes. By applying small voltages to these electrodes, McIver was able to trap ions for 1-2 ms (approximately 100 times that of the drift cell). These advantages led to a much greater dynamic range, sensitivity and mass resolution. More importantly, the extended trapping capability of the McIver cell was a prerequisite for the FTICR detection technique invented by Comisarow and Marshall later that decade. In the second half of the 1970s, Comisarow and Marshall adapted Fourier transform methods to ICR spectrometry and built the first FTICR-MS instrument. (Comisarow and Marshall 1974; Marshall, Comisarow et al. 1979) Since then, FTICR-MS has matured into a state-of-the-art high-resolution mass spectrometry instrument for the analysis of a wide variety of compounds (biological or not).

All FTICR-MS systems have in common five main components: a magnet (nowadays usually a superconducting magnet); analyser cell (placed in the strong magnetic field created by the magnet); ultra-high vacuum system, and ion source (Fig. 2); and a sophisticated data system (many of the components in the data system are similar to those used in NMR).

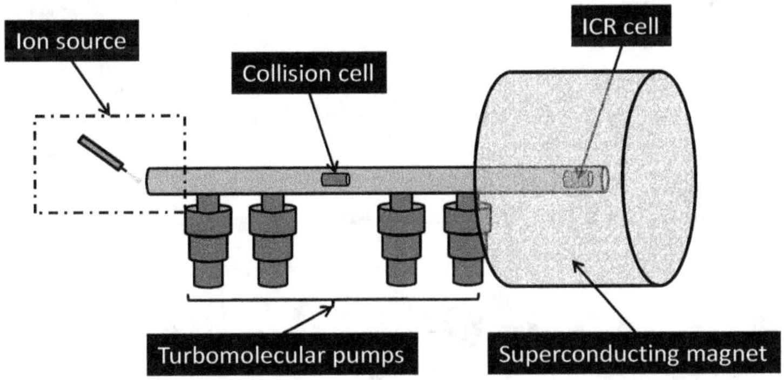

Fig. 2. Schematic representation of an FTICR mass spectrometer. Note that not all components are present in this scheme, for example, the ion optics is not presented, nor the rotary vacuum pumps that are needed for the proper functioning of the turbomolecular pumps

In this section, we shall not discuss the magnet, vacuum and data systems, and focus on the ICR cell, which is the heart of the FTICR-MS instrument. It is here that the ions are stored, mass analysed and detected.

Inside the FTICR cell an ion has three natural motions: the cyclotron, the trapping and the magnetron motions. The nature of each motion will be briefly explained in the following paragraphs.

Many fundamental aspects of FTICR can be understood from very simple idealised models:

- Ion cyclotron frequency, radius, velocity and energy, as a function of ion mass, ion charge and magnetic field strength, follow directly from the motion of an ion in a spatially uniform static magnetic field.
- Ion cyclotron motion may be rendered coherent (and thus observable) by the application of a spatially uniform RF electric field (excitation) at the same frequency as the ion cyclotron frequency. The ICR signal results from induction of an oscillating "image" charge on two conductive infinitely extended opposed parallel electrodes. A frequency-domain spectrum is obtained by Fourier transformation of the digitised ICR signal.
- Confinement of ions by application of a three-dimensional axial quadrupolar DC electric field shifts the ion cyclotron frequency, whereas excitation and detection remain essentially linear, but with a reduced proportionality constant.
- Collisions broaden the ICR signal in a simple way, and actually make it possible to cool and compress an ion packet for improved detection.
- Although FTICR-MS has been coupled to virtually every type of ion source, most ion sources work best outside the magnet. Thus, several methods have been developed to guide the externally generated ions into the ion trap inside a high-field magnet.
- The above features may be combined in various experimental "event sequences" to perform tandem-in-time mass spectrometry (MS/MS or MS^n).

Cyclotron motion

An ion moving in the presence of a uniform electric and magnetic fields, E and B, is subjected to a Lorentz force given by equation 1, where m, q and v are the mass, charge and velocity.

$$F = ma = m\frac{dv}{dt} = qE + qvB \tag{1}$$

Let us now consider only the presence of the magnetic field, B. If the ion maintains constant speed (i.e. no collisions), then the magnetic field bends the ion path into a circle of radius r, the cyclotron motion (Fig. 2).

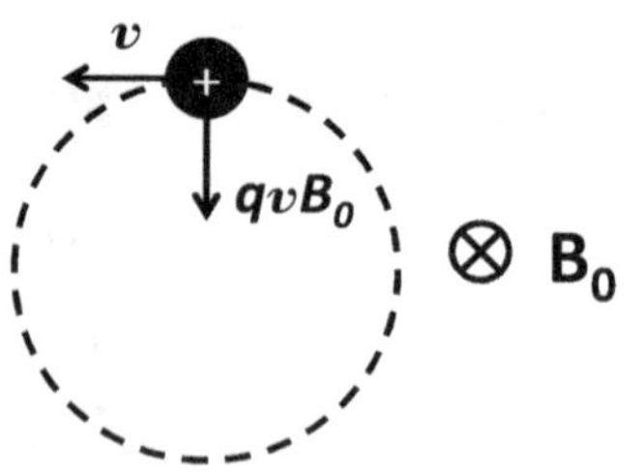

Fig. 2. Ion cyclotron motion for a positive or negative ion due to the presence of a magnetic field (B) perpendicular to the plane of the paper

The angular acceleration and angular velocity, ω, are defined by equations 2 and 3.

$$\frac{dv}{dt} = \frac{{v_{xy}}^2}{r} \tag{2}$$

$$\omega = \frac{v_{xy}}{r} \tag{3}$$

Substituting these terms in equation 1 we obtain[1]

$$\omega_c = \frac{qB}{m} \tag{4}$$

Equation 4 is the celebrated "cyclotron" equation that relates the cyclotron frequency, ω_c, with the mass and charge of the ion. It is clear that for a given *m/z* all ions have the same ICR frequency independent of their initial velocity; hence, energy focussing is not required to precisely determine the *m/z* ratio of an ion.

Several useful conclusions arise from the cyclotron equation (eq. 4). Considering an *m/z* range of 10 to 100 000, the frequencies lie between a few kHz and a few MHz (Fig. 3), which is a very convenient range for commercially available electronics.

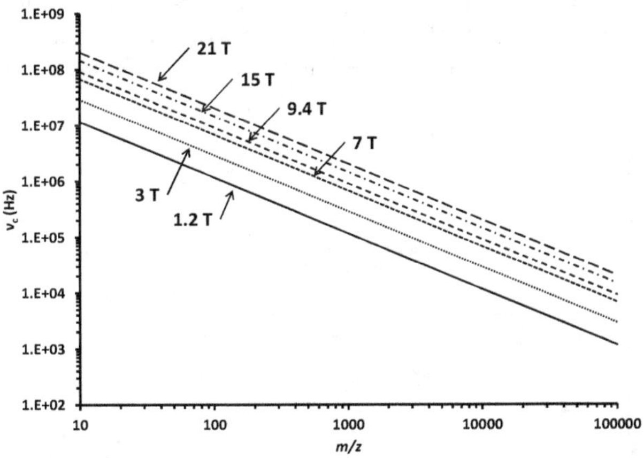

Fig. 3. ICR orbital frequency (for ions in the *m/z* range between 10 and 100 000) as a function of *m/z* at 1.2 T, 3 T, 7 T, 9.4 T, 15 T (highest magnetic field strength commercially available) and 21 T (highest magnetic field strength magnet under development at the National High Magnetic Field Laboratory in the USA)

The first derivative of the cyclotron equation with respect to *m* gives equation 5, which allows us to conclude that frequency resolution and mass resolution are essentially the same (apart from the minus sign).

---

[1] All of the equations will be presented in S.I. units. To convert, for example, m/q to m/z the reader should take into account that q = z×e, where z is in multiples of elementary charge (e = 1.602×10-19 C) and m is in atomic mass units (1u = 1.660×10-27 kg).

$$\frac{d\omega_c}{dm} = -\frac{qB}{m^2} = -\frac{\omega_c}{m} \Leftrightarrow \frac{\omega_c}{d\omega_c} = -\frac{m}{dm} \tag{5}$$

Finally, considering that the average kinetic energy of an ion with velocity v_{xy} in equilibrium with its surroundings at a temperature T (in K) is given by equation 6 (where k is the Boltzmann's constant) and that the ion cyclotron orbital radius, r, is given by equation 7 (derived from equation 3).

$$\frac{1}{2}m\langle v_{xy}{}^2\rangle \cong kT \Leftrightarrow v_{xy} = \sqrt{2kT/m} \tag{6}$$

$$r = \frac{mv_{xy}}{qB_0} \tag{7}$$

Substituting v_{xy} in equation 7 we obtain an expression that relates the ion cyclotron orbital radius as a function of m/z ratio (equation 8).

$$r = \frac{1}{qB_0}\sqrt{2mkT} \tag{8}$$

Considering T=298 K and various magnetic field strengths, we can construct a graphical representation of r as a function of m/z (Fig. 4). This representation allows us to conclude that even large ions are confined by the magnetic field to a small orbital radius. For example, a "modest" superconducting magnet of 3 T confines an ion with m/z 2000 to an orbit with a radius smaller than 0.5 mm.

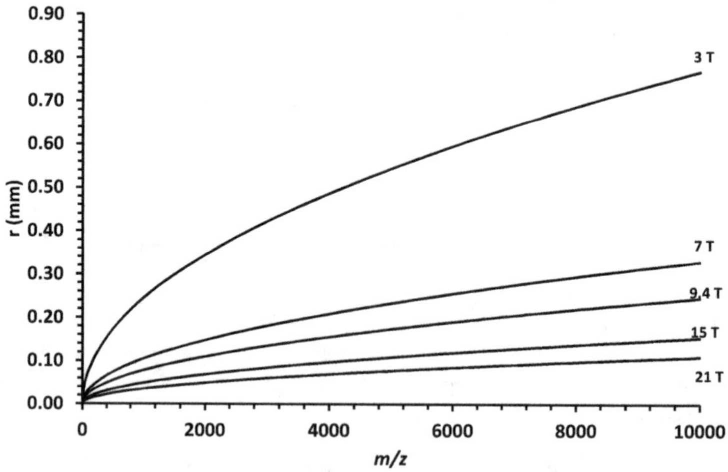

Fig. 4. ICR orbital radius, r, vs. m/z ratio at 298K and 3 T, 7 T, 9.4 T, 15 T and 21 T

It is also possible to relate the kinetic energy of a trapped ion with its orbital radius by rearranging equation 7 (equation 9). A graphical representation of the ion's kinetic energy as a function of the orbital radius, depicted in Fig. 5 for a singly charged ion at m/z 400, reveals that ions can be heated to high kinetic energies even in a relatively small container.

$$E_k = \frac{r^2 q^2 {B_0}^2}{2m} \tag{9}$$

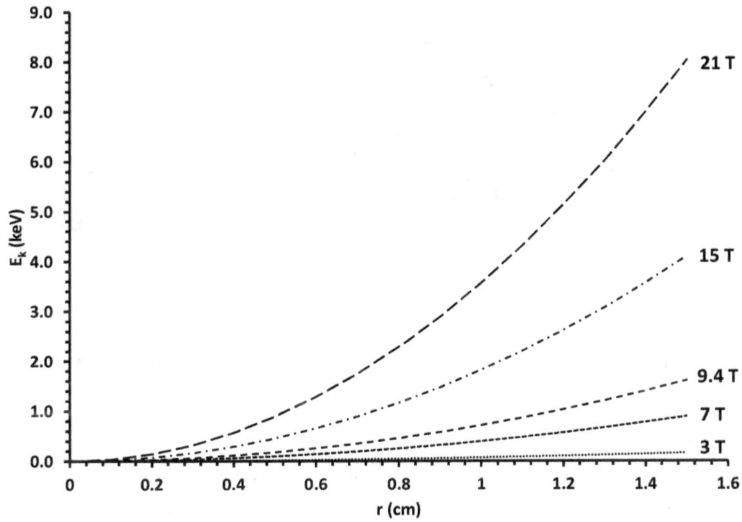

Fig. 5. Ion kinetic energy as a function of the ICR orbital radius (singly charged ion at *m/z* 400)

Trapping motion

The static magnetic field effectively confines ions in the xy plane; nevertheless, ions are still free to escape along the z-axis. To prevent this, a small electrostatic potential (equation 10), usually ≈ 1 V, is applied to the trapping electrodes (positioned at $z = \pm d/2$ from the center of the ion trap). (Vartanian, Anderson et al. 1995)

$$V(z) \cong \frac{V_t}{2} + \frac{k' z^2}{2} \tag{10}$$

where k′ is a constant.

The trapping electric field can be obtained by the negative gradient with respect to z of the electrostatic potential (equation 11).

$$E(z) = -\frac{dV(z)}{dz} = -k'z \tag{11}$$

This electric field subjects the trapped ion to a force, $F(z)$, given by equation 12

$$F(z) = -\frac{d^2 z}{dz^2} = -qk'z \tag{12}$$

Equation 12 resembles the harmonic oscillator equation; hence, ions trapped in a quadratic z-potential must oscillate back and forth along the z-direction at a natural trapping frequency, the so-called trapping motion (or trapping oscillation). The trapping frequency, ω_T, and the k' constant are defined by equations 13 and 14, respectively.

$$\omega_T = \sqrt{\frac{k'q}{m}} \tag{13}$$

$$k' = \frac{4V_T}{d^2} \tag{14}$$

Substituting k' in equation 13 we obtain

$$\omega_T = \frac{2}{d}\sqrt{\frac{qV_T}{m}} \tag{15}$$

In general, the trapping frequency is much smaller than the ICR frequency so it is not detected.

Magnetron motion

The combination of the magnetic field and the radial component of the electric field created by the trapping potential induce a third motion: the magnetron rotation. The magnetron frequency (equation 16), ω_m, is independent of both the mass and charge of the ion. Nevertheless, it is proportional to the trapping voltage (V_T) and the cell geometry factor (α) and inversely proportional to the cell edge length (a) and the magnetic field strength (B). (Vartanian, Anderson et al. 1995)

$$\omega_m = \frac{2\alpha V_T}{a^2 B} \tag{16}$$

Cyclotron + Trapping + Magnetron Motions

The three natural ion motions (cyclotron rotation, magnetron rotation and trapping oscillation) are depicted in Fig. 6. As mentioned earlier, the magnetron and trapping frequencies are usually much smaller than the cyclotron frequency and generally are not detected. Nevertheless, several reviews on the subject of ion trajectories inside the ICR cell and their influence on the cyclotron frequency were published over the years, e.g. the one by Vartanian *et al.* (Vartanian, Anderson et al. 1995)

Excitation and detection of an ICR signal

Ion cyclotron motion does not by itself generate an observable electric signal. When the ion packets enter the ICR cell, their ion cyclotron orbits are centred on the z-axis (i.e. they are too small to be detected) and must be made spatially coherent by moving them away from the centre of the cell. For that purpose excitation is needed and this is achieved by applying a spatially uniform electric field oscillating at or near the cyclotron frequency of ions of a particular *m/z* range.

Excitation is used in three ways in FTICR-MS:

- To accelerate ions coherently to a larger (and thus detectable) orbital radius – Fig. 7 a);

- To increase ion kinetic energy above the threshold for ion dissociation and/or ion-molecule reaction - Fig. 7 b);
- To accelerate ions to a cyclotron radius larger than the radius of the ion trap, so that ions are removed from the instrument - Fig. 7 c).

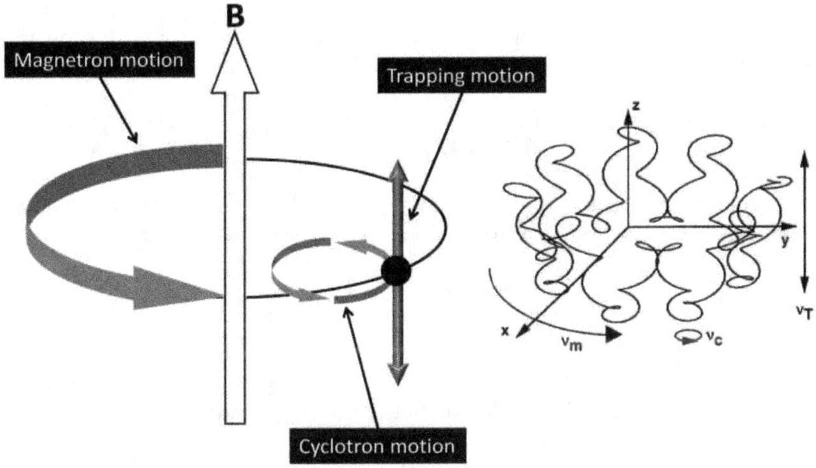

Fig. 6. Schematic representation of the three natural motions of an ion confined in an ICR cell and the resulting ion trajectory shape

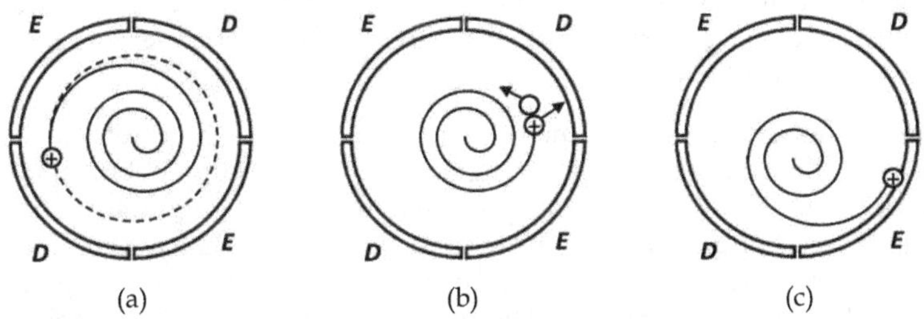

Fig. 7. Ion cyclotron excitation uses: a) acceleration of ions to larger orbital radius for detection; b) acceleration of ions to induce ion dissociation or ion-molecule reactions; c) removal of unwanted ions from the ICR cell. (*E* - Excitation electrode, *D* - Detection electrode)

In this chapter we will focus on the detection; the other two ways of using excitation (to increase the ion's kinetic energy and to remove undesired ions from the instrument) will not be addressed in this chapter. Nevertheless, more information can be obtained in a review by Marshall and co-workers. (Marshall, Hendrickson et al. 1998)

Spatial coherence is created by applying an oscillating resonant, $\omega = \omega_c$, phase-coherent electric field excitation $E(t)$ (equation 17).

$$E(t) = E_0 \cos \omega t\, j \tag{17}$$

This linearly-polarised electric field can be decomposed into two counter-rotating components, $E(t)=E_L(t)+E_R(t)$ (equations 18 and 19).

$$E_L(t) = \frac{E_0}{2}\cos\omega t\,\boldsymbol{j} + \frac{E_0}{2}\sin\omega t\,\boldsymbol{i} \tag{18}$$

$$E_R(t) = \frac{E_0}{2}\cos\omega t\,\boldsymbol{j} - \frac{E_0}{2}\sin\omega t\,\boldsymbol{i} \tag{19}$$

With the excitation the ion speeds up and its radius increases linearly with time and the rate of power absorption is given by equation 20. (Marshall, Hendrickson et al. 1998)

$$A(t_{excite}) = \frac{{E_0}^2 q^2 t_{excite}}{4m} \tag{20}$$

The integration of equation 20 from $t=0$ to $t=t_{excite}$ yields the total energy absorbed during the excitation period and assuming a complete conversion into kinetic energy we obtain

$$\frac{m{\omega_c}^2 r^2}{2} = \int_0^{t_{excite}} A(t)\,dt = \frac{{E_0}^2 q^2 (t_{excite})^2}{8m} \tag{21}$$

Substituting the cyclotron equation (equation 4) in equation 21 we obtain an expression that relates the radius with the excitation electric field and the excitation time.

$$r = \frac{E_0 t_{excite}}{2B_0} \tag{22}$$

An interesting conclusion arises from equation 22 in that the orbital radius of the excited ions is independent of the *m/z* ratio, which means that ions of different *m/z* ratios can be excited to the same ICR orbital radius.

The detection of the ions occurs as the ion packets pass two detector plates. As the ion packets have past these plates, charge moves within the detection circuit to counteract the proximity of the ions. The potential change (voltage) between the detection plates can be measured as a function of time and it is from here that the raw data is obtained. It should be noted that the ions repeatedly pass the detector plates for the duration of the acquisition, as non-destructive detection is employed. The magnitude of the signal is proportional to the total charge and to the proximity of the ions to the detection plates (orbital radius), and is independent of magnetic field strength. The raw data will represent the detections of all the ions at the same time, with their different cyclotron frequencies. It is therefore necessary to extract data concerning the different ion packets. This is done through the usage of a mathematical procedure known as Fourier transform (FT) where frequency information is obtained from time-domain data. Fig. 8 illustrates the process of obtaining a mass spectrum from the time-domain data through Fourier transform, conversion to *m/z* and calibration.

Unlike other mass spectrometers (e.g. sector instruments, time-of-flight, quadrupole) where mass analysis and ion detection are spatially separated events, in FTICR all analytical steps are made on the same spatial place but separated in time. Fig. 9 shows a typical sequence of events for a tandem mass spectrometry experiment performed in a FTICR mass spectrometer. Before ion introduction, the ICR cell is emptied with a quench pulse. After the ions have been introduced into the cell a significant amount of time is required for the ion selection, dissociation, excitation, detection, time-domain data storage and Fourier

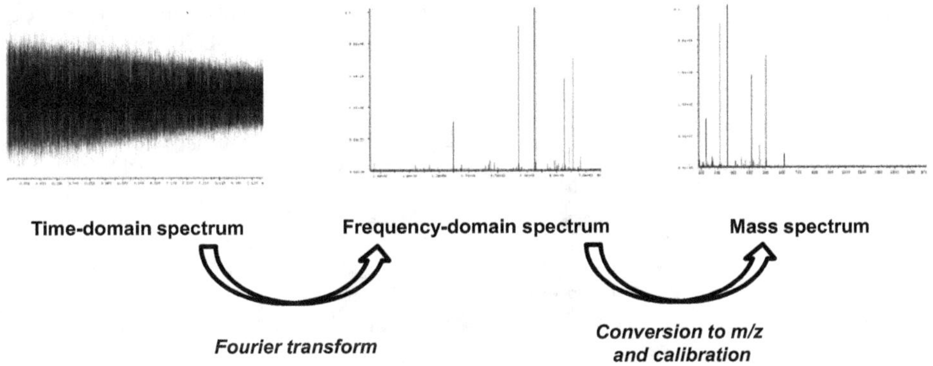

Fig. 8. Illustration of the processing of raw data. A Fourier transform is performed on time-domain data to convert it to the frequency-domain, and the resulting spectrum is then calibrated to *m/z* values

transformation events before the next experiment (i.e. sequence) is started. The time involved in the events that follow ion introduction, greatly depends on the instrument used and on the type of experiment (for example, the acquisition of a normal full scan mass spectrum will take less time than other mass spectra since the ion selection and dissociation steps will not be performed). (Heeren, Kleinnijenhuis et al. 2004)

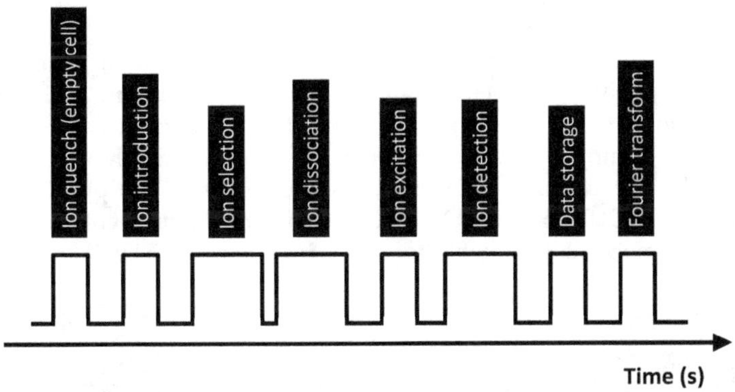

Fig. 9. Example for a tandem mass spectrometry sequence performed in a FTICR mass spectrometer. The sequence shows the order of the different time-separated analytical steps

Orbitrap mass spectrometry

The new millennium introduced a "new" mass analyser in the field of mass spectrometry that had not been stirred up since the first principles of FT-ICR. The *Orbitrap* (Fig. 10) is an ion trap that operates based only on an electrostatic field, by radially trapping ions about a central spindle electrode. An outer barrel-like electrode is coaxial with the inner spindle-like electrode and *m/z* values are measured from the frequency of harmonic ion oscillations along the axis of the electric field. This axial frequency is independent of the energy and spatial distribution of the ions. Ion frequencies are measured non-destructively by acquisition of time-domain image current transients, with subsequent fast Fourier transforms (FFT) being used to obtain the mass spectra. (Hu, Noll et al. 2005)

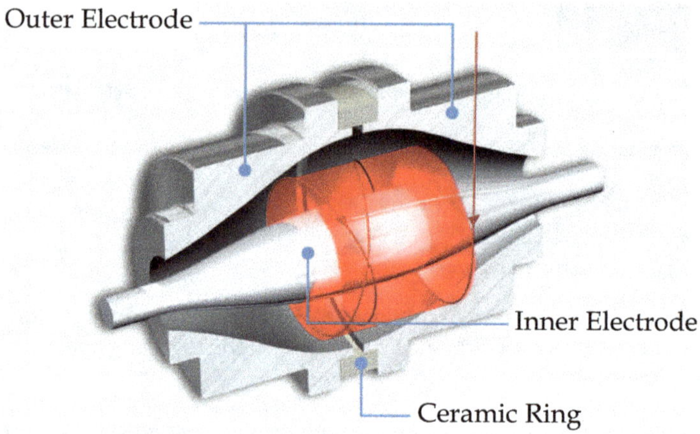

Fig. 10. *Orbitrap* mass analyser. Ions are captured in a quadrologarithmic electrostatic field. An outer electrode enclosing a central spindle electrode consists of two halves separated by a dielectric material. The image current of ions moving as concentric rings along the central electrode is picked up by the outer electrode sections. Image kindly supplied by Thermo Fisher Scientific

Features of the *Orbitrap* at its present stage of development include high mass resolution (up to 150 000), large space charge capacity, high mass accuracy (2–5 ppm), a mass/charge range of at least 6000, and dynamic range greater than 10^3. The current commercially available *Orbitrap* systems are equipped with several features that increase the range of applications. Fig. 11 depicts a hybrid ion trap-*Orbitrap* mass spectrometer equipped with an Electron Transfer Dissociation (ETD) module. ETD is the ion trap equivalent to Electron Capture Dissociation (a very popular fragmentation technique used in FTICR mass spectrometers)

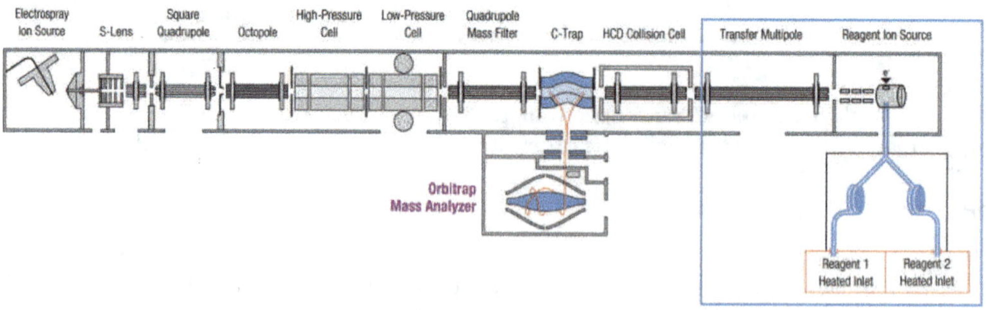

Fig. 11. Schematic representation of a hybrid ion trap-*Orbitrap* mass spectrometer. The main parts of a commercially available instrument are annotated on the diagram. The blue box highlights an optional ETD module which further extends the versatility of the system. Image kindly supplied by Thermo Fisher Scientific

that enables, for example, the probing of post-translational modifications of proteins. (Jung, Pasini et al. 2010)

Principles of operation of the *Orbitrap* mass analyser (overview)

The principles of operation of the *Orbitrap* can be summarised as follows:

- It exclusively uses an electrostatic field to trap the ions (no RF or magnetic fields).
- Moving ions are trapped around the central electrode with only electrostatic attraction being compensated by centrifugal force (from initial tangential velocity).
- The outer electrode confines ions axially.
- Control of frequencies of oscillation (axial in particular) by shape/geometry of electrodes.

In a very simple way, in this mass analyser, ions are injected tangentially into an electric field between specially shaped electrodes and trapped because their electrostatic attraction to the inner electrode is balanced by centrifugal forces. Thus, ions cycle around the central electrode in rings - rotational motion (elliptical orbiting). In addition, the ions also move back and forth along the axis of the central electrode - axial oscillation. Therefore, ions of a specific mass-to-charge ratio move in rings which oscillate along the central spindle-like electrode. The frequency of these harmonic oscillations is independent of the ion velocity and is inversely proportional to the square root of the mass-to-charge ratio (m/z). By sensing the ion oscillation in a manner similar to that used in FT-ICR (ion image current detection and FFT), the trap can be used as a mass analyser. There are three types of ion trapping linear, segmented and orbital trapping. The latter is the focus of our review.

Orbital trapping (History)

Orbital trapping was first implemented by Kingdon in 1923. (Kingdon 1923) In its classical shape, the Kingdon trap contains a wire stretched along the axis of an outer cylinder with flanges enclosing the trapping volume. When a voltage is applied between the wire and the cylinder, the strong field attracts ions to the wire. Only ions that have sufficient tangential velocity miss the wire and survive. Motion along the wire is restrained by the field curvature caused by the flanges of the outer cylinder. A more elaborate electrode shape has been developed by Knight ("ideal Kingdon trap"). (Knight 1981)

The modified Kingdon trap

Orbital trapping is known to be used in experiments on the spectroscopy of ions. In these applications, ions have been formed within the trap or injected tangentially prior to switching on the field of the trap. (Knight 1981; Lewis 1982; Lisheng and Church 1991; Sekioka, Terasawa et al. 1991)

Makarov (Makarov 2000) proposed the concept of orbital trapping with a fresh review for application to mass analysis. A "new" type of mass analyser was described which employs orbital trapping in an electrostatic field.

For didactic purposes the orbitrap can be considered as an enhanced 'Knight-style' ideal Kingdon trap, but with specially shaped electrodes. When a DC voltage is applied between the two axially symmetric electrodes it results in the electrostatic field with potential distribution described in equation 23.

$$U(r,z) = \frac{k}{2}\left(z^2 - \frac{r^2}{2}\right) + \frac{k}{2}(R_m)^2 \ln\left[\frac{r}{R_m}\right] + C \tag{23}$$

where r and z are cylindrical coordinates (z= 0 being the plane of the symmetry of the field), C is a constant, k is field curvature (or more recently known as axial restoring force determined by the exact shape of the electrodes and applied potential), and R_m is the characteristic radius. For a more detailed description, please refer to literature. (Gillig, Bluhm et al. 1996; Makarov 2000; Makarov, Denisov et al. 2006)

This field, shown in equation (22), is the sum of a quadrupole field of the ion trap (which confines ions axially) and a logarithmic field of a cylindrical capacitor (provides orbital ion trapping); therefore, it is also called a quadro-logarithmic field. It is clear from this expression that there are no cross-terms in r and z, thus motion in z is independent of r, φ-motion (where φ is the angular coordinate).

Geometry of the Orbitrap (axially symmetric electrodes and equations of motion)

The trap consists of an outer barrel-like electrode and a central spindle-like electrode along the axis (see Fig. 10). The geometrical shape of these axially symmetrical electrodes can be deduced from eq. 22:

$$z_{1,2}(r) = \sqrt{\frac{r^2}{2} - \frac{(R_{1,2})^2}{2} + (R_m)^2 \ln\left[\frac{R_{1,2}}{r}\right]} \tag{24}$$

where index 1 denotes the central electrode (spindle like), index 2 denotes the outer electrode (with flanges enclosing the trapping volume), z = 0 is the plane of symmetry, and $R_{1,2}$ are the maximum radii of the corresponding electrodes.

Stable ion trajectories involve both orbiting motion around (rotational) the central electrode (r, φ-motion, where φ is the angular coordinate) and simultaneous harmonic oscillations in the z-direction (coherent axial oscillations).

The Ion Trajectories (Equations of motion)

For the *Rotational Motion,* the r, φ–motion, although not used for mass analysis, is still important because ions must be trapped in the radial plane. Equilibrium is reached between attraction and centrifugal forces that act on the ions.

For the *Axial Motion,* the movement back and forth around the central electrode, along z, is (approximately) described by a simple harmonic oscillator. The angular frequency of axial oscillations in rad s^{-1} (ω) is defined by equation 25 in terms of the charge (q) and mass (m) of the ion, k the field constant, being essential for establishing the fundamental relation for mass analysis. (Makarov 2000; Hu, Noll et al. 2005; Makarov, Denisov et al. 2006)

$$\omega = \sqrt{(q/m)k} \tag{25}$$

Use of the axial angular frequency as opposed to rotational or radial frequency is critical because only this frequency is completely independent of energy and spatial spread of ions (see below). A mass analyser employing such electrostatic axially harmonic orbital trapping is referred hereinafter as the *Orbitrap*.

This axial frequency may be determined using image current detection and fast FT algorithms. (Dienes, Pastor et al. 1996) The detection of an ion image current due to motion along the *Orbitrap* axis is only possible as long as the ion packets retain their spatial coherence in the axial direction (*i.e.*, phase coherence and small spatial extent). The outer electrode is split in half at $z=0$ allowing the ion image current due to axial motion to be collected. The current is differentially amplified from each half of the outer electrode and then undergoes analog-to-digital conversion before processing and collection by customised control and acquisition software. Transient ion image current in the outer electrodes (split at $z=0$) is acquired as a time-domain transient and transformed to the spectra in the mass spectrometer by applying fast Fourier transform (FFT). The image current is amplified and processed exactly in the same way as in FT ICR; (see previous section) therefore, similar sensitivity and signal-to-noise ratios are expected. The resulting frequency spectrum however differs due to a much slower decrease of ion frequency with mass according to equation 25 when compared with equation 4.

As a conclusion, potential advantages of the *Orbitrap* include:

- high mass resolution (up to 100 000-200 000) since the quadro-logarithmic field (equation 22) may be defined with very high accuracy;
- increased space charge capacity at higher masses due to independence of trapping potential on mass-to-charge ratio and larger trapping volumes in contrast to FT ICR and ion trap;
- high mass accuracy (2-5 ppm internal and external calibration, respectively), dynamic range over which accurate masses can be determined, and upper mass limit;
- high sensitivity and stability; and
- maintenance free analyser (when compared to the FTICR mass analyser where cryogenic liquids or expensive cooling apparatus are needed to maintain the superconducting magnet).

Applications

The mass analysers described in this chapter are capable of performing high resolution and high accuracy mass measurements on a routine basis. It is worth mentioning that before such mass analysers were developed, accurate mass measurements were performed only on sector instruments with a great deal of work and time behind each analysis within a restricted mass range (basically the masses in the close vicinity of the measured mass). Nowadays, high resolution can be attained in a broad mass range and the spectra are acquired in a couple of minutes (usually the time spent to acquire an HR mass spectrum is well below 1 minute).

To demonstrate how useful high resolution is, electrospray ionisation (ESI) mass spectra will be shown that were acquired on two different instruments: an Apex Ultra FTICR mass spectrometer equipped with a 7 T superconducting magnet (Bruker Daltonics, Bremen, Germany) and a LCQ-Duo ion trap mass spectrometer (Thermo Fisher Scientific, Bremen Germany). The high- and low-resolution ESI mass spectra of ubiquitin are depicted in Fig. 12. Even though both systems provide information about charge state distribution, it is perfectly clear that the isotopic envelope is completely resolved when using an FTICR system (see insets in Fig. 12).

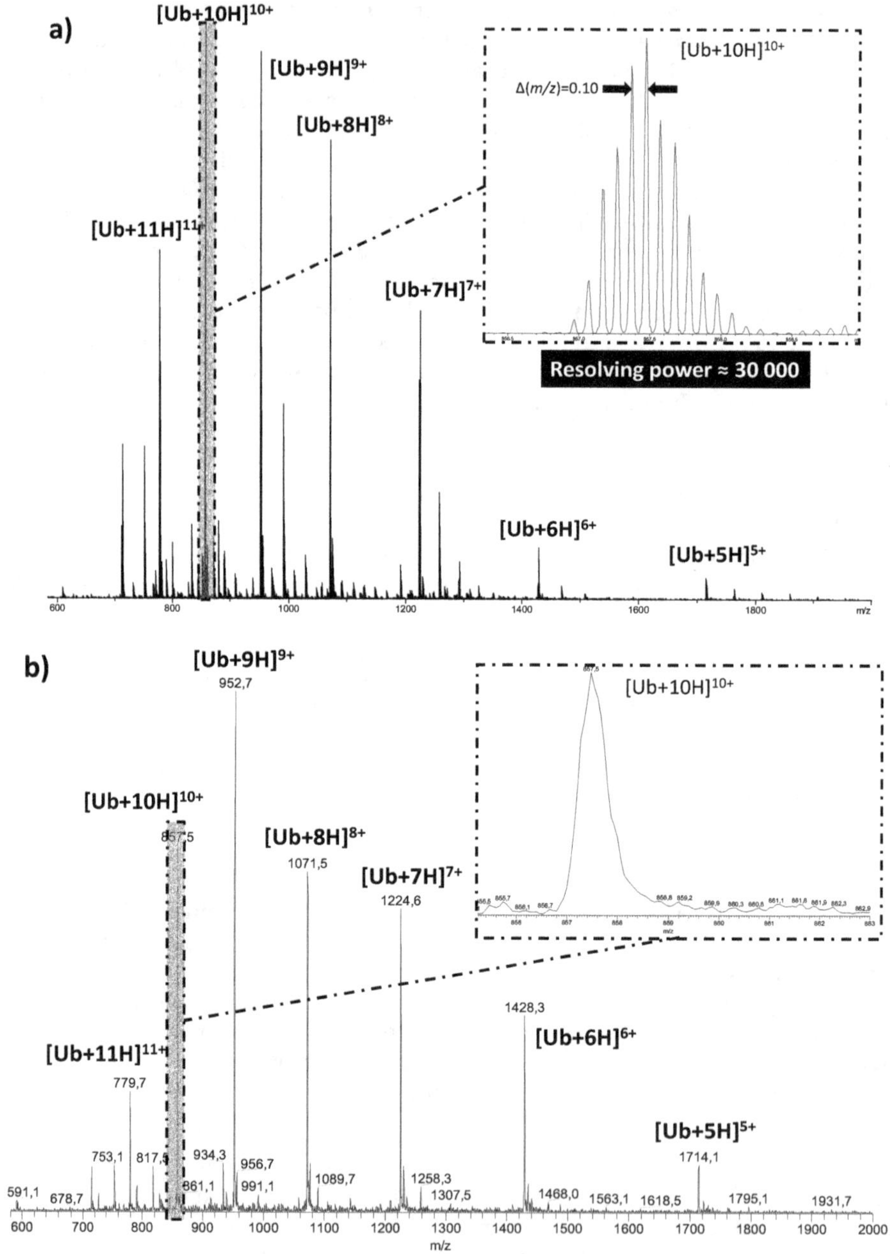

Fig. 12. High resolution (a) and low resolution (b) electrospray ionisation mass spectra of Ubiquitin (Ub). The insets show an expansion of the peak at *m/z* 857.5 attributed to $[Ub+10H]^{10+}$ ion. (High resolution system: 7 T Apex Ultra FTICR-MS from Bruker Daltonics; Low resolution system LCQ-Duo ion trap mass spectrometer from Thermo Fisher Scientific

There are basically two ways to determine the charge state of a given ion. One involves the analysis of the isotope envelope, implying that it can only be used in high-resolution mass spectrometry, while the other relates each peak with the subsequent one and can be used in high- and low-resolution mass spectrometry.

The mass difference between two isotopic peaks is related to the charge state by $z=1/\Delta$ (m/z), hence, for the ubiquitin ion at m/z 857.47 ($[Ub+10H]^{10+}$ ion of Fig. 12a) the charge state will be given by $z= (1/0.10) = 10$. It is worth mentioning that isotopic peaks are useful in the interpretation as long as there is no interference from other ions.

Considering that two adjacent peaks in the mass spectrum, for example, m/z 857.5 and m/z 957.7 of Fig. 12b, are due to the same entity (i.e. they reflect different charge states of ubiquitin), one can assume that $m = 857.5z'$ and $m = 957.7z$ (where $z' = z+1$), i.e. $857.5 (z+1) = 957.7z$, hence $z = 9$. As such the ion at m/z 957.7 corresponds to a 9+ ion while the ion at m/z 857.5 corresponds to a 10+ ion.

Acknowledgments

The authors acknowledge the funding provided by Fundação para a Ciência e a Tecnologia (PEst-OE/QUI/UI0612/2011 and REDE/1501/REM/2005). Pedro A. Alves acknowledges Fundação para a Ciência e a Tecnologia for the PhD grant (SFRH/BD/49069/2008). Thermo Fisher Scientific and Dr. Michaela Scigelova are gratefully acknowledged for supplying the *Orbitrap* images.

References

Aston, F. W. (1919). *LXXIV. A positive ray spectrograph*. Philosophical Magazine Series 6 38 (228): 707 - 714.

Bainbridge, K. T. (1932). *The isotopic weight of H_2*. Physical Review 42 (1): 1-10.

Bainbridge, K. T. and E. B. Jordan (1936). *Mass spectrum analysis 1. The mass spectrograph. 2. The existence of isobars of adjacent elements*. Physical Review 50 (4): 282-296.

Barber, M., R. S. Bordoli, et al. (1981). *Fast atom bombardment of solids (F.A.B.): a new ion source for mass spectrometry*. Journal of the Chemical Society, Chemical Communications (7): 325-327.

Comisarow, M. B. and A. G. Marshall (1974). *Fourier transform ion cyclotron resonance spectroscopy*. Chemical Physics Letters 25 (2): 282-283.

Covey, T. R., E. C. Huang, et al. (2002). *Structural characterization of protein tryptic peptides via liquid chromatography/mass spectrometry and collision-induced dissociation of their doubly charged molecular ions*. Analytical Chemistry 63 (13): 1193-1200.

Dass, C. (2007). *Fundamentals of Contemporary Mass Spectrometry*. Hoboken, New Jersey, John Wiley & Sons.

Dempster, A. J. (1918). *A new method of positive ray analysis*. Physical Review 11 (4): 316-325.

Dienes, T., S. J. Pastor, et al. (1996). *Fourier transform mass spectrometry - advancing years (1992–mid. 1996)*. Mass Spectrometry Reviews 15 (3): 163-211.

Fenn, J. B., M. Mann, et al. (1989). *Electrospray ionization for mass spectrometry of large biomolecules*. Science 246 (4926): 64-71.

Gillig, K. J., B. K. Bluhm, et al. (1996). *Ion motion in a Fourier transform ion cyclotron resonance wire ion guide cell*. International Journal of Mass Spectrometry and Ion Processes 157-158: 129-147.

Gross, J. H. (2004). *Mass Spectrometry: A Textbook*. Berlin, Springer-Verlag.

Hager, J. W. (2002). *A new linear ion trap mass spectrometer*. Rapid Communications in Mass Spectrometry 16 (6): 512-526.

Heeren, R. M. A., A. J. Kleinnijenhuis, et al. (2004). *A mini-review of mass spectrometry using high-performance FTICR-MS methods*. Analytical and Bioanalytical Chemistry 378 (4): 1048-1058.

Hoffmann, E. d. and V. Stroobant (2007). *Mass spectrometry: Principles and Applications*. Chichester, John Wiley & Sons.

Hu, Q., R. J. Noll, et al. (2005). *The Orbitrap: a new mass spectrometer*. Journal of Mass Spectrometry 40 (4): 430-443.

Jennings, K. R. (1968). *Collision-induced decompositions of aromatic molecular ions*. International Journal of Mass Spectrometry and Ion Physics 1 (3): 227-235.

Johnson, E. G. and A. O. Nier (1953). *Angular aberrations in sector shaped electromagnetic lenses for focusing beams of charged particles*. Physical Review 91 (1): 10-17.

Jung, H. R., D. Pasini, et al. (2010). *Quantitative mass spectrometry of Histones H3.2 and H3.3 in Suz12-deficient mouse embryonic stem cells reveals distinct, dynamic post-translational modifications at Lys-27 and Lys-36*. Molecular & Cellular Proteomics 9 (5): 838-850.

Karas, M. and F. Hillenkamp (2002). *Laser desorption ionization of proteins with molecular masses exceeding 10,000 daltons*. Analytical Chemistry 60 (20): 2299-2301.

Kingdon, K. H. (1923). *A method for the neutralization of electron space charge by positive ionization at very low gas pressures*. Physical Review 21 (4): 408-418.

Knight, R. D. (1981). *Storage of ions from laser-produced plasmas*. Applied Physics Letters 38 (4): 221-223.

Lewis, R. R. (1982). *Motion of ions in the Kingdon trap*. Journal of Applied Physics 53 (6): 3975-3980.

Lisheng, Y. and D. A. Church (1991). *Confinement of injected beam ions in a Kingdon trap*. Nuclear Instruments and Methods in Physics Research Section B: Beam Interactions with Materials and Atoms 56-57, Part 2: 1185-1187.

Makarov, A. (2000). *Electrostatic axially harmonic orbital trapping: a high-performance technique of mass analysis*. Analytical Chemistry 72 (6): 1156-1162.

Makarov, A., E. Denisov, et al. (2006). *Performance evaluation of a hybrid linear ion trap/Orbitrap mass spectrometer*. Analytical Chemistry 78 (7): 2113-2120.

Marshall, A. G., M. B. Comisarow, et al. (1979). *Relaxation and spectral line shape in Fourier transform ion resonance spectroscopy*. Journal of Chemical Physics 71 (11): 4434-4444.

Marshall, A. G., C. L. Hendrickson, et al. (1998). *Fourier transform ion cyclotron resonance mass spectrometry: A primer*. Mass Spectrometry Reviews 17 (1): 1-35.

Marshall, A. G. and F. R. Verdun (1990). *Fourier tansforms in NMR, optical, and mass spectrometry: a user's handbook*. New York, Elsevier.

Munson, M. S. B. and F. H. Field (1966). *Chemical ionization mass spectrometry. I. General Introduction*. Journal of the American Chemical Society 88 (12): 2621-2630.

Nier, A. O. (1940). *A mass Spectrometer for routine isotope abundance measurements*. Review of Scientific Instruments 11 (7): 212-216.

Paul, W. (1990). *Electromagnetic traps for charged and neutral particles*. Reviews of Modern Physics 62 (3): 531-540.

Perry, R. H., R. G. Cooks, et al. (2008). *Orbitrap mass spectrometry: Instrumentation, ion motion and applications*. Mass Spectrometry Reviews 27 (6): 661-699.

Rodrigues, J. A., A. M. Taylor, et al. (2007). *Mass Spectrometry of Carbohydrates: Newer Aspects,* In *Advances in Carbohydrate Chemistry and Biochemistry.* D. Horton. London, Academic Press. Volume 61: 59-141.

Ryhage, R. (2002). *Use of a mass spectrometer as a detector and analyzer for effluent emerging from high temperature gas liquid chromatography columns.* Analytical Chemistry 36 (4): 759-764.

Sekioka, T., M. Terasawa, et al. (1991). *Ion storage in Kingdon trap.* Radiation Effects and Defects in Solids 117 (1-3): 253-259.

Sommer, H., H. A. Thomas, et al. (1951). *The measurement of e/M by cyclotron resonance.* Physical Review 82 (5): 697-702.

Thomson, J. J. (1897). *Cathode rays.* Philosophical Magazine Series 5 44 (269): 293-316.

Thomson, J. J. (1907). *On rays of positive electricity.* Philosophical Magazine Series 6 13 (77): 561-575.

Vartanian, V. H., J. S. Anderson, et al. (1995). *Advances in trapped ion cells for Fourier transform ion cyclotron resonance mass spectrometry.* Mass Spectrometry Reviews 14 (1): 1-19.

Voyksner, R. D. (1997). *Combining Liquid Chromatography with Electrospray Mass Spectrometry,* In *Electrospray Ionization Mass Spectrometry: Fundamentals, Instrumentation, and Applications.* R. B. Cole. New York, John Wiley & Sons: 323-342.

Watson, J. T. and K. Biemann (2002). *Direct recording of high resolution mass spectra of gas chromatographic effluents.* Analytical Chemistry 37 (7): 844-851.

Whitehouse, C. M., R. N. Dreyer, et al. (2002). *Electrospray interface for liquid chromatographs and mass spectrometers.* Analytical Chemistry 57 (3): 675-679.

Wolff, M. M. and W. E. Stephens (1953). *A pulsed mass spectrometer with time dispersion.* Review of Scientific Instruments 24 (8): 616-617.

Zhang, L.-K., D. Rempel, et al. (2005). *Accurate mass measurements by Fourier transform mass spectrometry.* Mass Spectrometry Reviews 24 (2): 286-309.

Permissions

All chapters in this book were first published in FTMA, by InTech Open; hereby published with permission under the Creative Commons Attribution License or equivalent. Every chapter published in this book has been scrutinized by our experts. Their significance has been extensively debated. The topics covered herein carry significant findings which will fuel the growth of the discipline. They may even be implemented as practical applications or may be referred to as a beginning point for another development.

The contributors of this book come from diverse backgrounds, making this book a truly international effort. This book will bring forth new frontiers with its revolutionizing research information and detailed analysis of the nascent developments around the world.

We would like to thank all the contributing authors for lending their expertise to make the book truly unique. They have played a crucial role in the development of this book. Without their invaluable contributions this book wouldn't have been possible. They have made vital efforts to compile up to date information on the varied aspects of this subject to make this book a valuable addition to the collection of many professionals and students.

This book was conceptualized with the vision of imparting up-to-date information and advanced data in this field. To ensure the same, a matchless editorial board was set up. Every individual on the board went through rigorous rounds of assessment to prove their worth. After which they invested a large part of their time researching and compiling the most relevant data for our readers.

The editorial board has been involved in producing this book since its inception. They have spent rigorous hours researching and exploring the diverse topics which have resulted in the successful publishing of this book. They have passed on their knowledge of decades through this book. To expedite this challenging task, the publisher supported the team at every step. A small team of assistant editors was also appointed to further simplify the editing procedure and attain best results for the readers.

Apart from the editorial board, the designing team has also invested a significant amount of their time in understanding the subject and creating the most relevant covers. They scrutinized every image to scout for the most suitable representation of the subject and create an appropriate cover for the book.

The publishing team has been an ardent support to the editorial, designing and production team. Their endless efforts to recruit the best for this project, has resulted in the accomplishment of this book. They are a veteran in the field of academics and their pool of knowledge is as vast as their experience in printing. Their expertise and guidance has proved useful at every step. Their uncompromising quality standards have made this book an exceptional effort. Their encouragement from time to time has been an inspiration for everyone.

The publisher and the editorial board hope that this book will prove to be a valuable piece of knowledge for researchers, students, practitioners and scholars across the globe.

List of Contributors

Hassan Safouhi
Mathematical Section, Campus Saint-Jean, University of Alberta, Alberta, Canada

Sei Ueda
Osaka Institute of Technology, Japan

Florica Matei
University of Agricultural Science and Veterinary Medicine, Cluj-Napoca, Romania

Nicolae Aldea
National Institute for Research and Development of Isotopic and Molecular Technologies, Cluj-Napoca, Romania

F. X. Xin and T. J. Lu
State Key Laboratory for Mechanical Structure Strength and Vibration, School of Aerospace, Xi'an Jiaotong University, Xi'an, China

S. Abdullah and M. Z. Nuawi
Universiti Kebangsaan Malaysia, Malaysia

T. E. Putra
Universitas Syiah Kuala, Indonesia

Bingzheng Li
Taiyuan University of Science & Technology, P. R. China

E. R. Mejía, J. D. Ospina and M. A. Márquez
Materials Engineering School, Applied Mineralogy and Bio-process Group (GMAB), National University of Colombia, Medellín AA 1027, Colombia

A. L. Morales
Solid State Group, University Research Centre Antioquia University, Medellín AA 1226, Colombia

Chanel Fortier
United States Department of Agriculture, Agricultural Research Service, Southern Regional Research Center, New Orleans, LA, USA

Hassen Aroui
Laboratoire de Dynamique Moléculaire et Matériaux Photoniques, Université de Tunis, Ecole Supérieure des Sciences et Techniques de Tunis, Tunisia

Johannes Orphal
Institute for Meteorology and Climate Research (IMK), Karlsruhe Institute of Technology (KIT), Germany

Biao Huang
School of Material and Engineering, Fujian Agricultural and Forestry University, P. R. China

Fridolin Kwabia Tchana
Laboratoire Interuniversitaire des Systèmes Atmosphériques (LISA), UMR CNRS 7583 Universités Paris Est Créteil et Paris- Diderot, Institut Pierre Simon Laplace, France

Mizi Fan and Dasong Dai
Department of Civil Engineering, Brunel University, London, UB8 3PH, UK
School of Material and Engineering, Fujian Agricultural and Forestry University, P. R. China

Paulo J. Amorim Madeira, Pedro A. Alves and Carlos M. Borges
Centro de Química e Bioquímica, Departamento de Química e Bioquímica, Faculdade de Ciências da Universidade de Lisboa, Portugal

Index

www.ingramcontent.com/pod-product-compliance
Lightning Source LLC
LaVergne TN
LVHW051046220926
848485LV00001BA/28
* 9 7 8 1 6 3 9 8 7 2 3 9 8 *